Springer Series in
CHEMICAL PHYSICS 8

Springer-Verlag Berlin Heidelberg GmbH

Physics and Astronomy ONLINE LIBRARY

http://www.springer.de/phys/

Springer Series in
CHEMICAL PHYSICS

Series Editors: F. P. Schäfer J. P. Toennies W. Zinth

The purpose of this series is to provide comprehensive up-to-date monographs in both well established disciplines and emerging research areas within the broad fields of chemical physics and physical chemistry. The books deal with both fundamental science and applications, and may have either a theoretical or an experimental emphasis. They are aimed primarily at researchers and graduate students in chemical physics and related fields.

63 **Ultrafast Phenomena XI**
Editors: T. Elsaesser, J.G. Fujimoto, D.A. Wiersma, and W. Zinth

64 **Asymptotic Methods in Quantum Mechanics**
Application to Atoms, Molecules and Nuclei
By S.H. Patil and K.T. Tang

65 **Fluorescence Correlation Spectroscopy**
Theory and Applications
Editors: R. Rigler and E.S. Elson

66 **Ultrafast Phenomena XII**
Editors: T. Elsaesser, S. Mukamel, M.M. Murnane, and N.F. Scherer

67 **Single Molecule Spectroscopy**
Nobel Conference Lectures
Editors: R. Rigler, M. Orrit, T. Basché

68 **Nonequilibrium Nondissipative Thermodynamics**
With Application to Low-Pressure Diamond Synthesis
By J.-T. Wang

Series homepage – http://www.springer.de/phys/books/chemical-physics/

Volumes 1–62 are listed at the end of the book

Ji-Tao Wang

Nonequilibrium Nondissipative Thermodynamics

With Application to Low-Pressure Diamond Synthesis

With 114 Figures

Springer

Professor Ji-Tao Wang
Department of Microelectronics
Fudan University
Shanghai 200433, China

Series Editors:

Professor F.P. Schäfer
Max-Planck-Institut für Biophysikalische Chemie
37077 Göttingen-Nikolausberg, Germany

Professor J.P. Toennies
Max-Planck-Institut für Strömungsforschung
Bunsenstrasse 10
37073 Göttingen, Germany

Professor W. Zinth
Universität München,
Institut für Medizinische Optik
Öttingerstr. 67
80538 München, Germany

ISSN 0172-6218

ISBN 978-3-642-07674-9

Library of Congress Cataloging-in-Publication Data applied for.

Die Deutsche Bibliothek - CIP-Einheitsaufnahme
Wang, Ji-Tao:
Nonequilibrium nondissipative thermodynamics: with applications to
low-pressure diamond synthesis/Ji-Tao Wang.
(Springer series in chemical physics; Vol. 68)
ISBN 978-3-642-07674-9 ISBN 978-3-662-04829-0 (eBook)
DOI 10.1007/978-3-662-04829-0

http://www.springer.de

Originally published by Springer-Verlag Berlin Heidelberg New York in 2002
Softcover reprint of the hardcover 1st edition 2002

Typesetting: Camera-ready copy by the author
Cover concept: eStudio Calamar Steinen
Cover production: *design & production* GmbH, Heidelberg

Printed on acid-free paper SPIN: 10835588 57/3141/tr - 5 4 3 2 1 0

Preface

Albert Einstein (1879–1955) said, "*The most beautiful thing we can experience is the mysterious. It is the source of all true art and science.*" For the last 30 years or so, activated low pressure diamond synthesis has been regarded as "mysterious" or a "thermodynamic paradox" perhaps "violating the second law of thermodynamics". A new field of thermodynamics, nonequilibrium nondissipative thermodynamics — has recently arisen from the puzzle.

For complex systems including nonspontaneous reaction(s), the basic classification of equilibrium thermodynamics and nonequilibrium thermodynamics on the basis of the second law of thermodynamics has to be changed or overthrown. Maybe we should also cite another statement of Albert Einstein: "*A theory is more impressive the greater the simplicity of its premises is, the more different kinds of things it relates, and the more extended its area of applicability. Therefore the deep impression which classical thermodynamics made upon me. It is the only physical theory of universal content concerning which I am convinced that, within the framework of the applicability of its basic concepts, it will never be overthrown.*" and it might be thought that classical thermodynamics would never be overthrown. However, Einstein said, "it will never be overthrown" only "within the framework of the application of its basic concepts". This means that outside "the framework of the application of its basic concepts" some basic points of view from classical thermodynamics might be overthrown or changed.

In this book classical thermodynamics has to be changed into modern thermodynamics for certain complex systems in emerging science and technology. For instance, a slightly more complex isothermal and isobaric system consisting of a nonspontaneous reaction and another simultaneous spontaneous reaction is under consideration. The Gibbs free energy expression of the second law of thermodynamics, i.e., $(\mathrm{d}G)_{T,p} \leq 0$, should be used as the criterion. The situation where $[(\mathrm{d}G_1)_{T,p} > 0$, $(\mathrm{d}G_2)_{T,p} < 0$ and $(\mathrm{d}G)_{T,p} \leq 0]$ is possible, and is known as reaction coupling. $(\mathrm{d}G_1)_{T,p}$, $(\mathrm{d}G_2)_{T,p}$ and $(\mathrm{d}G)_{T,p} = (\mathrm{d}G_1)_{T,p} + (\mathrm{d}G_2)_{T,p}$ are Gibbs free energy changes of reaction 1, reaction 2 and the whole system, respectively. Therefore, both situations of equilibrium $[(\mathrm{d}G_1)_{T,p} = 0$, $(\mathrm{d}G_2)_{T,p} = 0$ and $(\mathrm{d}G)_{T,p} = 0]$ and nonequilibrium $[(\mathrm{d}G_1)_{T,p} > 0$, $(\mathrm{d}G_2)_{T,p} < 0$ and $(\mathrm{d}G)_{T,p} = 0]$ correspond to the equality of the second law of thermodynamics. Therefore, the systematization of modern thermodynamics for isothermal and isobaric system has been drastically changed, as shown in the following figure.

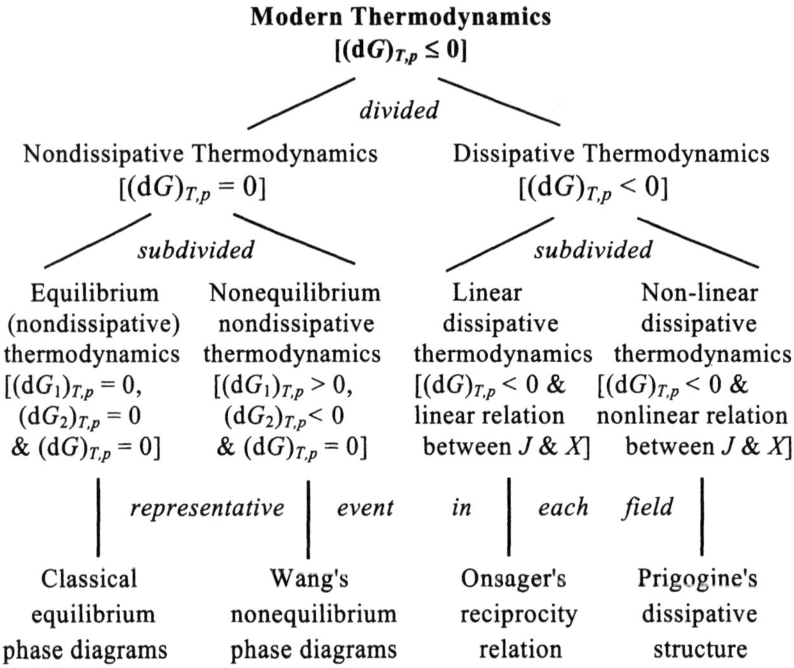

X and J represent thermodynamic force and thermodynamic flow for irreversible process, respectively. $(dG_1)_{T,p}$, $(dG_2)_{T,p}$, and $(dG)_{T,p} = [(dG_1)_{T,p} + (dG_2)_{T,p}]$ are Gibbs free energy changes of reaction 1, reaction 2 and the whole system, respectively.

In general, the figure above should be drawn on the basis of modern formulation of the second law of thermodynamics. It will be discussed in detail in the book.

Obviously, the equality of the second law is no longer a necessary and sufficient condition of a system to be in equilibrium. Some related concepts in classical thermodynamics should also be overthrown for complex systems. However, most of the basic concepts of classical thermodynamics are still correct and useful as the foundation for developing modern thermodynamics, as mentioned in Chap. 1 of the book.

In Chap. 2 it can be seen that not many substantial premises are needed, but the area of application of thermodynamics has been greatly extended. Meanwhile, a new field of nonequilibrium nondissipative thermodynamics has been introduced by the author into modern thermodynamics. That is the title of this book and will be the main topic of this book, presented in depth in Chaps. 3–8. The calculation of nonequilibrium phase diagrams and their quantitative agreement with a lot of experimental data for activated low-pressure diamond synthesis reported in the literature are the most important achievements in this new field in the last 10 years. Because the criterion of reaction coupling, such as $[(\mathrm{d}G_1)_{T,p} > 0$, $(\mathrm{d}G_2)_{T,p} < 0$ and $(\mathrm{d}G_1)_{T,p} \leq 0]$, includes both equality and inequality, some applications of reaction coupling discussed in Chap. 9 and elsewhere in this book will be concerned not only with nonequilibrium nondissipative thermodynamics but also with dissipative thermodynamics.

Those who have also contributed to or worked on the projects concerned with the book are Jan-Otto Carlsson of Uppsala University of Sweden, Pei-Ju Zheng of the Research Center for Analysis and Measurement of Fudan University, Zhong-Qiang Huang of the Applied Mathematics Department of Tonji University, David Wei Zhang, Hao Wang, Chuan-Bao Cao, Qing-He Yang, Yong-Zhong Wan, Zhi-Jie Liu, Jian-Yun Zhang, Shi-Jin Ding and Peng-Fei Wang of the Electronic Engineering Department of Fudan University.

The author is grateful to the Swedish Board for Technical Development (Prof. J.-O. Carlsson being in charge of the project for my visit to Sweden), the National Natural Science Foundation of China (NSFC), the Advanced Materials Committee of the High Technology Research and Development Program (863 Program) of China, the Science and Technology Department and the Foreign Affairs Department of the Chinese Education Ministry, and the Fundamental Research Department of the Chinese Ministry of Science and Technology for financial support of projects related to the book.

The author is also grateful to the Springer editors Dr. Claus Ascheron and Dr. Angela Lahee for their discussions and suggestions on the arrangement of the content and the title of the book. The first version of this manuscript was submitted for peer reviewing on July 28th 2000 and the present book is a revised version thereof.

Shanghai
November 2001

Prof. Ji-Tao Wang

Contents

1. Classical Thermodynamics

1.1 Basic Laws of Classical Thermodynamics

The brilliant French engineer Sadi Carnot (1796–1832) published his scientific publication *Reflections on the Motive Force of Fire and on the Machines Fitted to Develop that Power* in 1924. He said: "Wherever there exists a difference of temperature, motive force can be produced." He argued that the reversible cyclic engine must produce the maximum work (motive force), but he used the caloric theory of heat, according to which heat is an indestructible substance. Carnot's theorem might be taken as the beginning of thermodynamics. However, as an academic discipline classical thermodynamics was set up on the basis of two thermodynamic laws during the middle of the 19th century.

In about 1850, the mechanical equivalent of heat was proved by James Prescott Joule (1818–1889), and was later generalized into the law of conservation of energy. That is, the total energy of a system and its surroundings must remain constant, although it may be changed from one form to another. Whenever a quantity of one kind of energy disappears an exactly equivalent amount of other kinds must be produced. In other words, energy cannot be created or destroyed inside a system. That is the first law of thermodynamics.

The first law of thermodynamics has very important practical implication. At that time many researchers tried to produce a perpetual motion machine, which would continuously produce mechanical work without supplying an equivalent amount energy from other sources. That could be called a perpetual motion machine of the first kind. That is, of course, against the first law of thermodynamics, so another statement for the first law of thermodynamics is the impossibility of constructing a perpetual motion machine of the first kind. Up to now, nobody has found any fact contrary to this empirical rule from the daily activities of human beings. That is the most powerful demonstration for the first law of thermodynamics.

Mathematically, the first law can be formulated for a closed system as: the change of internal energy of a system $\mathrm{d}U$ (positive for increase, and negative for decrease; this will be the same for other state functions below) equals the heat absorbed by the system δQ minus the work done by the system δW:

$$\mathrm{d}U = \delta Q - \delta W. \tag{1.1}$$

There is no exchange of heat and work between the system and its surroundings for an isolated system, and no exchange of matter either. Therefore, for an isolated system

$$(\mathrm{d}U)_{\mathrm{iso}} = 0. \tag{1.2}$$

In the other words, the internal energy of an isolated system always keeps constant.

After the impossibility of constructing a perpetual motion machine of the first kind was known, some still hoped to find another kind of machine which could produce mechanical work by supplying an equivalent quantity of heat from a single heat reservoir. If that could be realized, one would be able to get a machine, which could do work from one heat reservoir of ocean. In this way, all the machines in the world might work for hundreds of years with a negligible decrease of the temperature of ocean. It would be a perpetual motion machine of the second kind. This suggestion is not contrary to the first law of thermodynamics, but it has never been realized. From a lot of failures another basic rule had been obtained as the second law of thermodynamics.

A simple statement of the second law of thermodynamics is that it is impossible to construct a perpetual motion machine of the second kind. More specifically, it is impossible to construct a machine functioning in cycles, which can convert heat completely into an equivalent amount of work without producing changes elsewhere. The term "functioning in cycles" is inserted to indicate that the machine must return exactly to its original state at regular stages, so that it can operate continuously. In 1850 Rudolf Clausius (1822–1888) said "heat cannot of itself, without the intervention of any external agency, pass from a colder to a hotter body." Lord Kelvin (William Thomson, 1824–1907) stated in 1851 that "it is impossible to obtain work by cooling a body below the lowest temperature of the system". All of these statements are equivalent for the second law of thermodynamics.

The mathematical formulation of the second law can be written for isolated systems (or adiabatic systems) as

$$(\mathrm{d}S)_{\mathrm{iso}} = 0 \text{ (for reversible processes)} \tag{1.3}$$

$$(\mathrm{d}S)_{\mathrm{iso}} > 0 \text{ (for irreversible processes)} \tag{1.4}$$

or

$$(\mathrm{d}S)_{\mathrm{iso}} \geq 0. \tag{1.5}$$

That is, the entropy of any isolated system will never decrease. Therefore, the second law of thermodynamics is also called the entropy increase principle. Here, the entropy change of a system $\mathrm{d}S$ is defined as the heat absorbed by the system in a reversible process δQ_R divided by the absolute temperature T,

$$\mathrm{d}S = \frac{\delta Q_R}{T}. \tag{1.6}$$

It should be kept in mind that any practical process is irreversible. The heat absorbed by the system in a practical process δQ divided by the temperature T does not equal the entropy change of the system $\mathrm{d}S$. Only in a reversible process can the equality sign be used, that is,

$$\mathrm{d}S \geq \frac{\delta Q}{T}. \tag{1.7}$$

Equation (1.7) is also a formulation of the second law of thermodynamics.

At the beginning of the 20th century H.W. Nernst (1862–1941) and others proposed the third law of thermodynamics, i.e., the impossibility of reaching zero degree in the absolute temperature scale (0 K), that is the impossibility cooling a body to 0 K by any finite number of steps or procedures. Therefore, the absolute temperature T should always be positive. Comparison of (1.7) with (1.6) shows that the heat absorbed by the system in a reversible process δQ_R is always greater than that in any irreversible process δQ, if both processes start at the same initial state and end at the same final state. According to the conservation of energy the maximum work will be done by the system only in reversible processes.

Thermodynamics studies changes of macroscopic properties of a system consisting of a large number of particles (such as of order 6.023×10^{23}). It also studies the conversion relationship among heat, work and other forms of energy, and the relationship between physical quantities and states during the change of the system. In thermodynamics, functions used for describing the states of the system are called functions of states, or state parameters. When the system is in a definite state, the properties of the system are dependent only on the state of the system, and independent of its history. A series of properties of the system will be changed during the change of the state of the system. The amount of the change is dependent only on the initial state and the final state, but is independent of the path taken by the change. State functions can be divided into two classes. One class of state functions, such as temperature and pressure, which are independent of the quantity of material in the system, are called intensive properties. The other kind of state functions, such as volume, mass and internal energy, which are dependent on the quantity of material in the system, are called extensive properties.

As mentioned above, the basic laws of classical thermodynamics are the first, second and third laws of thermodynamics. The mathematical formulations for the first and second laws are usually as follows.

The first law for closed systems is

$$\mathrm{d}U = \delta Q - \delta W$$

and the second law for isolated systems (or adiabatic systems) is

$$(\mathrm{d}S)_{\mathrm{iso}} = 0 \text{ (for reversible processes)}$$

$$(\mathrm{d}S)_{\mathrm{iso}} > 0 \text{ (for irreversible processes)}$$

or

$$(\mathrm{d}S)_{\mathrm{iso}} \geq 0.$$

Here,

$$\mathrm{d}S = \frac{\delta Q_R}{T}.$$

According to the third law of thermodynamics, T will never be negative. In other words, the entropy of a pure crystalline substance may be taken to be zero at the absolute zero of thermodynamic temperature, i.e., 0 K. It may also be shown that the entropy of pure crystalline substances at absolute zero is not a function of pressure, i.e., $\left(\frac{\partial S}{\partial p}\right)_{T\to 0} = 0$. At temperatures above absolute zero, however, the entropy of a substance is a function of the pressure. Because of this the tabulated values of the absolute entropy, relative to a value of zero at a temperature of absolute zero, are usually given at the standard reference pressure of 101.325 kPa. This reference state is usually denoted by the superscript $(^0)$ on S, i.e., S^0. The absolute entropy of a substance can be evaluated by two distinct methods. One of these requires specific heat and latent heat data, while the other is statistical in origin and requires molecular information. The third law is more important for materials under very low temperature, so it will not be discussed in detail in this book.

1.2 Classical Equilibrium Thermodynamics

From the middle of the 19th century to the beginning of the 20th century, based on the first and second laws of thermodynamics a rather huge, rigorous, and systematic equilibrium thermodynamics had been deduced, so equilibrium thermodynamics is often called classical equilibrium thermodynamics. It was very easy to think that classical thermodynamics might be equivalent to equilibrium thermodynamics, or that equilibrium thermodynamics or classical thermodynamics might represent the whole of thermodynamics. However, it should be pointed out that the second law of thermodynamics consists of two parts: equality and inequality. Equality can be used for equilibrium systems or systems with only reversible processes, while inequality can be used for nonequilibrium systems or systems with irreversible processes. Therefore, classical thermodynamics, even at its initial stage, already included the two parts of equilibrium thermodynamics and nonequilibrium thermodynamics. The second law of equilibrium thermodynamics is the equality $\mathrm{d}S = \frac{\delta Q}{T}$, so it is easy to combine this with the equality of the first law such as $\mathrm{d}U = \delta Q - \delta W$ for a closed system, and to deduce a whole systematic classical equilibrium thermodynamics. On the other hand, the second law for nonequilibrium thermodynamics is the inequality $\mathrm{d}S > \frac{\delta Q}{T}$, which is more difficult to use in a quantitative calculation and deduction. In this way, the development of nonequilibrium thermodynamics had been at a low ebb and had not attracted much attention for a long time.

It should be pointed out that the systematic development of classical equilibrium thermodynamics has also provided a firm foundation for nonequilibrium thermodynamics. A lot of concepts, functions and relationships among state functions in nonequilibrium thermodynamics are still based on equilibrium thermodynamics. Therefore, before a discussion of nonequilibrium thermodynamics, some basic aspects of equilibrium thermodynamics and an overview of classical thermodynamics should be discussed first.

For a detailed discussion about an object, a part of the whole world called a system, has to be separated from other connecting parts called its surroundings. If there is no exchange of either matter or energy between the system and its surroundings, then the system is called an isolated system. If there is only an exchange of energy between the system and its surroundings, then the system is called a closed system. If there are exchanges of both matter and energy between the system and its surroundings, then the system is called an open system.

The basic parameters or state functions in thermodynamics are temperature T, pressure p, volume V, internal energy U and entropy S. For convenience of applications under different conditions, the enthalpy H was also introduced. Later, H.L.P. van Helmholtz (1821–1894) and J. Willard Gibbs (1839–1903) introduced two other state functions, i.e., the Helmholtz free energy F, and the Gibbs free energy G. The definitions of these three derivative functions are

$$H = U + pV \tag{1.8}$$

$$F = U - TS \tag{1.9}$$

and

$$G = H - TS. \tag{1.10}$$

To remember these, the relationships among them can be illustrated as in Fig. 1.1.

For closed systems, the formulation of the first law is $\mathrm{d}U = \delta Q - \delta W$. To distinguish between expansion work $p\mathrm{d}V$ and other forms of work $\delta W'$, (1.1) can be rewritten as

$$\mathrm{d}U = \delta Q - p\mathrm{d}V - \delta W'. \tag{1.11}$$

According to the definition of H, this can be changed into

$$\mathrm{d}H = \delta Q + V\mathrm{d}p - \delta W'. \tag{1.12}$$

The physical meanings of U and H are clearly shown in (1.11) and (1.12). In isochoric processes with no work done by the system, the increase of the internal energy $\mathrm{d}U$ equals the heat absorbed by the system δQ during the process.

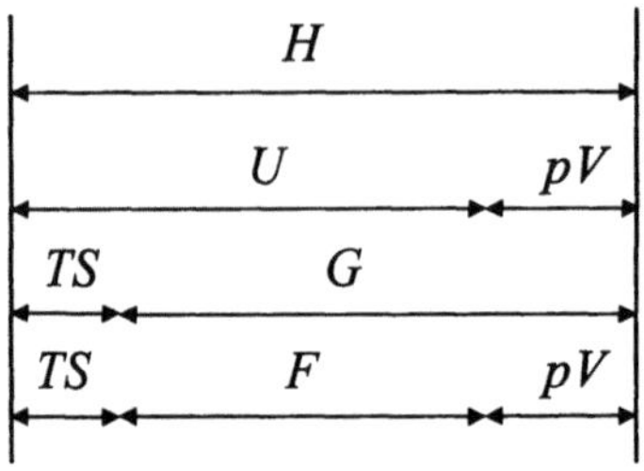

Fig. 1.1. Relations among thermodynamic functions

In isobaric processes with only expansion work, the increase of enthalpy $\mathrm{d}H$ equals the heat absorbed by the system δQ during the process.

For an equilibrium system or a system with only reversible processes, the formulation of the second law is

$$T\mathrm{d}S = \delta Q. \tag{1.13}$$

Therefore, based on the definitions of F and G the combined formulation of the first and second laws can also be written as

$$\mathrm{d}F = -S\mathrm{d}T - p\mathrm{d}V - \delta W' \tag{1.14}$$

and

$$\mathrm{d}G = -S\mathrm{d}T + V\mathrm{d}p - \delta W'. \tag{1.15}$$

The physical meanings of F and G can be found in (1.14) and (1.15), that is, in isothermal and isochoric processes, the decrease of the Helmholtz free energy of the system $-\mathrm{d}F$ equals the work done by the system in a reversible process $\delta W'$, and in isothermal and isobaric processes, the decrease of the Gibbs free energy of the system $-\mathrm{d}G$ equals the nonexpansion work done by the system in the reversible process $\delta W'$.

For closed systems during reversible processes with only expansion work, the set of equations (1.11), (1.12), (1.14) and (1.15) can be simplified to

$$\mathrm{d}U = T\mathrm{d}S - p\mathrm{d}V \tag{1.16}$$

$$\mathrm{d}H = T\mathrm{d}S + V\mathrm{d}p \tag{1.17}$$

$$\mathrm{d}F = -S\mathrm{d}T - p\mathrm{d}V \tag{1.18}$$

and

$$\mathrm{d}G = -S\mathrm{d}T + V\mathrm{d}p. \tag{1.19}$$

It is shown that the dimensions of the functions U, H, F and G are all energy. In fact, all of them are energy factors of criteria for determining the direction of change under different conditions, and energy factors of the heat given up or the maximum nonexpansion work done by the system.

Equations (1.16)–(1.19) are formulations of complete differentials, so a series of useful partial differential equations can be written as

$$T = \left(\frac{\partial U}{\partial S}\right)_V = \left(\frac{\partial H}{\partial S}\right)_p \tag{1.20}$$

$$p = -\left(\frac{\partial U}{\partial V}\right)_S = -\left(\frac{\partial F}{\partial V}\right)_T \tag{1.21}$$

$$V = \left(\frac{\partial H}{\partial p}\right)_S = \left(\frac{\partial G}{\partial p}\right)_T \tag{1.22}$$

and

$$S = -\left(\frac{\partial F}{\partial T}\right)_V = -\left(\frac{\partial G}{\partial T}\right)_p . \tag{1.23}$$

Based on the properties of complete differentials of state functions, we obtain:

$$\left(\frac{\partial T}{\partial V}\right)_S = -\left(\frac{\partial p}{\partial S}\right)_V \tag{1.24}$$

$$\left(\frac{\partial T}{\partial p}\right)_S = \left(\frac{\partial V}{\partial S}\right)_p \tag{1.25}$$

$$\left(\frac{\partial S}{\partial V}\right)_T = \left(\frac{\partial p}{\partial T}\right)_V \tag{1.26}$$

and

$$\left(\frac{\partial S}{\partial p}\right)_T = -\left(\frac{\partial V}{\partial T}\right)_p . \tag{1.27}$$

The above four equations are usually called the Maxwell relationships, and they show the relationship among certain thermodynamic functions. They can be used to get a partial differentials that are difficult to measure in terms of easily measurable partial differentials.

The most difficult thing may be how to find the physical meaning of entropy among all the thermodynamic functions. It seems very difficult to find a physical quantity directly corresponding to the entropy. Roughly speaking, the value of the entropy may be regarded as a measure of the microscopic "randomness" or "state of chaos" in a given system. All spontaneous processes in an isolated system are accompanied by an increase of the entropy, and therefore lead to an increase in the random distribution. For an ideal gas consisting of simple ball-type particles the value of the entropy is quantitatively proportional to the logarithm of the microscopic thermodynamic probability of the system, $S = k \ln \Omega$. k is called the Boltzmann constant, and the formula was first proposed by Ludwig Boltzmann (1844–1906) in 1896. The second law of thermodynamics is the entropy increase principle, that is, the microscopic state spontaneously tends to the most probable distribution state with the maximum thermodynamic probability. The entropy is a macroscopic physical quantity, while the thermodynamic probability is a microscopic variable of the system. Therefore, the Boltzmann formula is an important connection between a macroscopic quantity and a microscopic quantity, and it becomes the foundation of statistical thermodynamics.

The English sentence "Good physicist Has Studied Under Very Fine Teacher" with two arrows has been recommended for remembering the relationships among the state functions of thermodynamics, as shown in Fig. 1.2. Here, the first letter of each word represents a main state function of thermodynamics. Comparison between the figure and the series of equations mentioned above enables one to see how to use it.

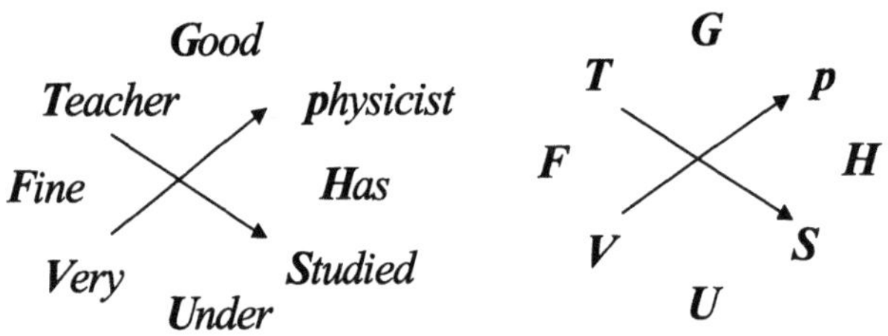

Fig. 1.2. An English sentence for remembering the relationships among thermodynamic functions

For example, the four parameters G, H, U and F with dimensions of energy are located at the top, right, bottom and left, respectively. The conditions corresponding to them are located nearby. G is located between T and p; that means the function G is a criterion for the spontaneous reaction direction at constant temperature and pressure. Similarly, F is located between T and V; that means the function F is a criterion for the spontaneous reaction direction at constant temperature and volume.

1.3 Nonequilibrium Thermodynamics in Classical Thermodynamics

All spontaneous processes are irreversible. From the inequality of the second law of thermodynamics, the entropy in any isolated system with an irreversible process does always increase, i.e., $(\mathrm{d}S)_{\mathrm{iso}} > 0$ (for irreversible processes). For closed systems

$$T\mathrm{d}S > \delta Q \text{ (for irreversible processes)} \tag{1.28}$$

where δQ is the heat absorbed by the system in a practical process, and T is the temperature of surroundings. The corresponding combination formulation of the first and second laws will be

$$dU < TdS - pdV - \delta W' \tag{1.29}$$

$$dH < TdS + Vdp - \delta W' \tag{1.30}$$

$$dF < -SdT - pdV - \delta W' \tag{1.31}$$

and

$$dG < -SdT + Vdp - \delta W'. \tag{1.32}$$

For closed systems with only expansion work, these formulae can be used as criteria for the direction of spontaneous processes in a given system. The suitable conditions for each formula are different. For instance, the spontaneous direction should be $dF < 0$ at constant temperature and volume, while the spontaneous direction should be $dG < 0$ at constant temperature and pressure, that is,

$$(dU)_{S,V} < 0 \tag{1.33}$$

$$(dH)_{S,p} < 0 \tag{1.34}$$

$$(dF)_{T,V} < 0 \tag{1.35}$$

and

$$(dG)_{T,p} < 0. \tag{1.36}$$

Because these equations are all inequalities, they can only be used as criteria for the direction of spontaneous processes. For a long time it was difficult to make further progress. Therefore, classical equilibrium thermodynamics had been regarded as the whole of classical thermodynamics or even as the whole of thermodynamics. That was a problem for the traditional point of view of thermodynamics [1.1, 1.2], which had led to some mistakes or misunderstandings. For example, there had been a misunderstanding that diamond synthesis from graphite at low pressure had been regarded as thermodynamically impossible. In fact, the conclusion was an unsuitable statement coming from the classical equilibrium thermodynamics, which is only suitable for equilibrium systems. The correct statement should be that diamond could not spontaneously form from graphite at low pressure. The word "spontaneously" in the statement should never be omitted. That is to say, in the case of supplying external energy, the synthesis of diamond from graphite at low pressure is not contrary to the fundamental principles of thermodynamics. Activated low-pressure diamond synthesis was achieved by former Soviet Union scientists in about 1970, but the new technique was generally accepted only after the middle of the 1980s. The reason of this delay may have been due to the unsuitable or incomplete conclusions of classical equilibrium thermodynamics and the out-of-date traditional concept of thermodynamics. In fact, there are only three general statements about impossibilities, i.e., the first, second and third laws of thermodynamics. That is, a perpetual motion machine of the first kind, a perpetual motion machine of the second kind and the absolute zero of temperature are all impossible to achieve. All other impossibilities with general thermodynamic meanings in traditional concepts should be abandoned. In other words, all other impossibilities may be valid only under special conditions. If the conditions are changed, they may become possible.

1.4 Criteria of Change Direction and Limitation

The second law of thermodynamics itself is the most generalized criterion of the direction and limitation of reactions or processes. The classical form of the second law is suitable for isolated systems. The equality $(\mathrm{d}S)_{\text{iso}} = 0$ is suitable for reversible processes, while the inequality $(\mathrm{d}S)_{\text{iso}} > 0$ is suitable for irreversible processes. In isolated systems all spontaneous processes must be irreversible; i.e., any spontaneous process in isolated systems tends to increase the entropy of the system, i.e., $(\mathrm{d}S)_{\text{iso}} > 0$. Spontaneous changes result in equilibrium states of the system. After the system reaches its equilibrium state, no macroscopic irreversible change will take place. The entropy will not increase in equilibrium systems, i.e., $(\mathrm{d}S)_{\text{iso}} = 0$. The entropy criterion can be written as $(\mathrm{d}S)_{\text{iso}} \geq 0$. If the system is not an isolated system, the criterion $(\mathrm{d}S)_{\text{iso}} \geq 0$ can still be used, but the entropy change should include both the entropy change of system and the entropy change of its surroundings, i.e., $(\mathrm{d}S_{\text{sys}} + \mathrm{d}S_{\text{surr}})_{\text{iso}} \geq 0$. In practical usage, this is not convenient. For processes at constant temperature and volume (or at constant temperature and pressure), it will be more convenient to introduce the state function F (or G) as the criterion.

At constant temperature and volume without work done, the most convenient way is to take the function F as a criterion for a closed system. In such cases, any spontaneous process must result in a decrease of F, i.e., $(\mathrm{d}F)_{T,V} < 0$. The value of F decreases until equilibrium, i.e., $(\mathrm{d}F)_{T,V} = 0$. Therefore,

$$(\mathrm{d}F)_{T,V} \leq 0. \tag{1.37}$$

In practice, the most general conditions are constant temperature and pressure conditions, so the Gibbs free energy, the function of G, is the most generally adopted criterion for a closed system.

$$(\mathrm{d}G)_{T,p} \leq 0. \tag{1.38}$$

Any spontaneous processes at constant temperature and pressure with only expansion work done must be accompanied with a decrease of the Gibbs free energy, i.e., $(\mathrm{d}G)_{T,p} < 0$. In the other words, the equilibrium state at constant temperature and pressure with only expansion work done must be accompanied by the minimum of the Gibbs free energy of the system. That is the basic principle of the minimization of the Gibbs free energy for the calculation of phase diagrams (CALPHAD) in classical equilibrium thermodynamics. It should be emphasized that all these thermodynamic criteria are only concerned with the direction of spontaneous processes. It only indicates the spontaneous direction of the process, the opposite direction not being spontaneous, but it does not mean that the opposite direction is forbidden in any particular case. As mentioned above, the transformation from graphite to diamond at low pressure is a process with $(\mathrm{d}G)_{T,p} > 0$. This process will not be able to take place spontaneously, but under artificial conditions with the input of external energy the transformation from graphite to diamond at low pressure is still possible to achieve. Thermodynamic criteria indicate only the possibilities. Thermodynamics will not be concerned with the microscopic mechanism, the reaction rates, and so on for the realization of thermodynamic possibilities. Due to the strictness of thermodynamics, up to now no phenomenon contrary to any real thermodynamic conclusion has been found in any field relevant to thermodynamics.

1.5 Relationship Between Gibbs Free Energy and Temperature or Pressure

In production, or in research work, there are a lot of reactions or processes running at constant temperature and pressure. More detailed discussions should be done on the character of the Gibbs free energy, such as the relationship between the function of G and T or p. For a pure substance, 1 mole is usually taken as a unit. The relationship between its G value and temperature can be derived as follows. According to the basic thermodynamic formula, we know

$$\left(\frac{\partial G}{\partial T}\right)_p = -S. \tag{1.39}$$

Substitution in the definition of G, $G = H - TS$, gives

$$\left(\frac{\partial G}{\partial T}\right)_p = \frac{G-H}{T} \tag{1.40}$$

or

$$\frac{1}{T}\left(\frac{\partial G}{\partial T}\right)_p - \frac{G}{T^2} = -\frac{H}{T^2} \tag{1.41}$$

that is,

$$\left[\frac{\partial\left(\frac{G}{T}\right)}{\partial T}\right]_p = -\frac{H}{T^2}. \tag{1.42}$$

After integration,

$$\int \mathrm{d}\left(\frac{G}{T}\right) = \int \left(-\frac{H}{T^2}\right) \mathrm{d}T \tag{1.43}$$

$$\frac{G}{T} = -\int \frac{H}{T^2} \mathrm{d}T + I. \tag{1.44}$$

Here, I is the integration constant.

If the heat capacity at constant pressure, C_p, can be expressed as a function of temperature, such as $C_p = A + BT + CT^{-2} + DT^2$, then

$$H = H_0 + \int C_p \mathrm{d}T = H_0 + AT + \frac{1}{2}BT^2 - CT^{-1} + \frac{1}{3}DT^3. \tag{1.45}$$

Here, H_0 is the integration constant. By substitution of G/T in (1.44) and integration

$$\frac{G}{T} = \frac{H_0}{T} - A \ln T - \frac{1}{2}BT - \frac{1}{2}CT^{-2} - \frac{1}{6}DT^2 + I \tag{1.46}$$

that is,

$$G = H_0 - AT \ln T - \frac{1}{2}BT^2 - \frac{1}{2}CT^{-1} - \frac{1}{6}DT^3 + IT. \tag{1.47}$$

From thermodynamic data books, the relation of C_p–T for a given substance and the standard values of H^0 and G^0 (or S^0) at the standard pressure and 298.15 K can be found. From these values and the formulae mentioned above, I and H_0 can be obtained first, and then in the valid range of the C_p–T relation all values of H^0, G^0 and S^0 at different temperature T and at standard pressure can be calculated. The superscript (^{0}) here represents only the condition of standard pressure, and does not denote the temperature. In this book the enthalpy H^0 of any stable pure substance at 101.325 kPa and 298.15 K is relatively taken as zero.

According to the basic formula $\mathrm{d}G = -S\mathrm{d}T + V\mathrm{d}p$, the relationship between the function G and the pressure for pure substances can be derived from

$$\left(\frac{\partial G}{\partial p}\right)_T = V. \tag{1.48}$$

After integration,

$$G(p_2, T) - G(p_1, T) = \int_{p_1}^{p_2} V \mathrm{d}p. \tag{1.49}$$

If p_1 is the standard pressure, $p_1 = p^0$, then the G value at pressure p will be

$$G(p, T) = G^0(p^0, T) + \int_{p^0}^{p} V \mathrm{d}p. \tag{1.50}$$

If the substance is an ideal gas and the amount equals 1 mole, then

$$G_{\mathrm{m}}(p, T) = G^0_{\mathrm{m}}(p^0, T) + RT \ln \left(\frac{p}{p^0}\right). \tag{1.51}$$

Therefore, if the $G^0(p^0, T)$ value at T and at standard pressure is known, then the $G(p, T)$ value at other pressures can be calculated. The subscript "m" here represents 1 mole.

Suppose that 1 mole of a solid pure substance, e.g., graphite, undergoes a pressure change from an initial pressure of 1 atmosphere to 100 atmospheres. The change of G value can be calculated by the following steps.

$$V_{\mathrm{m}} = \frac{M}{\rho} = \frac{12.011 \text{ g mol}^{-1}}{2.260 \text{ g cm}^{-3}} \tag{1.52}$$

where M is the atomic weight of carbon. It is assumed that the density ρ is a constant in the relevant pressure range. Therefore,

$$\begin{aligned} \Delta G_{\mathrm{m}} &= (12.011/2.260) \times (100 - 1) \times 101.325 \text{ kPa cm}^3 \text{ mol}^{-1} \\ &= 53 \text{ J mol}^{-1}. \end{aligned} \tag{1.53}$$

The G value for 1 mole of a pure substance is usually of the order of kJ mol^{-1}, so such a small value of 53 J mol^{-1} is negligible. Usually, chemical reactions run at about 1 atmosphere, so the influence of a pressure change on the G value of a solid substance is negligible. However, in high-pressure physics the pressure change may be of the order of GPa. In such conditions, the influence of the pressure on the solid should be considered. This will be discussed in Chap. 3.

1.6 Chemical Potential and Its Relationship with Temperature and Pressure

For multicomponent systems, the composition of the system, i.e., the number of moles of each component, may change. That may be due to the influence of matter flow in an open system, or due to the reaction or phase transformation in a closed system. Gibbs and G.N. Lewis (1875–1946) introduced the concept of chemical potential for such systems.

In a multicomponent system its state can be represented by any two independent variables of the thermodynamic functions p, T, V, U, S, ... , and the amount of each component. For instance, for the Gibbs free energy, the G value, T, p, n_1, n_2, ... , n_k can be selected as independent variables, i.e., $G = G(T, p, n_1, n_2, ..., n_k)$. If there are k components in the system, and the number of moles of each component are n_1, n_2, ... , n_k, respectively, then the total differential of G can be written as

$$\begin{aligned} \mathrm{d}G &= \left(\frac{\partial G}{\partial T}\right)_{p,n_j} \mathrm{d}T + \left(\frac{\partial G}{\partial p}\right)_{T,n_j} \mathrm{d}p + \sum_{i=1}^{k} \left(\frac{\partial G}{\partial n_i}\right)_{T,p,n_j} \mathrm{d}n_i \\ &= -S\mathrm{d}T + V\mathrm{d}p + \sum_{i=1}^{k} \left(\frac{\partial G}{\partial n_i}\right)_{T,p,n_j} \mathrm{d}n_i \end{aligned} \tag{1.54}$$

where n_j represents the number of moles of all other components except n_i. The definition for the last term can be given as

$$\mu_i = \left(\frac{\partial G}{\partial n_i}\right)_{T,p,n_j} . \tag{1.55}$$

μ_i is called the chemical potential of component i. When the temperature T, pressure p and all other components n_j are kept constant, the increase of the Gibbs free energy of a huge system due to the addition of 1 mole of substance i is called the chemical potential of substance μ_i. The system should be so huge that concentrations of all components remain unchanged during the addition.

It should be noted that the appropriate independent parameters should be selected for different thermodynamic functions. The following four partial differentials can be called chemical potentials, if the respective subscripts remain constant, that is

$$\begin{aligned} \mu_i &= \left(\frac{\partial U}{\partial n_i}\right)_{S,V,n_j} = \left(\frac{\partial H}{\partial n_i}\right)_{S,p,n_j} \\ &= \left(\frac{\partial F}{\partial n_i}\right)_{T,V,n_j} = \left(\frac{\partial G}{\partial n_i}\right)_{T,p,n_j} . \end{aligned} \tag{1.56}$$

It should also be carefully noted that not all partial differentials could be called as chemical potentials. Therefore, for multicomponent and composition variable systems, the four basic formulae of thermodynamics (1.16)–(1.19) should be rewritten as

$$\mathrm{d}U = T\mathrm{d}S - p\mathrm{d}V + \sum_i \mu_i \mathrm{d}n_i \tag{1.57}$$

$$\mathrm{d}H = T\mathrm{d}S + V\mathrm{d}p + \sum_i \mu_i \mathrm{d}n_i \tag{1.58}$$

$$\mathrm{d}F = -S\mathrm{d}T - p\mathrm{d}V + \sum_i \mu_i \mathrm{d}n_i \tag{1.59}$$

and

$$\mathrm{d}G = -S\mathrm{d}T + V\mathrm{d}p + \sum_i \mu_i \mathrm{d}n_i . \tag{1.60}$$

The most general condition for running reactions or processes is at constant temperature and pressure, so $\mu_i = \left(\frac{\partial G}{\partial n_i}\right)_{T,p,n_j}$ is the most general form being used. If there is no special mention, the chemical potential usually represents the chemical potential expression of the Gibbs free energy. The subscripts T, p and n_j of $\left(\frac{\partial G}{\partial n_i}\right)_{T,p,n_j}$ mean that the reaction or process is running at constant temperature, pressure, and amounts of other components. It should be noted that only $\left(\frac{\partial G}{\partial n_i}\right)_{T,p,n_j}$ can also be called the partial molar differential of the Gibbs free energy, but the other expressions of the chemical potential cannot be called the partial molar differentials, because the other expressions are not suitable at constant temperature and pressure.

It is very convenient to solve problems concerning multiphase equilibrium by the introduction of the concept of chemical potential, for instance, that there are phase α and phase β in a system. If a samll amount of substance i transforms from phase α to phase β then the change of the total Gibbs free energy of the system is

$$\mathrm{d}G = \mathrm{d}G^{\alpha} + \mathrm{d}G^{\beta} = \mu_i^{\alpha}\mathrm{d}n_i^{\alpha} + \mu_i^{\beta}\mathrm{d}n_i^{\beta}. \tag{1.61}$$

Here, $\mathrm{d}n_i^{\alpha} = -\mathrm{d}n_i^{\beta}$. If phase equilibrium has already been reached in the system, then $\mathrm{d}G = 0$ and $\mu_i^{\alpha}\mathrm{d}n_i^{\alpha} = \mu_i^{\beta}\mathrm{d}n_i^{\alpha}$, i.e., $\mu_i^{\alpha} = \mu_i^{\beta}$. This means that the chemical potentials of substance i in the two phases must be equal in equilibrium. The conclusion can be extensively used for any cases of multiphase systems. A similar conclusion can also be used for reaction systems. If a chemical reaction at constant temperature and pressure is

$$a\mathrm{A} + b\mathrm{B} + \ldots \rightarrow f\mathrm{F} + g\mathrm{G} + \ldots . \tag{1.62}$$

It can be shown from (1.60) that

$$\mathrm{d}G = \sum_i \mu_i \mathrm{d}n_i. \tag{1.63}$$

All amounts of individual substances change simultaneously and relatively in a chemical reaction. If the concept of the extent of reaction ξ is introduced, then

$$\mathrm{d}\xi = \frac{\mathrm{d}n_{\mathrm{F}}}{f} = \frac{\mathrm{d}n_{\mathrm{G}}}{g} = \ldots = -\frac{\mathrm{d}n_{\mathrm{A}}}{a} = -\frac{\mathrm{d}n_{\mathrm{B}}}{b} \tag{1.64}$$

$$\mathrm{d}\xi = \frac{\mathrm{d}n_i}{\nu_i} \text{ or } n_i = \nu_i \mathrm{d}\xi. \tag{1.65}$$

Therefore,

$$\mathrm{d}G = \sum_i \nu_i \mu_i \mathrm{d}\xi \tag{1.66}$$

where, ν_i represents the stoichiometric coefficient of component i. Suppose that the system is huge, only 1 mole of the reaction equation has taken place [i.e., a moles of A react with b moles of B (and other reactants) in (1.61)] and concentrations of all components remain constant. Then

$$\Delta G_{\mathrm{m}} = \sum_i \nu_i \mu_i. \tag{1.67}$$

ΔG_{m} represents the change of the Gibbs free energy of the system, after 1 mole of the reaction equation has taken place. In chemical equilibrium, the sum of the chemical potentials of all productions equals the sum of the chemical potentials of all reactants. This can be written as

$$\Delta G_m = \sum_i \nu_i \mu_i = 0. \tag{1.68}$$

The relationship between chemical potential and temperature can be deduced as follows:

$$\begin{aligned}
\left(\frac{\partial \mu_i}{\partial T}\right)_{p,n_i,n_j} &= \left[\frac{\partial}{\partial T}\left(\frac{\partial G}{\partial n_i}\right)_{T,p,n_j}\right]_{p,n_i,n_j} \\
&= \left[\frac{\partial}{\partial n_i}\left(\frac{\partial G}{\partial T}\right)_{p,n_i,n_j}\right]_{T,p,n_j} \\
&= \left[\frac{\partial}{\partial n_i}(-S)\right]_{T,p,n_j} = -\left(S_{\mathrm{m}}\right)_i
\end{aligned} \tag{1.69}$$

where $(S_{\mathrm{m}})_i$ represents the partial differential molar entropy of substance i. By partial differentiation of $G = H - TS$ with respect to n_i at constant temperature and pressure,

$$\left(\frac{\partial G}{\partial n_i}\right)_{T,p,n_j} = \left(\frac{\partial H}{\partial n_i}\right)_{T,p,n_j} - T\left(\frac{\partial S}{\partial n_i}\right)_{T,p,n_j} \tag{1.70}$$

or

$$\mu_i = \left(H_{\mathrm{m}}\right)_i - T\left(S_{\mathrm{m}}\right)_i . \tag{1.71}$$

Note that $\left(\frac{\partial H}{\partial n_i}\right)_{T,p,n_j} = (H_{\mathrm{m}})_i$, $\left(\frac{\partial S}{\partial n_i}\right)_{T,p,n_j} = (S_{\mathrm{m}})_i$ are partial molar quantities at constant temperature and pressure, but not chemical potentials. Only $\left(\frac{\partial G}{\partial n_i}\right)_{T,p,n_j} = \mu_i$ is the chemical potential and partial molar quantity. Correspondingly,

$$\begin{aligned}
\left[\frac{\partial\left(\frac{\mu_i}{T}\right)}{\partial T}\right] &= \frac{T\left(\frac{\partial \mu_i}{\partial T}\right)_{p,n_i,n_j} - \mu_i}{T^2} \\
&= \frac{-\left[T\left(S_{\mathrm{m}}\right)_i + \mu_i\right]}{T^2} = -\frac{\left(H_{\mathrm{m}}\right)_i}{T}.
\end{aligned} \tag{1.72}$$

The relation between chemical potential and pressure is

$$\begin{aligned}
\left(\frac{\partial \mu}{\partial p}\right)_{T,n_i,n_j} &= \left[\frac{\partial}{\partial p}\left(\frac{\partial G}{\partial n_i}\right)_{T,p,n_j}\right]_{T,n_i,n_j} \\
&= \left[\frac{\partial}{\partial n_i}\left(\frac{\partial G}{\partial p}\right)_{T,n_i,n_j}\right]_{T,p,n_j} = \left(\frac{\partial V}{\partial n_i}\right)_{T,p,n_j} = \left(V_{\mathrm{m}}\right)_i .
\end{aligned} \tag{1.73}$$

For an ideal gas,

$$\mu\left(T,p\right)=\mu^{0}\left(T\right)+RT\ln\left(\frac{p}{p^{0}}\right) \tag{1.74}$$

or for substance i in an ideal gaseous mixture,

$$\mu_{i}\left(T,p\right)=\mu_{i}^{0}\left(T\right)+RT\ln\left(\frac{p_{i}}{p^{0}}\right). \tag{1.75}$$

The corresponding pressure of $\mu^{0}\left(T\right)$ or $\mu_{i}^{0}\left(T\right)$ is the standard pressure, so both $\mu^{0}\left(T\right)$ and $\mu_{i}^{0}\left(T\right)$ are only a function of temperature.

To sum up, the starting points for getting the relation between chemical potential and temperature or pressure are

$$\left(\frac{\partial\mu_{i}}{\partial T}\right)_{p,n_{i},n_{j}}=-\left(S_{\mathrm{m}}\right)_{i}$$

and

$$\left(\frac{\partial\mu_{i}}{\partial p}\right)_{T,n_{i},n_{j}}=\left(V_{\mathrm{m}}\right)_{i}$$

respectively.

In comparison with the corresponding equations for the pure substance, it can be found that the thermodynamic equations for both the multicomponent system and the system of a pure substance are similar in form. The distinction between them is that the molar quantities of pure substances have been replaced by partial molar differentials. In other words, for a pure substance its partial molar differentials equal its molar quantities. That is to say, the chemical potential of a pure substance is just the same as its 1 mole Gibbs free energy G_{m}.

1.7 Change of Gibbs Free Energy in Chemical Reactions

In the last two sections the relationship between the chemical potential of a pure substance or one substance in a multicomponent system and temperature or pressure has been discussed. If a chemical reaction takes place inside a system at constant temperature and pressure, then the change of the Gibbs free energy during the reaction should be taken as a thermodynamic criterion for the spontaneous direction of the reaction. Therefore, the change of the Gibbs free energy after one mole amount of the reaction has taken place is an important parameter, the negative value of which was defined by Th. De Donder as the chemical affinity A:

$$A=-\Delta G_{\mathrm{m}}=-\sum_{i}\nu_{i}\mu_{i}. \tag{1.76}$$

The chemical affinity is usually called the driving force of a reaction. If $A>0$, the spontaneous direction is forward to the right side. If $A<0$, the spontaneous direction is backward to the left side. If $A=0$, the reaction may be in equilibrium. Obviously, the relationship ΔG or ΔG_{m} of the reaction with temperature or pressure should be clearly understood. Similarly to the last two sections, from the starting point of $\left(\frac{\partial \Delta G}{\partial T}\right)_p = -\Delta S$ and $\Delta G = \Delta H - T\Delta S$, the relationship of the Gibbs free energy change of reaction with temperature similar to that of a pure substance can be deduced. That is, the following equations can be written:

$$C_p = \Delta A + \Delta BT + \Delta CT^{-2} + \Delta DT^2 \tag{1.77}$$

$$\begin{aligned}\Delta H &= \Delta H_0 + \int \Delta C_p dT \\ &= \Delta H_0 + \Delta AT + \frac{1}{2}\Delta BT^2 - \Delta CT^{-1} + \frac{1}{3}\Delta DT^3\end{aligned} \tag{1.78}$$

and

$$\begin{aligned}\Delta G = \Delta H_0 - \Delta AT\ln T - \frac{1}{2}\Delta BT^2. \\ -\frac{1}{2}\Delta CT^{-1} - \frac{1}{6}\Delta DT^3 + \Delta IT.\end{aligned} \tag{1.79}$$

If the values ΔH, ΔG at 298.15K and the relation C_p–T are known, the two integration constants of ΔH_0 and ΔI can be obtained from the equations listed above, and then all the values of ΔH, ΔG and ΔS (based on $\Delta H = \Delta G - T\Delta S$) at different temperatures in the valid range of C_p–T can be obtained by substituting T. If all reactants and products in the reaction are individually at the standard pressure (101.325 kPa), the symbols ΔH^0 and ΔG^0 will be used instead ΔH and ΔG.

Consider again the chemical reaction (1.62), i.e., $a\mathrm{A} + b\mathrm{B} + \ldots \rightarrow f\mathrm{F} + g\mathrm{G} + \ldots$. Suppose that all reactants and products are ideal gases. By substituting (1.75) in (1.67),

$$\Delta G_{\mathrm{m}} = \sum_i \nu_i \mu_i = \sum_i \nu_i \mu_i^0(T) + \sum_i \nu_i RT \ln\left(\frac{p_i}{p^0}\right). \tag{1.80}$$

If all reactants and products are at standard pressure, ΔG_{m}^0 should be used instead of ΔG_{m}:

$$\Delta G_{\mathrm{m}}^0 = \sum_i \nu_i \mu_i^0(T) \tag{1.81}$$

$$\Delta G_{\mathrm{m}} = \Delta G_{\mathrm{m}}^0 + RT\ln\frac{\left(\frac{p_{\mathrm{F}}}{p^0}\right)^f \left(\frac{p_{\mathrm{G}}}{p^0}\right)^g \cdots}{\left(\frac{p_{\mathrm{A}}}{p^0}\right)^a \left(\frac{p_{\mathrm{B}}}{p^0}\right)^b \cdots}, \tag{1.82}$$

where p_i represents the partial pressure of component i. When the system is in equilibrium, $\Delta G_{\mathrm{m}} = 0$, so

$$\Delta G^0_{\mathrm{m}} = -RT \ln \frac{\left(\frac{p^e_{\mathrm{F}}}{p^0}\right)^f \left(\frac{p^e_{\mathrm{G}}}{p^0}\right)^g \cdots}{\left(\frac{p^e_{\mathrm{A}}}{p^0}\right)^a \left(\frac{p^e_{\mathrm{B}}}{p^0}\right)^b \cdots}. \tag{1.83}$$

K^o_p is defined as a number without dimensions, and is called the thermodynamic equilibrium constant:

$$K^o_p = \frac{\left(\frac{p^e_{\mathrm{F}}}{p^0}\right)^f \left(\frac{p^e_{\mathrm{G}}}{p^0}\right)^g \cdots}{\left(\frac{p^e_{\mathrm{A}}}{p^0}\right)^a \left(\frac{p^e_{\mathrm{B}}}{p^0}\right)^b \cdots} \tag{1.84}$$

so

$$\Delta G^0_m = -RT \ln K^o_p \textbf{ or } \Delta G^0 = -nRT \ln K^o_p. \tag{1.85}$$

Here, p^0 represents the standard pressure (100.325 kPa), and p^e_i represents the equilibrium partial pressure of component i. It is defined so that

$$Q_p = \frac{\left(\frac{p_{\mathrm{F}}}{p^0}\right)^f \left(\frac{p_{\mathrm{G}}}{p^0}\right)^g \cdots}{\left(\frac{p_{\mathrm{A}}}{p^0}\right)^a \left(\frac{p_{\mathrm{B}}}{p^0}\right)^b \cdots}. \tag{1.86}$$

Therefore,

$$\Delta G_{\mathrm{m}} = -RT \ln K^o_p + RT \ln Q_p \tag{1.87}$$

$$\Delta G = -nRT \ln K^o_p + nRT \ln Q_p \tag{1.88}$$

$$\Delta G_{\mathrm{m}} = \Delta G^0_{\mathrm{m}} + RT \ln Q_p \tag{1.89}$$

or

$$\Delta G = \Delta G^0 + nRT \ln Q_p. \tag{1.90}$$

This is the complete relationship between the eqilibrium constant and ΔG of a practical chemical reaction, which was first deduced by J.H. vant Hoff (1886) and called the "reaction isotherm" or the "isotherm of chemical reactions".

It should be noted that

1. The thermodynamic equilibrium constant K^o_p is a number without dimensions, which is different from the eqilibrium constants K_p in many textbooks of chemistry. The ordinary equilibrium constants K_p are usually defined as

$$K_p = \frac{(p^e_{\mathrm{F}})^f (p^e_{\mathrm{G}})^g \cdots}{(p^e_{\mathrm{A}})^a (p^e_{\mathrm{B}})^b \cdots}, \tag{1.91}$$

where K_p is a number with dimensions. The value of K_p changes with the unit used for pressure, while the thermodynamic equilibrium constant K^o_p is independent of the pressure unit.

2. The superscript (^{0}) of the thermodynamic equilibrium constant K_p^o does not represent the standard pressure here, because from (1.80) and (1.84) K_p^o, being only a temperature relative function and independent of pressure can be found.

3. p_A, p_B, ... , p_F, p_G, ... are practical partial pressures of the reactants and products. ΔG^0 represents all pressures of reactants and products being individually at standard pressure.

It should be emphasized that ΔG^0 should never be used as a strict criterion of any practical reaction direction instead of ΔG. The key difference between ΔG^0 and ΔG is their dependence on practical partial pressures (or practical concentrations). For the former the value of ΔG^0 is independent of practical partial pressures and does not connect with the degree of advancement of the reaction, while for the latter the value of ΔG is closely dependent on the practical partial pressures and changes with the degree of advancement of the reaction. When ΔG equals zero, the system may be in equilibrium.

From the equations $\Delta G_m^0 = -RT \ln K_p^o$ and $\left[\frac{\mathrm{d}\left(\frac{\Delta G_m^0}{T}\right)}{\mathrm{d}T}\right] = -\frac{\Delta H_m^0}{T^2}$, it is easy to write

$$\frac{\mathrm{d} \ln K_p^o}{\mathrm{d}T} = \frac{\Delta H_m^0}{RT^2} \tag{1.92}$$

and

$$\ln K_p^o = -\frac{\Delta H_0}{RT} + \frac{\Delta A}{R} \ln T + \frac{\Delta B}{2R} T + \frac{\Delta C}{2R} T^{-2} + \frac{\Delta D}{6R} T^2 + \frac{\Delta I}{R}. \tag{1.93}$$

If ΔH_m^0 and K_p^o at a definite temperature, and the relationship C_p–T in a definite range are known, then from these equations the two integration constants ΔH_0 and ΔI can be obtained, and then the value of K_p^o at different temperature can be calculated.

For a reaction system consisting of ideal gases, in the derivation of the reaction eqilibrium constant via chemical potentials one finds

$$RT \ln K_p^o = \sum_i \nu_i \mu_i^0 (T) = \Delta G_m^0. \tag{1.94}$$

$\mu_i^0 (T)$ is only a function of temperature, so

$$\left[\frac{\partial \ln K_p^o}{\partial p}\right]_T = 0. \tag{1.95}$$

For condensed systems, the starting point for the derivation of the relationship between ΔG_m and pressure is the following equation:

$$\left[\frac{\partial (\Delta G_m)}{\partial p}\right]_T = \Delta V_m. \tag{1.96}$$

After integration from the standard pressure to an arbitrary pressure,

$$\Delta G_{\mathrm{m}} = \Delta G_{\mathrm{m}}^{0} + \int_{p^0}^{p} \Delta V_{\mathrm{m}} \mathrm{d}p. \tag{1.97}$$

Suppose that in condensed systems the densities of the reactants and products does not change with pressure, so

$$\Delta G_{\mathrm{m}} = \Delta G_{\mathrm{m}}^{0} + \Delta V_{\mathrm{m}} \left(p - p^{0}\right). \tag{1.98}$$

Based on this relation, the density data of diamond and graphite, and the value of $\Delta G_{\mathrm{m}}^{0}$, it can be estimated how high a pressure is needed to achieve the transformation from graphite to diamond. This will be discussed later.

1.8 Limitation to the Applicability of Classical Thermodynamics

Albert Einstein in 1949 wrote: "A theory is more impressive the greater the simplicity of its premises is, the more different kinds of things it relates, and the more extended its area of applicability. Therefore the deep impression which classical thermodynamics made upon me. It is the only physical theory of universal content concerning which I am convinced that, within the framework of the applicability of its basic concepts, it will never be overthrown." [1.3] His opinion had often been quoted as an emphasis of the authority and the correctness of classical thermodynamics. However, the last words about "within the framework of the applicability of its basic concepts" did not attracted enough attention of many scientists.

Of course, it was very difficult to find a clear answer for what is the "framework of the applicability" of classical thermodynamics at that time. It is well known that it is no problem to use classical thermodynamics for simpler systems, such as systems with only an adiabatic expansion, isothermal expansion, adiabatic compression or isothermal compression. It is also well known that it is very difficult to use classical thermodynamics for complex systems with multiple simultaneous processes in the life sciences. Maybe some believe that there is no possibility of using thermodynamics in living systems due to the "mystique" of life, but most scientists do not believe that.

After recent research on thermodynamics for activated low-pressure diamond growth from the vapor phase it is clearer that classical thermodynamics should be used only for simple systems. In general, some concepts in classical thermodynamics are not suitable for more complex systems with multiple simultaneous reactions or processes. For instance, $(\Delta G)_{T,p} = 0$, an equality of the second law of classical thermodynamics, has been intuitively taken as a necessary and sufficient condition for the last 150 years for an isothermal and isobaric system being in equilibrium. Such a basic concept of classical thermodynamics is really suitable for simpler systems within the framework of the applicability of its basic concepts, but is not valid for complex systems outside the framework of the applicability of its basic conceptions. There is never a general demonstration for such a basic concept.

In general, one should say that $(\Delta G)_{T,p} = 0$ is a necessary condition but not a sufficient condition for a system being in equilibrium. $(\Delta G)_{T,p} = 0$ means only that the system has no dissipation (or nondissipation) of the Gibbs free energy. The system may be in equilibrium and may be in nonequilibrium. That is one of essential ideas of this book and will be discussed in detail later.

2. Fundamentals of Modern Thermodynamics

In 1931, Lars Onsager (1903–1976) proposed the first general relation in nonequilibrium thermodynamics: the reciprocity relation. The term "classical thermodynamics" also emerged no later than 1931 in the literature. Therefore, 1931 or even earlier (at the beginning of the 20th century) could be regarded as the starting point of modern thermodynamics. In 1959 the term "modern thermodynamics" had already been used in the summer program at MIT [2.1]. Important progress, such as Onsager's reciprocity relation and Ilya Prigogine's (1917–) dissipative structure, should also be regarded as the outstanding achievements of modern thermodynamics. However, modern thermodynamics has still been in its developing stage throughout the 20th century, some basic points of view and even the term "modern thermodynamics" had not been accepted in general.

As pointed in the brief introduction of the book *Modern Thermodynamics — from heat engines to dissipative structures* by D. Kondepudi & I. Prigogine (1998) "Thermodynamics is a core part of science. Most books present thermodynamics very much the way it was presented in the 19th century." [2.2] That is a severe comment on the real situation about thermodynamics. It means that most scientists had received such undergraduate and/or graduate education in classical thermodynamics (even only classical equilibrium thermodynamics) in the last 150 years or so. Maybe it is a double-edged sword, so it also means a severe situation of modern thermodynamics itself that after about 70–100 year development modern thermodynamics itself had not been well accepted. This may be due to some basic problems or discrepancies inside "modern thermodynamics". The situation from both sides seriously had hindered the application of thermodynamics in modern science and technology and the development of modern thermodynamics itself.

In this chapter a new systematization of modern thermodynamics will be proposed by the author on the basis of our ten-year research on thermodynamics for activated low-pressure diamond growth from the vapor phase. On the other hand, this chapter will provide a theoretical basis of the book for nonequilibrium nondissipative thermodynamics and its application to activated low-pressure diamond synthesis from the vapor phase. Detailed nonequilibrium (stationary) phase diagrams theorem and the thermodynamic reaction coupling model for the activated low-pressure diamond process will be discussed in the following chapters.

2.1 Modern Formulation of Basic Thermodynamic Laws

For their contributions to nonequilibrium thermodynamics L. Onsager of Yale University and I. Prigogine of Universite Libre de Bruxelles (ULB) received Nobel prizes for chemistry in 1968 and 1977, respectively. These are important milestones in the history of nonequilibrium thermodynamics and modern thermodynamics.

As mentioned in Chap. 1, for thermodynamic research it is necessary to separate from the whole world a macroscopic part with a well-defined macroscopic geometrical volume, which is called a system. The boundary of the system is a surface or interface, which separates the outside world (the exterior), called the surroundings, from the system. Classical thermodynamics had been concerned mainly with the study of isolated and closed systems. A striking achievement of recent developments has been to withdraw this limitation so as to generalize the methods of thermodynamics to open systems which are of great importance for biological thermodynamics as well as for many other fields such as meteorology, geology and so on. Nonequilibrium thermodynamics is often necessary for studying systems in these fields, in which open systems are usually concerned. It is useful to classify thermodynamic systems according to the exchanges of energy and matter through their boundaries. Systems are usually separated into three classes, i.e., isolated systems which can exchange neither energy nor matter, closed systems which exchange energy but no matter, and open systems which exchange both energy and matter with their surroundings. For convenience of study of different systems, the basic laws of thermodynamics had better be revised into a new form. The revised form of the basic thermodynamic laws can be called the modern formulation of thermodynamic laws, while the original form can be called the classical formulation of thermodynamic laws.

The change of internal energy in the first law of thermodynamics can be split into two parts [2.3, 2.4]. One part is connected with the change from inside which is denoted by the symbol $\mathrm{d}_i U$. Another part is connected with the exchange of energy (for open systems the exchange of matter should be under consideration as well) between a system and its surroundings, which is denoted by the symbol $\mathrm{d}_e U$. Therefore, the total change of internal energy of the system can be written as

$$\mathrm{d}U = \mathrm{d}_e U + \mathrm{d}_i U. \tag{2.1}$$

According to the conservation of energy, the first law of thermodynamics can be rewritten as

$$\mathrm{d}_i U = 0 \tag{2.2}$$

and for closed systems

$$\mathrm{d}U = \mathrm{d}_e U = \delta Q - \delta W. \tag{2.3}$$

For the second law of thermodynamics entropy is the most important state function. Entropy in Greek is $\varepsilon\nu\ \tau\rho\omega\pi\eta$ meaning "evolution". The entropy of the system is an extensive property. If a system consists of several parts, the total entropy equals the sum of the entropies of each part. Similar to the modern formulation of the first law, the entropy change of a system can be split into two parts [2.2–2.7]. One is the entropy production denoted by $\mathrm{d}_i S$ due to the irreversible processes inside the system. Another is the entropy flow denoted by $\mathrm{d}_e S$ due to interactions with the surroundings, such as exchanges of energy and/or matter. Therefore, the total entropy change of the system equals

$$\mathrm{d}S = \mathrm{d}_e S + \mathrm{d}_i S. \tag{2.4}$$

entropy flow of a system $\mathrm{d}_e S$ can be positive or negative, but entropy production of any system $\mathrm{d}_i S$ is never negative; that is,

$$\mathrm{d}_i S \geq 0. \tag{2.5}$$

For the system with only reversible processes inside, $\mathrm{d}_i S = 0$. For the system with at least one irreversible process inside, then maybe $\mathrm{d}_i S > 0$. For an isolated system, $\mathrm{d}_e S = 0$, so that $(\mathrm{d}S)_{\text{iso}} = \mathrm{d}_i S \geq 0$; that is to say, for any isolated system, the modern formulation results in the same conclusion as the classical formulation of the second law. However, the modern formulation of the second law, $\mathrm{d}_i S \geq 0$, can be used not only for isolated systems, but also for closed systems and open systems. Therefore, (2.5) is a more general form of the second law of thermodynamics.

As mentioned above, the modern formulation of the second law, $\mathrm{d}_i S \geq 0$, is suitable for any macroscopic isolated, closed or open system. Moreover, the modern formulation is also suitable for each macroscopic part of any system. Therefore, if an isolated system is split into two macroscopic parts I and II, then $\mathrm{d}_i S^{\mathrm{I}} \geq 0$ and $\mathrm{d}_i S^{\mathrm{II}} \geq 0$. $\mathrm{d}_i S^{\mathrm{I}}$ and $\mathrm{d}_i S^{\mathrm{II}}$ represent the entropy productions of part I and part II, respectively. The physical situation such that

$$\mathrm{d}_i S^{\mathrm{I}} < 0,\ \mathrm{d}_i S^{\mathrm{II}} > 0 \text{ and } \left(\mathrm{d}_i S^{\mathrm{I}} + \mathrm{d}_i S^{\mathrm{II}}\right) \geq 0 \tag{2.6}$$

is excluded. We can, therefore, say that the "absorption" of entropy in one part, compensated by a sufficient "production" in another part of the system is prohibited. This formulation implies that in every macroscopic region of the system the entropy production due to irreversible processes is positive. The term "macroscopic region" refers to any region containing a number of molecules sufficiently large for microscopic fluctuations to be negligible.

Another physical situation must be strictly distinguished: If there are two different processes taking place simultaneously in the same system (or in the same macroscopic region of a system), and the entropy productions due to process 1 and process 2 are denoted by $\mathrm{d}_i S_1$ and $\mathrm{d}_i S_2$ respectively, then it is possible to have:

$$\mathrm{d}_i S_1 < 0, \ \mathrm{d}_i S_2 > 0 \text{ and } (\mathrm{d}_i S_1 + \mathrm{d}_i S_2) \geq 0. \tag{2.7}$$

Because only the sum $\mathrm{d}_i S_1 + \mathrm{d}_i S_2$ represents the total entropy production $\mathrm{d}_i S$ of the macroscopic system, $\mathrm{d}_i S = (\mathrm{d}_i S_1 + \mathrm{d}_i S_2) \geq 0$ completely accords with the modern formulation of the second law of thermodynamics. Such a situation is called thermodynamic coupling between irreversible processes, or just thermodynamic coupling. If process 1 or/and 2 is a chemical reaction, then it can be called nonequilibrium thermodynamic reaction coupling or reaction coupling for short. The introduction of the concept of reaction coupling is one of the most important events in modern thermodynamics, but the concept had been bogged down in controversy for about 70 years and had not been well quantitatively demonstrated before the success of the reaction coupling model for the activated low-pressure diamond process. The details will be discussed later. The term "reaction" may sometimes have extended meanings.

For a clearer understanding of the meanings of the entropy flow $\mathrm{d}_e S$ and the entropy production $\mathrm{d}_i S$, the calculation of the entropy flow in a closed system may be used as a starting point for discussion. The entropy flow in a closed isochoric system can be quantitatively calculated based on the heat absorbed or given up by the system. The calculation must be on a basis of a reversible process; that is.

$$\mathrm{d}_e S = \frac{\delta Q_R}{T}. \tag{2.8}$$

For open systems, the entropy flow should include the contribution due to the exchange of matter as well. That is the distinction between closed and open systems. If no irreversible process takes place inside the system, then $\mathrm{d}_i S = 0$. For systems with irreversible processes taking place the qualitative result $\mathrm{d}_i S$ ¿ 0 can usually be obtained. The detailed calculation of $\mathrm{d}_e S$ and $\mathrm{d}_i S$ will be further discussed later.

2.2 Some Basic Concepts in Nonequilibrium Thermodynamics

As mentioned above, a series of basic terms and concepts of nonequilibrium thermodynamics have already been introduced by the entropy and the entropy increase principle in classical thermodynamics. For instance, the irreversibility of change with time, the equilibrium state and nonequilibrium state, reversible processes and irreversible processes, and so on, are used. We now begin our discussion on these.

2.2.1 Irreversibility of Change with Time

The introduction of the concept of the entropy and the entropy increase principle is a very important event in classical thermodynamics. The increase of the total entropy for isolated systems is irreversible with respect to the change of time. This has been called in some thermodynamics textbooks the connection with "evolution" or the "arrow of time". Such irreversibility with respect to time is rarely found in other field of physics. For instance, in classical mechanics, a system consisting of n simple particles can be described by a phase space consisting of n coordinate parameters and n motion parameters. Inside the phase space the different orbits formed from different starting points will never cross each other. All states on an orbital are equivalent and reversible. In classical mechanics, motion can be treated in a four-dimensional position and time space. The coordinate of time can be positive or negative. However, in thermodynamic systems, the change of time follows a single way going forward. The direction of change for an isolated system is always toward the increase of the entropy, and the opposite direction is forbidden, i.e., irreversible. In any isolated system, which includes spontaneous processes, the entropy of the system will increase, because spontaneous processes are irreversible. When the total entropy of an isolated system comes to its maximum value, it reaches its equilibrium state. Moreover, once an isolated system has reached its equilibrium state, it will never return to any nonequilibrium states even in which the system has stayed. However, it is possible that the same equilibrium state could result from different initial nonequilibrium states.

2.2.2 Equilibrium State, Stationary State and Nonequilibrium State

Intuitively, if a system is under a definite condition with a strained limitation (such as a fixed boundary condition, a concentration limitation, ...), after a certain time the system will reach a stationary state being independent of time. Such a state is called a stationary nonequilibrium state or simply a stationary state. If the system is isolated, i.e., the system has no any limitations from the exterior, the system will develop into a fixed final state without any macroscopic processes. That is a special stationary state — an equilibrium state. Without any constrained limitation, closed systems and open systems may also reach their equilibrium states. Once the equilibrium state has been reached, there will never be any macroscopic spontaneous process.

A macroscopic system in its stationary state will not change with time, but the macroscopic process is still continuously going on. The stationary state is independent of time. That is only due to the total result of internal changes and external exchanges. Therefore, the stationary state must be distinguished from the equilibrium state. The most general states of systems are nonequilibrium states. When a system is in its nonequilibrium state, the inner macroscopic state changes continuously with time. A stationary state is a special state among nonequilibrium states, while an equilibrium state is a special state among stationary states. The stationary state is of special importance in nonequilibrium states. For explanation, a simple practical example is considered. A metallic rod is connected with a high-temperature heat reservoir on one side, while it is connected with a low-temperature heat reservoir on the other. After a certain interval of time, the temperature at each point keeps its constant value independent of time. Therefore, the rod is in a stationary state, but not in an equilibrium state, because the temperature at every point does not equal each other. Meanwhile, there is a constant heat flow through the metallic rod macroscopically. A living body can also be taken as another example. When the living body lives in a certain stage, the system may be in a macroscopically stationary state: eating food continuously is accompanied with exhausting the waste, and the metabolism is taking place continuously. Such a living body can also be taken as a stationary state.

2.2.3 Reversible and Irreversible Processes

Practical macroscopic processes are irreversible. The spontaneous process taking place in an isolated system is irreversible, and results in an increase of the entropy of the system. However, in thermodynamics there is the important concept of a reversible process. In general, a process commencing from an initial equilibrium state is called reversible if at any time during the process both the system and environment can be returned to their initial states. The concept of reversibility by its definition requires restorability for both the system and its surroundings. It is the nature of a reversible process that all heat and work interactions, which occurred across the boundaries during the original (forward) process, are equal in magnitude but reversed in direction during the reverse process. Thus no net history is left in the surroundings when the system regains its initial states. The reversible process is an ideal one, which can only be approached by a quasistatic process, which consists of a series of infinitesimally slow steps, so that the system is always in equilibrium internally and with its surroundings.

It should be noted that a clear understanding of these terms or concepts is very important. Sometimes these terms may be substituted for each other, and regarded as synonyms, but on the other hand these terms may be completely different from each other. For instance, a spontaneous process must be an irreversible process, but an irreversible process does not necessarily mean a spontaneous process. Practical electrolysis with external power is an irreversible process, but it is obviously not a spontaneous process.

2.3 Assumption of Local Equilibrium

The huge achievement of classical thermodynamics is well known. A rather complete academic field of classical thermodynamics has been strictly deduced from only a few laws, principles or assumptions arising from experiences. In particular, a lot of formulations have been quantitatively deduced in classical equilibrium thermodynamics. The conclusions and predictions can be checked by different kinds of experiments or observations. However, classical thermodynamics has met a series of unusual difficulties during the treatment of practical questions concerning nonequilibrium states or irreversible processes. Perhaps the first difficulty was that the formulation of the second law for irreversible processes is usually an inequality. It was difficult to use for quantitative deductions and derivations. Besides, there were more basic difficulties of whether these classical thermodynamic functions and formulations were still valid for systems in nonequilibrium states or for systems with irreversible processes. Perhaps the first problem was how to characterize the research system in a nonequilibrium state.

For example, suppose that there is a system with a unified temperature gradient in it. The left end of the system is connected with a heat reservoir of 400 K, while the right end is connected with a heat reservoir of 300 K. What temperature is at the middle of the system? Maybe one can quickly make an intuitive answer: "It is 350 K." In fact, this is a rather complex question. It should be concerned with the definition of temperature. In thermodynamics, the temperature of a body is defined and measured by the zeroth law and the third law of thermodynamics. The zeroth law requires that heat equilibrium should be reached first between the thermometer and the body for measurement. In a nonuniform heat field the body (consisting of a large number of particles) itself is not yet in equilibrium. What can be said about the temperature? Of course in practice, the intuitive answer, 350 K, may still be acceptable. However, in academic discussion, temperature, pressure, and all state functions or parameters in equilibrium thermodynamics should be defined again for nonequilibrium systems. The simplest way is to accept the assumption of local equilibrium [2.3, 2.4].

The key points of the assumption of local equilibrium are:

1. The system in its nonequilibrium state is split into a lot of units with small volume. From a macroscopic point of view, each volume unit is so small that the property of any point inside the unit can be used without any obvious error instead of the property of the unit. From a microscopic point of view, there are still a lot of particles (such as molecules, atoms, ...) inside each volume unit, so that the statistical requirements can still be satisfied and the statistical properties (temperature, pressure, entropy and so on) can still be represented.

2. From the viewpoint of time, it is supposed that at time t these small volume units are isolated from their surroundings. At time $t + \delta t$ the original nonequilibrium state at time t has already been in equilibrium, so that the system at time $t + \delta t$ can be treated by equilibrium thermodynamics. If the time interval δt is still very small in comparison with the time scale of the macroscopic change of the whole system, the thermodynamic functions or parameters of each small volume unit at time t can be represented by the corresponding thermodynamic functions or parameters after equilibrium has been reached at time $t + \delta t$.

3. In addition, it should be supposed that the relationship among the obtained approximate thermodynamic functions or parameters is the same as the relationship among the corresponding functions or parameters in equilibrium thermodynamics.

After the assumption of local equilibrium has been made, the system in its nonequilibrium state can be described and treated by the method similar to that in classical equilibrium thermodynamics. It is obvious that the assumption of local equilibrium is not deduced from the basic laws of thermodynamics. Its suitable range is of a certain limitation. Overall, in such nonequilibrium systems the gradient of the thermodynamic functions or parameters should not be too big. For instance, for a dilute gas the change of thermodynamic functions or parameters in the range of the mean free path of molecules l should be much smaller than the mean value of the corresponding functions or parameters inside the range. For temperature and pressure, the requirements are

$$\frac{\Delta T}{T} \approx \frac{l\,|\nabla T|}{T} \ll 1 \tag{2.9}$$

and

$$\frac{\Delta p}{p} \approx \frac{l\,|\nabla p|}{p} \ll 1. \tag{2.10}$$

ΔT and Δp represent the changes of temperature and pressure in the range of l, respectively. ∇T and ∇p represent the gradients of temperature and pressure at this place, respectively. Besides the assumption of local equilibrium, the attenuation rate of fluctuation in each volume unit should be much faster than the rate of macroscopic change in the system, so that the motion of a large number of particles can be surely subordinated to statistics laws.

The assumption of local equilibrium is an important basic assumption in nonequilibrium thermodynamics. It is useful in nearly all research to be transplanted from equilibrium systems to nonequilibrium systems. Besides temperature, pressure, and thermodynamic functions and parameters, the concept of the phase of a substance can also be taken as an example. In equilibrium thermodynamics a "phase" can be defined as a homogeneous and physically distinct part of a system which is separated from other parts of the system by definite bounding surfaces. In other words, the complete physically and chemically homogeneous part inside a system is called a "phase", there are obvious boundaries between phase and phase, and on the boundary there are sudden and discontinuous changes in its nature. Obviously, such a definition of phase could not be used directly for nonequilibrium systems. There are usually temperature, pressure and/or concentration gradients in a nonequilibrium system. However, after the acceptance of the assumption of local equilibrium a gaseous system with temperature, pressure or concentration gradients can still be treated approximately as a single phase. Because any local region of a gaseous system with temperature, pressure and/or concentration gradients is treated as part of a continuous single phase, the whole gaseous system can be approximately treated as a single phase system. This approximation coincides with the ordinary concept. Otherwise, it would not be accepted by most people that a gaseous system with only a small temperature gradient were treated as a system with an infinite number of "phases". In this way on the basis of the assumption of local equilibrium, a multiphase system is still distinguishable from a single-phase system in nonequilibrium thermodynamics. Therefore, the existence of nonequilibrium phase diagrams and corresponding phase lines are possible and necessary. Discussions on the multiphase system of stationary nonequilibrium states, the corresponding nonequilibrium phase diagrams, the thermodynamic coupling between irreversible processes in a multiphase nonequilibrium system, and so on, are rarely found in other books concerning the nonequilibrium thermodynamics. In this book discussions on these topics will recur.

2.4 Calculation of Entropy

According to classical thermodynamics, for isolated systems an irreversible process will result in a positive entropy change of the system $[(\mathrm{d}S)_{\mathrm{iso}} > 0]$, but for closed and open systems the entropy change of the system may be positive or negative. There is no general relationship between the irreversible process and the entropy change of the system. However, based on modern thermodynamics, there is a direct relationship between the irreversible process and the entropy production of the system, i.e., any practical irreversible process will result in a positive entropy production $(\mathrm{d}_i S > 0)$ for any isolated system, closed system or open system. Therefore, the first task in modern thermodynamics or in nonequilibrium thermodynamics was how to find an equality or equalities for calculating the entropy production of irreversible processes. Such work has already been done [2.3, 2.4]. For the purposes in this book, some resulting formulations will be listed here. The details on the sources of these formulae can be found in the references.

For nonequilibrium thermodynamic systems or thermodynamic systems with irreversible processes the total change of entropy with time can be written as

$$\frac{\mathrm{d}S}{\mathrm{d}t} = \frac{\mathrm{d}}{\mathrm{d}t}\int_V s\mathrm{d}V = \int_V \frac{\partial s}{\partial t}\mathrm{d}V = \int_V \mathrm{d}V\left(-\nabla\cdot\boldsymbol{J_s}+\sigma\right), \tag{2.11}$$

where, s represents the entropy value per unit volume. $\boldsymbol{J_s}$ represents the entropy flow going out through unit area per unit time, and is called the entropy flux. $\nabla\cdot\boldsymbol{J_s}$ is the divergence of the entropy flux, which represents the entropy flowing out from unit volume per unit time, and is called the velocity of the entropy flow. σ is the entropy production in unit volume per unit time, and called the velocity of the entropy production or the strength of the entropy source. Comparison with $\mathrm{d}S = \mathrm{d}_eS + \mathrm{d}_iS$ gives

$$\frac{\partial s}{\partial t} = -\nabla\cdot\boldsymbol{J_s}+\sigma. \tag{2.12}$$

The first term on the right side is the velocity of the entropy coming from the surroundings and flowing into a definite volume unit. The second term is the velocity of the entropy production due to the irreversible process inside the definite volume unit of the system. According to nonequilibrium thermodynamics the detailed formulations of $\boldsymbol{J_s}$ and σ can be written as

$$\boldsymbol{J_s} = s\boldsymbol{u} + \frac{\boldsymbol{j_q}}{T} - \sum_i \frac{\mu_i}{T}\boldsymbol{j_i} \tag{2.13}$$

and

$$\sigma = \boldsymbol{j_q}\cdot\nabla\left(\frac{1}{T}\right)+\sum_i \boldsymbol{j_i}\cdot\left[-\nabla\frac{\mu_i}{T}+\frac{M_i\boldsymbol{F_i}}{T}\right]-\frac{1}{T}\boldsymbol{\Pi}:\nabla\boldsymbol{u}+\sum_\rho\frac{A_\rho}{T}r_\rho. \tag{2.14}$$

Here,

$$A_\rho = -\sum_i \nu_{i\rho}\mu_i. \tag{2.15}$$

M_i is the molecular weight of component i, $\boldsymbol{F_i}$ is the force on unit amount of component i due to the external field, $\boldsymbol{\Pi}$ is the tensor of the viscosity stress. A_ρ is the affinity of the ρth reaction, and r_ρ is the reaction rate of the ρth reaction.

Each term in both (2.13) for $\boldsymbol{J_s}$ and (2.14) for σ has clear physical meanings. The first term in the formula for $\boldsymbol{J_s}$ is the entropy flux due to convection, the second term is the entropy flux due to conduction, and the third term is the entropy flux due to diffusion. Here, $\boldsymbol{u}$ is the velocity of the mass center of a fluid volume unit; $\boldsymbol{j_q}$ is the heat through unit area per unit time, i.e., the flux of the heat flow; $\boldsymbol{j_i}$ is the flux of the mass flow of component i through unit area per unit time. The first term on the right in the formula for σ is concerned with heat conduction, the second term is concerned with diffusion process (including natural diffusion and diffusion under external field), the third term is concerned with viscous flow, and the fourth term is concerned with the chemical reaction. It can readily be shown in the formula for σ that each term on the right side is always a product of two factors. The factor, that is related to the velocity of irreversible processes can in general be called the thermodynamic flow, denoted by the symbol J_k. Another factor, which is related to the driving force to form the irreversible thermodynamic flow, can in general be called the thermodynamic force driving the irreversible process denoted by X_k. Now the formula for σ can simply be written in its general form:

$$\sigma = \sum_k J_k X_k. \tag{2.16}$$

Table 2.1. Flows $\boldsymbol{J_k}$ and forces $\boldsymbol{X_k}$ of irreversible processes

Irreversible process	J_k	X_k
Diffusion	Mass flow $\boldsymbol{j}_i$	$-\nabla \frac{\mu_i}{T} + \frac{M_i \boldsymbol{F}_i}{T}$
Heat conduction	Heat flow $\boldsymbol{j}_q$	$\nabla \left(\frac{1}{T}\right)$
Viscose flow	Tensor of viscosity stress $\boldsymbol{\Pi}/T$	$-\nabla \boldsymbol{u}$
Chemical reaction	Reaction rate r_ρ	$\frac{A_\rho}{T}$

It should be pointed out that the forces and the flows of the irreversible processes defined and listed in Table 2.1 are not a unique set of the combination between J_k and X_k. There are some arbitrary sets of combination to be selected. However, once a set of expressions of forces or flows (e.g., forces at first) has been decided, then another set of expressions of flows or forces (e.g., flows) in form are also decided in company. Any decisions should accord with the following requirements:

1. The product of the force and the flow in any set should have the dimension of entropy production.

2. For the same system and the same set of irreversible processes, the sum of the products should remain constant with respect to the expression changes of forces and flows, that is to say, if a new set of expressions of forces X'_k are defined, the new set should be a linear combination of the original expressions of forces X_k, and the entropy productions should remain constant for the selection of a new set of expressions of flows J'_k, that is

$$\sum_k J_k X_k = \sum_k J'_k X'_k. \tag{2.17}$$

It may be found from different references that the driving force (or called the affinity) of chemical reactions is sometime written as A, and sometimes A/T for the reasons discussed. It should be noted that the definition of affinity is sometimes written as $A = -\frac{1}{T}\sum_\rho \nu_{i\rho}\mu_i$, and sometime the definition is written as $A = -\sum_\rho \nu_{i\rho}\mu_i$. It is important that the dimension of the product of a force and a flow must be the same as the dimension of entropy production.

The introduction of the entropy production $\mathrm{d}_i S$ or the velocity of the entropy production σ is very important for developing modern thermodynamics. In some respects its role is similar to the role of the entropy function in classical thermodynamics. For a clearer understanding the function Φ can also be adopted. Here, $\Phi = T\sigma$, and Φ has the dimension of a free energy in unit volume per unit time. It can be used to measure the dissipation rate of the free energy due to irreversible processes, so it can be called the dissipation function. Based on the detailed expression of the velocity of the entropy production σ, further quantitative research on entropy production and free energy dissipation is continuing.

If there are two reactions inside the system, the velocity of the entropy production will be

$$\frac{\mathrm{d}_i S}{\mathrm{d}t} = \frac{1}{T}\left(A_1 r_1 + A_2 r_2\right). \tag{2.18}$$

Therefore, the criterion of the reaction coupling or coupled reaction (2.7), i.e., $\mathrm{d}_i S_1 < 0$, $\mathrm{d}_i S_2 > 0$ and $(\mathrm{d}_i S_1 + \mathrm{d}_i S_2) \geq 0$, can be rewritten as

$$A_1 r_1 < 0,\ A_2 r_2 > 0 \text{ and } (A_1 r_1 + A_2 r_2) \geq 0 \tag{2.19}$$

This kind of formulation for the reaction coupling or coupled reaction was proposed by De Donder in 1936 [2.8]. In comparison with (1.76) in Chap. 1, i.e., $A = -\Delta G_{\mathrm{m}}$, the criterion of the reaction coupling at constant temperature and pressure can also be rewritten as

$$(\Delta G_1)_{T,p} > 0,\ (\Delta G_2)_{T,p} < 0 \text{ and } \left[(\Delta G_1)_{T,p} + \chi\,(\Delta G_2)_{T,p}\right] \leq 0 \tag{2.20}$$

where, χ is the rate ratio of two reactions, i.e., $\chi = \frac{r_2}{r_1}$. $(\Delta G_1)_{T,p}$, $(\Delta G_2)_{T,p}$, and $\left[(\Delta G_1)_{T,p} + \chi\,(\Delta G_2)_{T,p}\right]$ are the molar Gibbs free energy changes of reaction 1, reaction 2 and the whole reaction (or system) at constant temperature and pressure, respectively. Similarly, for a multireaction system at constant temperature and constant volume conditions, the molar free energy formulation for the reaction coupling or coupled reaction

$$(\Delta F_1)_{T,V} > 0,\ (\Delta F_2)_{T,V} < 0 \text{ and } \left[(\Delta F_1)_{T,V} + \chi\,(\Delta F_2)_{T,V}\right] \leq 0 \tag{2.21}$$

had already practically been introduced as early as 1931 by Dean Burk, except $\chi = 1$ in his case [2.9]. He maybe first introduced the concept of coupled reaction directly and simply on the basis of the molar free energy change formulation of the second law of thermodynamics.

It is known that for an isothermal isobaric chemical reaction in a closed system $\Delta S = \frac{\Delta H - \Delta G}{T}$. ΔH in an isobaric and isothermal reaction equals the heat absorbed by the system. In comparison with (2.4), $\Delta S = \Delta_e S + \Delta_i S$, $\frac{\Delta H}{T}$ is the term of the entropy flow $\Delta_e S$ from the surroundings, while $-\frac{\Delta G}{T}$ equals the entropy production $\Delta_i S$ due to the irreversible process; that is

$$\Delta_i S = -\frac{\Delta G}{T}. \tag{2.22}$$

This equality is very useful for calculating the entropy production of reactions taking place at constant temperature and pressure.

Of course, (2.20), (2.21) and (2.22) can be rewritten into their differential forms, i.e.,

$$(\mathrm{d}G_1)_{T,p} > 0,\ (\mathrm{d}G_2)_{T,p} < 0 \text{ and } \left[(\mathrm{d}G_1)_{T,p} + (\mathrm{d}G_2)_{T,p}\right] \leq 0 \tag{2.23}$$

$$(\mathrm{d}F_1)_{T,V} > 0,\ (\mathrm{d}F_2)_{T,V} < 0 \text{ and } \left[(\mathrm{d}F_1)_{T,V} + (\mathrm{d}F_2)_{T,V}\right] \leq 0 \tag{2.24}$$

$$\mathrm{d}_i S = -\frac{\mathrm{d}G}{T}. \tag{2.25}$$

Equation (2.7) can also be rewritten into its integrated form, i.e.,

$$\Delta_i S_1 < 0,\ \Delta_i S_2 > 0 \text{ and } (\Delta_i S_1 + \chi \Delta_i S_2) \geq 0. \tag{2.26}$$

All (2.7), (2.19), (2.20), (2.21), (2.23), (2.24) and (2.26) are equivalent as criteria of reaction coupling.

If there are more reactions inside the system, $\mathrm{d}_i S_1$ will represent the sum of the entropy productions of all nonspontaneous reactions, while $\mathrm{d}_i S_2$ will represent the sum of the entropy productions of all spontaneous reactions.

2.5 Systematization of Modern Thermodynamics

Since the setup of the second law of thermodynamics, $(\mathrm{d}S)_{\mathrm{iso}} \geq 0$, it had already been shown that classical thermodynamics consisted of an equality part and an inequality part. The equality was regarded as the criterion of a system in its equilibrium state, while the inequality was regarded as the criterion of a system in its nonequilibrium state. That is, classical thermodynamics was divided into equilibrium thermodynamics and nonequilibrium thermodynamics. The combination of the equality part of the second law $(\mathrm{d}S)_{\mathrm{iso}} = 0$ and the equality of the first law $\mathrm{d}U = \delta Q - \delta W$ produced a huge and rather strict academic field of classical equilibrium thermodynamics. The nonequilibrium thermodynamics of classical thermodynamics was considered to be "drowned" by the achievement of equilibrium thermodynamics. However, the existence of the $(\mathrm{d}S)_{\mathrm{iso}} > 0$ part was never denied. That might be due to a lot of difficulties arising from the inequality form in developing nonequilibrium thermodynamics. Due to the effort over about a hundred years, much progress in nonequilibrium thermodynamics has been achieved. Therefore, nonequilibrium thermodynamics could be subdivided into linear nonequilibrium thermodynamics and nonlinear nonequilibrium thermodynamics. The linear and nonlinear relations between thermodynamic forces and thermodynamic flows of the nonequilibrium system were taken as the criteria of linear nonequilibrium thermodynamics and nonlinear nonequilibrium thermodynamics, respectively. In this way, the whole field of traditional thermodynamics had been divided into three branches, i.e., equilibrium thermodynamics, linear nonequilibrium thermodynamics and nonlinear nonequilibrium thermodynamics [2.3, 2.4].

1. Equilibrium thermodynamics (or thermodynamics for reversible processes). There is no irreversible process in the corresponding systems. That is to say, both thermodynamic force and thermodynamic flow are equal to zero in the system. Classical equilibrium thermodynamics had been well developed, has been playing a very important role in physics, chemistry, engineering science and industrial manufacturing and will continue to play an important role in the future. However, classical thermodynamics is not suitable for some complex systems including living systems, and so is mainly used for simpler isolated and closed systems. For an open system both entropy change of the system, $(\mathrm{d}S)_{\mathrm{syst}}$, and entropy change of its surroundings, $(\mathrm{d}S)_{\mathrm{surr}}$, should be considered, i.e., $(\mathrm{d}S)_{\mathrm{iso}} = \left[(\mathrm{d}S)_{\mathrm{syst}} + (\mathrm{d}S)_{\mathrm{surr}}\right]$. In fact, $(\mathrm{d}S)_{\mathrm{surr}}$ is more difficult to calculate.

2. Linear nonequilibrium thermodynamics (or thermodynamics for linear irreversible processes). The character of nonequilibrium thermodynamics is its study of systems that includes irreversible processes. Due to the irreversible process taking place in the system the thermodynamic force and the thermodynamic flow are not equal to zero. The thermodynamic force is the cause of the thermodynamic flow, so it can be considered that there is a functional relation between them. Supposing that the functional relation is continuous and it is possible to express it as a Taylor function at the reference point of an equilibrium state in which the force and the flow both equal zero, then for a simple process

$$J = J(X) = J_0(X_0) + \left(\frac{\partial J}{\partial X}\right)_0 (X - X_0) + \frac{1}{2}\left(\frac{\partial^2 J}{\partial X^2}\right)(X - X_0)^2 + \dots . \tag{2.27}$$

In the equilibrium state, $X_0 = 0$, $J_0 = 0$, so

$$J = LX + \frac{1}{2}\left(\frac{\partial^2 J}{\partial X^2}\right)_0 X^2 + \dots \tag{2.28}$$

where, $L = \left(\frac{\partial J}{\partial X}\right)_0$. When the deviation from the equilibrium state is so small that the higher-power terms are much smaller and can be neglected, then

$$J = LX. \tag{2.29}$$

From experimental observations, this form of relationship has in fact been obtained. For instance, the magnitude of the heat flow is usually directly proportional to the temperature gradient $\left(\frac{\partial T}{\partial r}\right)$. The magnitude of the diffusion flow is also directly proportional to the concentration gradient. These are Fourier's law and Fick's law for heat conduction and for diffusion, respectively. The proportional coefficients L are called phenomenological coefficients. If L is basically independent of X (i.e., L is approximately a constant with respect to the change of X), then such an irreversible process can be called an irreversible process in its linear regime. The corresponding field for thermodynamic research was called linear nonequilibrium thermodynamics or thermodynamics for linear irreversible processes.

3. Nonlinear nonequilibrium thermodynamics (or thermodynamics for nonlinear irreversible processes). When the thermodynamic force is not too small in the system, the relation formula will not be linear between the thermodynamic force and the thermodynamic flow of the irreversible process. The higher-power terms in the relation formula between the thermodynamic flow J and the thermodynamic force X must be considered. Usually, such a system deviates rather far from its equilibrium state. According to research done by Prigogine et al., some macroscopically ordered phenomena may appear in far-from-equilibrium systems. Prigogine called them self-organization phenomena. Some scientists called them synergetical phenomena. The appearance and maintenance of macroscopic phenomena of ordered self-organization must depend on the input of external energy. Therefore, Prigogine called the special nonequilibrium macroscopic ordered state of the system a dissipative structure.

Figure 2.1 shows the systematization of traditional thermodynamics. The development of nonequilibrium thermodynamics took place mainly in the 20th century. Sometimes such a systematization had also been based on the modern formulation of the second law of thermodynamics, i.e., $d_iS \geq 0$ for any system, as also shown in Fig. 2.1. Because the systemization of thermodynamics did not change too much by the substitution of $[\mathrm{d}_iS \geq 0]$ for $[(\mathrm{d}S)_{\mathrm{iso}} = 0]$, the difference between $[(\mathrm{d}S)_{\mathrm{iso}} = 0]$ and $[\mathrm{d}_iS \geq 0]$ had seemed to be only formal and of no substantial meaning. That may also be an important reason why the modern formulation of the second law was not well accepted and why "Most books present thermodynamics very much the way it was presented in the 19th century." written in the brief introduction of the book *Modern Thermodynamics — from heat engines to dissipative structures* as mentioned before.

After reaction coupling [i.e., $\mathrm{d}_iS_1 < 0$, $\mathrm{d}_iS_2 > 0$ and $\mathrm{d}_iS = (\mathrm{d}_iS_1 + \mathrm{d}_iS_2) \geq 0$] had been well quantitatively verified in activated low-pressure diamond growth from the vapor phase (see the following chapters), the systematization was drastically changed [2.6,2.7,2.10–2.14]. A new systematization of modern thermodynamics is illustrated in Fig. 2.2. All the criteria for the different fields are listed in square brackets [ ].

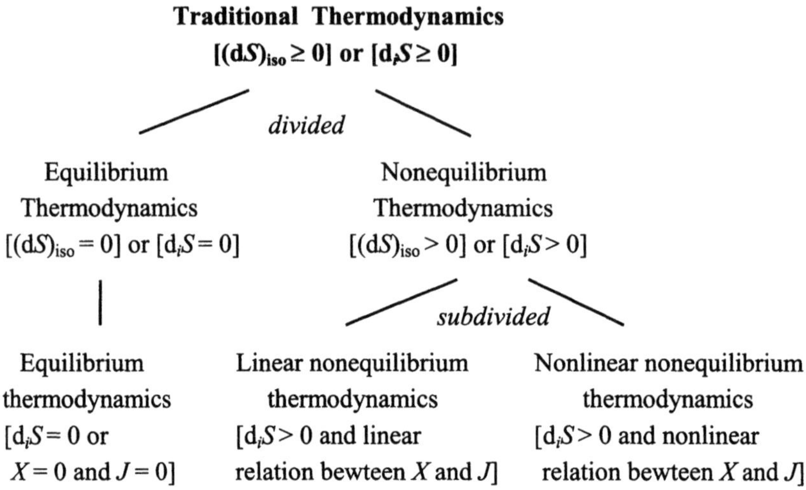

Fig. 2.1. Traditional systematization of thermodynamics on the basis of classical or modern formulation of the second law of thermodynamics. X and J represent the thermodynamic force and thermodynamic flow for an irreversible process, respectively

It is obvious that the equality part of the second law of modern thermodynamics, i.e., nondissipative thermodynamics (or nonentropy-production thermodynamics) in Fig. 2.2, could not be called equilibrium thermodynamics, because it includes both equilibrium and nonequilibrium parts. The inequality part of the second law of modern thermodynamics, i.e., dissipative thermodynamics (or entropy-production thermodynamics) should not be called nonequilibrium thermodynamics either, because it does not include nonequilibrium nondissipative thermodynamics. Although equilibrium nondissipative thermodynamics, linear dissipative thermodynamics and nonlinear dissipative thermodynamics in modern thermodynamics are nearly the same as equilibrium thermodynamics, linear nonequilibrium thermodynamics and nonlinear nonequilibrium thermodynamics in traditional thermodynamics, respectively, the new field (or new branch field) of nonequilibrium nondissipative thermodynamics emerged in modern thermodynamics.

Nonequilibrium nondissipative thermodynamics is only for more complex systems (especially for open systems) including reaction coupling between simultaneous reactions (or processes), but not for simpler systems without reaction coupling. That is the reason why there was no such corresponding field in classical thermodynamics. Maybe activated low-pressure diamond process is the simplest of such complex systems which includes reaction coupling, so it can be quantitatively verified (see the following chapters). The whole book will be mainly devoted to the new field of modern thermodynamics, especially to nonequilibrium phase diagrams for activated low-pressure diamond

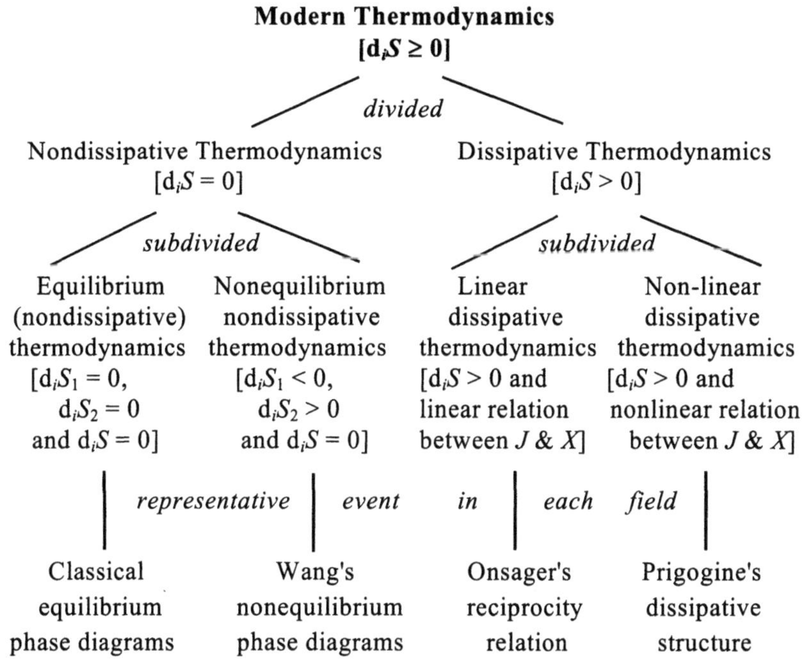

Fig. 2.2. A new systematization of modern thermodynamics on the basis of the modern formulation of the second law of thermodynamics. X and J represent the thermodynamic force and thermodynamic flow for an irreversible process, respectively. d_iS_1, d_iS_2 and $d_iS = (d_iS_1 + d_iS_2)$ are entropy productions of reaction 1, reaction 2 and the whole system, respectively

growth from the vapor phase, which quantitatively agree with a lot of experimental data reported in the literature. In general, nonequilibrium phase diagrams depend on time, and should be displayed by video or computer on the screen. However, for a system in its stationary states phase diagrams are independent of time and can be illustrated on paper. Such phase diagrams are called nonequilibrium stationary phase diagrams. There are no big gaps between nonequilibrium phase diagrams and nonequilibrium stationary phase diagrams in the theoretical treatment, so the term "nonequilibrium (stationary) phase diagrams" is often used to emphasize that all nonequilibrium phase diagrams drawn on paper are nonequilibrium stationary phase diagrams. In particular, the system for activated low-pressure diamond synthesis discussed in this book is in fact often running for several hours or several days in its stationary states.

The situation [$d_iS_1 < 0$, $d_iS_2 > 0$ and $d_iS = (d_iS_1 + d_iS_2) = 0$] is an ideal one and an extreme for nonequilibrium (stationary) phase diagram calculation, which can be called the nonequilibrium nondissipation principle. Any practical process, including activated low-pressure diamond growth, could only occur under the situation [$\mathrm{d}_i S_1 < 0$, $\mathrm{d}_i S_2 > 0$ and $\mathrm{d}_i S = (\mathrm{d}_i S_1 + \mathrm{d}_i S_2)$ a little greater than zero], and something about linear dissipative thermodynamics and even nonlinear dissipative thermodynamics should also be included in the following part of this chapter and later.

2.6 Onsager's Reciprocity Relation

According to experiments, the existence of certain linear phenomenological relationships in irreversible processes had already been observed many years ago. For instance, Fourier's law of heat conduction shows that heat flow in an isotropic medium is directly proportional to the gradient of the temperature (or the gradient of the reciprocal of the temperature), that is

$$\Theta = L\nabla\frac{1}{T} = -\frac{L}{T^2}\left(\frac{\partial T}{\partial r}\right) = -\lambda\left(\frac{\partial T}{\partial r}\right). \tag{2.30}$$

Here, $\lambda = \frac{L}{T^2}$ is called the coefficient of heat conduction, and Θ is the heat flow. As another example, Fick's law of diffusion in binary systems also coincides with this form of linear relationship.

Meanwhile, it should be noted that the discussion of these simple linear relationships is concerned with a system including only one irreversible process. In fact, many experiments show that if there are several irreversible processes in a system, a flow of one irreversible process will depend not only on the corresponding thermodynamic force of the process, but also on the influences of other irreversible processes, that is to say, there are some kinds of coupling between these irreversible processes. For instance, in a binary component gas system, a temperature gradient can lead to not only a heat flow but also a diffusion flow of matter. This is the well-known thermodiffusion phenomenon. Besides, a concentration gradient in a nonuniform system can lead to not only a flow of matter but also a flow of heat. Therefore, a thermodynamic flow J_k is a function of all other thermodynamic forces $\{X_l\}$ in the system:

$$J_k = J_k\left(\{X_l\}\right). \tag{2.31}$$

Similarly, it can be expressed as a function at the reference point of an equilibrium state denoted by subscript zero:

$$J_k\left(\{X_l\}\right) = J_k\left(\{X_{l,0}\}\right) + \sum_l \left(\frac{\partial J_k}{\partial X_l}\right)_0 X_l + \frac{1}{2}\sum_{l,m}\left(\frac{\partial^2 J_k}{\partial X_l \partial X_m}\right)_0 X_l X_m + \dots . \tag{2.32}$$

If the system is close to its equilibrium state, in which the thermodynamic forces of all these irreversible processes (X_l, X_k, ...) are so small that the high power terms are negligible, only the linear term need be considered, and $J_k(\{X_{l,0}\}) = 0$; then a linear phenomenological relation can be obtained as

$$J_k = \sum_l L_{kl} X_l \tag{2.33}$$

where, $L_{kl} = \left(\frac{\partial J_k}{\partial X_l}\right)_0$. The phenomenological coefficients L_{kk} represent the relation between the thermodynamic forces X_k and the corresponding thermodynamic flow J_k, while the phenomenological coefficients L_{kl} ($k \neq l$) represent the coupling relationship between different irreversible processes. The phenomenological coefficients L_{kl} might be concerned with pressure, temperature, concentrations, ... but not related to the change rate of these parameters.

As mentioned above, the linear phenomenological relationship is observed in research labs, but is not deduced from the basic laws of thermodynamics. Even by a backward step, it might be introduced as an assumption: if once the phenomenological laws are accepted some properties of the phenomenological coefficients L_{kl} can be obtained by thermodynamic methods from other natural principles.

If a system including two simultaneous irreversible processes is under consideration, then

$$J_1 = L_{11}X_1 + L_{12}X_2 \tag{2.34}$$

$$J_2 = L_{21}X_1 + L_{22}X_2. \tag{2.35}$$

The entropy production of the system will be

$$\begin{aligned}\sigma &= \sum_k J_k X_k = J_1 X_1 + J_2 X_2 \\ &= L_{11}X_1^2 + (L_{12} + L_{21})\, X_1 X_2 + L_{22} X_2^2.\end{aligned} \tag{2.36}$$

According to the second law of thermodynamics, $\sigma \geq 0$ must be satisfied, and that is independent of the selected values of X_1 and X_2. Apart from $X_1 = X_2 = 0$ and $\sigma = 0$, the selected values of X_1 and X_2 must satisfy the following sufficient and necessary conditions required in algebra:

$$L_{11} > 0,\ L_{22} > 0 \text{ and } (L_{12} + L_{21})^2 < 4L_{11}L_{22} \tag{2.37}$$

that is to say, these phenomenological coefficients are not arbitrary, but subordinate to certain limitations and principles.

Besides the second law of thermodynamics, these phenomenological coefficients are also limited by the Curie's space symmetry principle. The Curie's space symmetry principle shows that the symmetric elements of a macroscopic cause are always less than that of the produced effect in an isotropic medium. Prigogine introduced it into the treatment of irreversible processes, and concluded that the symmetric elements of a thermodynamic force must not be more than that of the thermodynamic flow which is coupled with it. That is called Currie–Prigogine's principle. For instance, a chemical reaction will not be able to drive a flow of matter or a flow of heat. It should be noted that the premise of the conclusion is for reactions in an isotropic system. For reactions taking place on a film or on a surface the limitation will not be valid. Several active diffusions (i.e., diffusions contrary to the concentration gradient) taking place in biological membranes are examples without such a limitations.

In addition, the phenomenological coefficients also subordinate the limitation of the principle of detailed balance, so that the Onsager's reciprocity relations could be deduced. That is, there is a symmetrical character between the linear phenomenological coefficients. Its mathematical formulation is

$$L_{kk'} = L_{k'k}. \tag{2.38}$$

Its written statement is that if the flow of the kth irreversible process J_k is influenced by the force of the k'th irreversible process $X_{k'}$, then the flow of the k'th irreversible process $J_{k'}$ must also simultaneously be influenced by the force of the kth irreversible process, that is, the two coupling coefficients characterizing the influences on each other between these two irreversible processes must be equal.

A chemical reaction will be discussed as an example as follows [2.4, 2.15]. If there is a reaction cycle in a reaction system shown in Fig. 2.3, then the formulations of the three reaction rates and affinities are

$$r_1 = k_1(\mathrm{A}) - k_{-1}(\mathrm{B}), \; A_1 = \mu_\mathrm{A} - \mu_\mathrm{B} \tag{2.39}$$

$$r_2 = k_2(\mathrm{B}) - k_{-2}(\mathrm{C}), \; A_2 = \mu_\mathrm{B} - \mu_\mathrm{C} \tag{2.40}$$

and

$$r_3 = k_3(\mathrm{C}) - k_{-3}(\mathrm{A}), \; A_3 = \mu_\mathrm{C} - \mu_\mathrm{A}. \tag{2.41}$$

(A), (B) and (C) represent the concentrations of substances A, B and C, respectively. Obviously only two of the formulations are independent among the three expressions for the affinities listed above, that is

$$A_3 = -(A_1 + A_2). \tag{2.42}$$

Therefore, the entropy production of the whole system can be written as

$$\sigma = \frac{1}{T}\sum_{k=1}^{3} A_k r_k = \frac{1}{T}\left[A_1\left(r_1 - r_3\right) + A_2\left(r_2 - r_3\right)\right]. \tag{2.43}$$

If A_1/T and A_2/T in the expression are taken as two independent thermodynamic forces, X_1 and X_2, respectively, then two corresponding thermodynamic flows are J_1 and J_2, respectively:

$$X_1 = A_1/T, \; J_1 = r_1 - r_3 \tag{2.44}$$

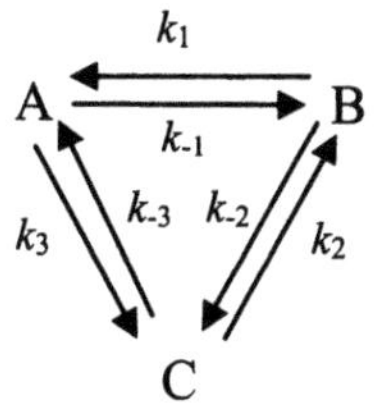

Fig. 2.3. A cycle reaction system

and

$$X_2 = A_2/T,\ J_2 = r_2 - r_3. \tag{2.45}$$

When the system is in equilibrium,

$$\mu_{A,0} = \mu_{B,0} = \mu_{C,0}. \tag{2.46}$$

Therefore,

$$A_{1,0} = A_{2,0} = A_{3,0} = 0 \tag{2.47}$$

$$r_{1,0} = r_{2,0} = r_{3,0} = 0. \tag{2.48}$$

This is the principle of detailed balance, which is also the result of the microscopic reversibility principle in statistical mechanics. According to the microscopic reversibility principle any forward and backward mean rates in a microscopic process during equilibrium must equal each other, that is,

$$k_1(\mathrm{A})_0 = k_{-1}(\mathrm{B})_0,\ k_2(\mathrm{B})_0 = k_{-2}(\mathrm{C})_0 \text{ and } k_3(\mathrm{C})_0 = k_{-3}(\mathrm{A})_0 \tag{2.49}$$

where, $(\mathrm{A})_0$, $(\mathrm{B})_0$ and $(\mathrm{C})_0$ represent the concentrations of the three components in equilibrium, respectively.

If now the system deviates a little from its equilibrium state, suppose that the concentrations of the three components are

$$(\mathrm{A}) = (\mathrm{A})_0 + x,\ (\mathrm{B}) = (\mathrm{B})_0 + y \text{ and } (\mathrm{C}) = (\mathrm{C})_0 + z \tag{2.50}$$

respectively, and

$$\frac{|x|}{(\mathrm{A})_0} \ll 1,\ \frac{|y|}{(\mathrm{B})_0} \ll 1,\ \frac{|z|}{(\mathrm{C})_0} \ll 1. \tag{2.51}$$

Then

$$r_1 = k_1 x - k_{-1} y,\ r_2 = k_2 y - k_{-2} z,\ r_3 = k_3 z - k_{-3} x \tag{2.52}$$

and

$$\begin{aligned} A_1 = \mu_A - \mu_B &= A_{1,0} + RT \ln\left[1 + \frac{x}{(\mathrm{A})_0}\right] - RT \ln\left[1 + \frac{y}{(\mathrm{B})_0}\right] \\ &\approx RT\left[\frac{x}{(\mathrm{A})_0} - \frac{y}{(\mathrm{B})_0}\right] = \frac{RT}{k_1(\mathrm{A})_0}(k_1 x - k_{-1} y). \end{aligned} \tag{2.53}$$

Therefore,

$$r_1 = \frac{k_1 \left(\mathrm{A}\right)_0}{RT} A_1. \tag{2.54}$$

By a similar derivation it can be found that

$$r_2 = \frac{k_2 \left(\mathrm{B}\right)_0}{RT} A_2 \tag{2.55}$$

and

$$r_3 = \frac{k_3 \left(\mathrm{C}\right)_0}{RT} A_3 = -\frac{k_3 \left(\mathrm{C}\right)_0}{RT} \left(A_1 + A_2\right). \tag{2.56}$$

Therefore, the phenomenological relation is

$$\begin{aligned} J_1 = r_1 - r_3 &= \frac{k_1 \left(\mathrm{A}\right)_0 + k_3 \left(\mathrm{C}\right)_0}{R} \cdot \frac{A_1}{T} + \frac{k_3 \left(\mathrm{C}\right)_0}{R} \cdot \frac{A_2}{T} \\ &= L_{11} X_1 + L_{12} X_2 \end{aligned} \tag{2.57}$$

and

$$\begin{aligned} J_2 = r_2 - r_3 &= \frac{k_3 \left(\mathrm{C}\right)_0}{R} \cdot \frac{A_1}{T} + \frac{k_2 \left(\mathrm{B}\right)_0 + k_3 \left(\mathrm{C}\right)_0}{R} \cdot \frac{A_2}{T} \\ &= L_{21} X_1 + L_{22} X_2 \end{aligned} \tag{2.58}$$

where the phenomenological coefficients are

$$L_{11} = \frac{k_1 \left(\Lambda\right)_0 + k_3 \left(\mathrm{C}\right)_0}{R} \tag{2.59}$$

$$L_{12} = \frac{k_3 \left(\mathrm{C}\right)_0}{R} = L_{21} \tag{2.60}$$

and

$$L_{22} = \frac{k_2 \left(\mathrm{B}\right)_0 + k_3 \left(\mathrm{C}\right)_0}{R}. \tag{2.61}$$

In this way, it has been confirmed that the Onsager's reciprocity relation is valid in the system. During the derivation of the relation the principle of detailed balance, i.e., (2.49), and the conditions of a slight deviation of the system from equilibrium, i.e., (2.51), have been adopted. Therefore, that is also a suitable range for the Onsager's reciprocity relations. The relation was first deduced by Onsager in 1931. This is an important foundation for linear dissipative thermodynamics.

However, if the discussed reaction cycle system, as shown in Fig. 2.3, were rewritten as in Fig. 2.4, that would be a single way reaction cycle. If such a system were in equilibrium, there would be nothing violating the thermodynamic laws, but such a reaction does violate the microscopic reversibility principle. Therefore, it does not exist. Onsager had carefully pointed out that the microscopic reversibility principle was a principle independent of thermodynamics. That is, when any macroscopic system reaches its equilibrium state, the forward reaction rate of each elemental reaction must equal its backward reaction rate.

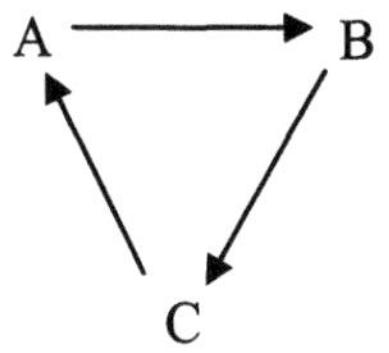

Fig. 2.4. A single way cycle reaction system

2.7 Entropy Production Minimization Principle

Another important foundation of linear dissipative thermodynamics is the entropy production minimization principle proposed by Prigogine in 1945, that is, in a close to equilibrium system the stationary state with respect to external constrained conditions is a state with a minimum of entropy production [2.3, 2.4, 2.16].

For any system without external constraint (such as any isolated system), free development will result in an equilibrium state. If the state function of the entropy is taken as a criterion, the entropy of the system will reach its maximum value. For an isothermal and isobaric change of a closed system, it is more convenient to use the Gibbs free energy as a criterion. The system has minimum Gibbs free energy during equilibrium. The Gibbs free energy minimization principle is the most useful under general constant temperature and pressure conditions. In equilibrium states no macroscopic process and no irreversible process takes place, so the entropy production of the equilibrium system equals zero.

Under an external constrained condition, the system cannot reach an equilibrium state but a stationary state independent of time. For instance, if both sides of a binary gaseous system are connected with heat reservoirs at different temperatures, respectively, there will be a definite temperature gradient in the system. At the same time a concentration gradient will be set up in the system. This is the well-known thermodiffusion phenomenon. In the final state of the system there are a steady heat flow and a fixed concentration gradient, but the mass flow of diffusion will be zero. In the stationary state system the entropy production must have a minimum value. For example, a simple demonstration can be deduced as follows.

According to the general formulation the entropy production can be written as

$$\sigma = J_1 X_1 + J_2 X_2. \tag{2.62}$$

The subscripts 1 and 2 represent heat conduction and diffusion processes, respectively. It is assumed that the heat conduction and the diffusion are satisfied by the linear relation, and that the Onsager's reciprocity relation is valid, so that

$$J_1 = L_{11}X_1 + L_{12}X_2 \tag{2.63}$$

$$J_2 = L_{21}X_1 + L_{22}X_2 \tag{2.64}$$

$L_{12} = L_{21}$,

$$\sigma = L_{11}X_1^2 + 2L_{21}X_1X_2 + L_{22}X_2^2. \tag{2.65}$$

During the free development of the diffusion force under the constrained condition of constant heat conduction force X_1, the change of the entropy production can be obtained by the partial differential of σ with respect to X_2:

$$\frac{\partial \sigma}{\partial X_2} = 2\left(L_{21}X_1 + L_{22}X_2\right) = 2J_2. \tag{2.66}$$

When a stationary state is reached, the diffusion flow equals zero, i.e., $J_2 = 0$, so

$$\frac{\partial \sigma}{\partial X_2} = 0 \tag{2.67}$$

that is, for a stationary state the entropy production should be an extreme value with respect to the change of the diffusion force, and $\frac{\partial^2 \sigma}{\partial X_2^2} = 2L_{22} > 0$, so the extreme value must be a minimum. The general demonstration of this principle on the basis of the variational method will not be further discussed here. It should be noted that during the demonstration of the entropy production minimization principle not only the Onsager's reciprocity relation but also the assumption of constant phenomenological coefficients has been used, so that the domain of validity of the Prigogine entropy production minimization principle is the close to equilibrium system of stationary states.

The entropy production minimization principle does not mean that the entropy production equals zero. The entropy production equals zero only for some special cases, such as in the equilibrium state. It shows that the general tendency of development is the minimization of the entropy production in the linear region of nonequilibrium states. Because the equilibrium state has been prevented from being reached, the system has to choose an entropy production minimization state, i.e., a stationary state. An equilibrium state at constant temperature and pressure corresponds to a Gibbs free energy minimization state, and there is no dissipation of the Gibbs free energy. However, in general for a system in its stationary state there is a dissipation of the Gibbs free energy, but it is a dissipation minimization state under a definite external constrained condition. Therefore, it can be found that the change from an equilibrium state into a stationary state is continuous. The equilibrium state (or called equilibrium nondissipative state) together with nonequilibrium nondissipative states are special stationary states with zero entropy production or nondissipation of the Gibbs free energy. The nonequilibrium nondissipation principle, i.e., $[\mathrm{d}_iS_1 < 0,\ \mathrm{d}_iS_2 > 0$ and $\mathrm{d}_iS = (\mathrm{d}_iS_1 + \mathrm{d}_iS_2) = 0]$, is the important theoretical basis for calculation of nonequilibrium phase diagrams. Further discussions of that will be given later.

2.8 Prigogine's Dissipative Structures

Traditional nonlinear nonequilibrium thermodynamics should actually be called nonlinear dissipative thermodynamics in modern thermodynamics. In this regime, systems are far from equilibrium, and there are no general extremum principles that predict the state to which the system will evolve. The lack of extremum principles that uniquely predict the state to which a far-from-equilibrium system will evolve is a fundamental aspect of nonlinear dissipative systems. In contrast to near-equilibrium systems, which evolve to a state that minimizes a free energy, nonlinear dissipative systems can evolve unpredictably, and their states may not always be unique. That is because, for a given set of nonequilibrium conditions, it is often possible to have more than one state. Which one of these states a particular system will evolve to is, in general, not predictable. The new states thus attained may often be "ordered states" that possess spatiotemporal organization. Patterns in fluid flow, inhomogeneities in concentrations exhibiting geometrical patterns with great symmetry, or periodic variations of concentrations are examples of such ordered states. This phenomenon is usually called "self-organization", and since it is a result of fluctuation is referred to as "order through fluctuations".

In nonlinear dissipative systems, oscillating concentrations and geometrical concentration patterns can be a result of chemical reaction and diffusion, the same dissipative processes that, in a closed or isolated system, wipe out inhomogeneities and drive the system to a stationary, timeless homogeneous state of equilibrium. Since the creation and maintenance of organized nonequilibrium states are due to dissipative processes, Prigogine calls these "dissipative structures" [2.3, 2.4].

Some examples of "self-organization" or "dissipative structures" are given in the following sections.

2.8.1 Bénard's Pattern

In 1900 the famous experiment "Bénard's Pattern" experiment was performed by Bénard and reported in his doctoral thesis [2.4, 2.17]. In this experiment, an adverse temperature gradient is maintained in a thin horizontal layer of fluid by heating the underside. Due to thermal expansion, the fluid at the bottom will be lighter than the fluid at the top. This top-heavy arrangement is potentially unstable. There will be a tendency on the part of the fluid to redistribute itself and overcome the unstability in its arrangement. However, this natural tendency is inhibited by its own viscosity. When the adverse temperature gradient exceeds a certain threshold value, the fluid starts to organize itself to show some kind of macroscopic regular patterns depending on various boundary conditions. For instance, if the upper surface of fluid is (a) free in the air, (b) covered with a solid plate in a round container or (c) covered with a solid plate but in a square container, then the corresponding Bénard patterns are a regular hexagonal cellular pattern (Fig. 2.5a), a concentric circle pattern (Fig. 2.5b), or a parallel line pattern (Fig. 2.5c), respectively. The phenomenon can also be called "self-organization". Figure 2.6 shows a cross section of the fluid at the first moment of such a "self-organization". The exact time and the exact place for starting self-organization may be ambiguous due to fluctuation. However, once the first cell —"Bénard's roll" forms, the dissection of the layer into cells follows. This is the key factor in forming a dissipative structure, and is called the positive feedback effect (or self-catalysis effect in chemical reactions). There are also motions inside the cells: of ascension at the center, and of descent at the boundaries with the adjoining cells. The detailed hydrodynamic equations can be written out, but they involve mainly hydrodynamics or kinetics and will not be discussed in detail here.

2.8.2 Laser Emission

The laser (light amplification by stimulated emission of radiation) appeared in the 1960s. It is also a typical example of a macroscopically ordered dissipative structure [2.21]. The amplification effect (i.e., positive feedback effect) of stimulated emission is the key factor to produce lasing. Here, the light intensity I will be taken as a parameter for discussion. Under normal conditions, atoms emit and absorb photons at random, so $I \approx 0$. During stimulated emission $I \neq 0$, and some factors related to $\frac{\mathrm{d}I}{\mathrm{d}t}$ should be considered.

1. The light intensity of stimulated emission is directly proportional to the light intensity of incidence. That is, the increase of light intensity due to stimulation emission will be directly proportional to

$$DI$$

D is called a gain coefficient. Obviously, D is related with the atom population number at the excited level N_i. Increasing N_i will cause D to increase as well.

2. Atom stimulation needs absorption of photons, and results in a decrease of the light intensity I and a negative value of $\frac{\mathrm{d}I}{\mathrm{d}t}$. That is, the increase of light intensity due to atom stimulation will be also directly proportional to

$$-A_c I$$

A_c is called a wear coefficient (and includes other effects that decrease the light intensity). Combining these two factors, the equation for the change in light intensity $\frac{\mathrm{d}I}{\mathrm{d}t}$ can be written as

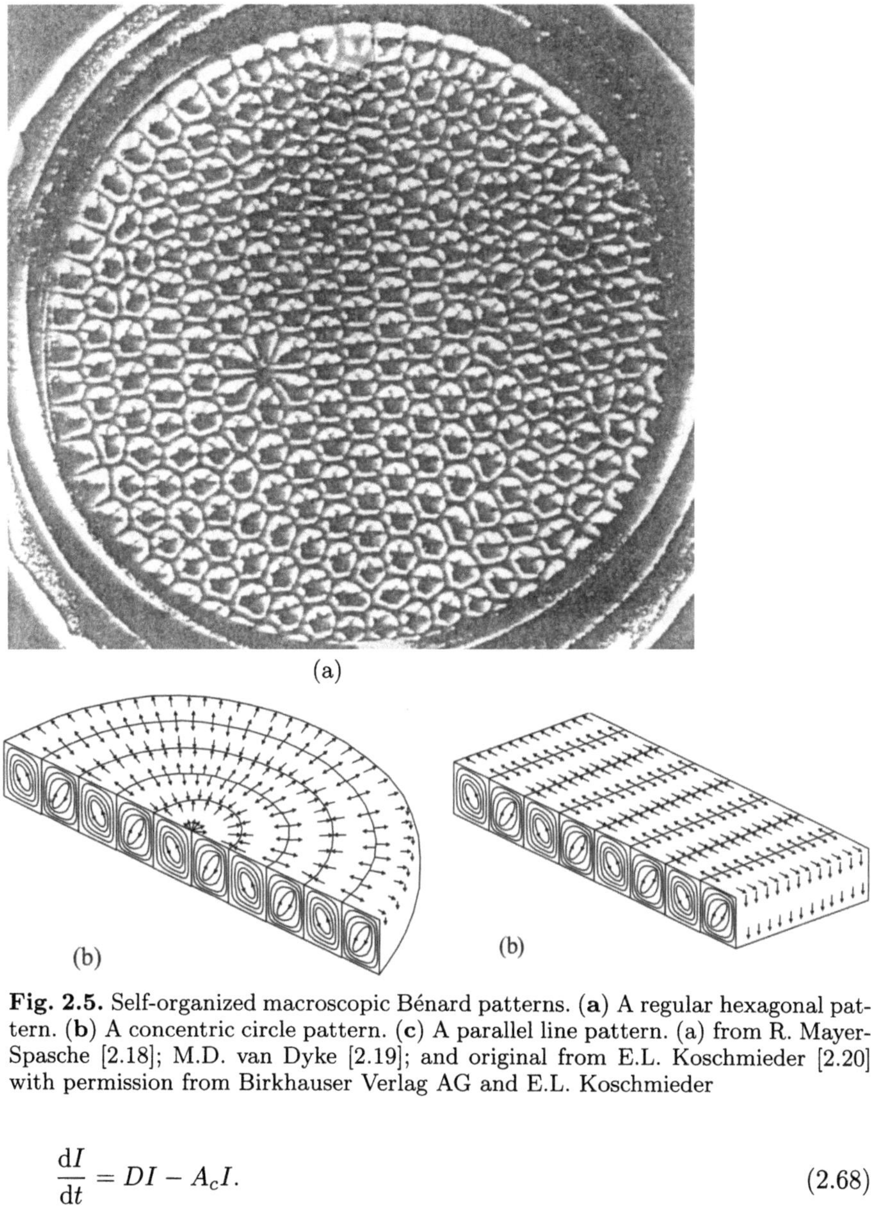

Fig. 2.5. Self-organized macroscopic Bénard patterns. (**a**) A regular hexagonal pattern. (**b**) A concentric circle pattern. (**c**) A parallel line pattern. (a) from R. Mayer-Spasche [2.18]; M.D. van Dyke [2.19]; and original from E.L. Koschmieder [2.20] with permission from Birkhauser Verlag AG and E.L. Koschmieder

$$\frac{\mathrm{d}I}{\mathrm{d}t} = DI - A_c I. \tag{2.68}$$

This is a linear equation. If $D > A_c$, the light intensity would increase infinitely. This is physically impossible to realize. It is prevented by the limitation of the atom population number at the excited level. It can be called the saturation effect, and is nonlinear. Suppose that the nonlinear effect can be represented by the simple nonlinear term

$$-\beta I^2$$

Fig. 2.6. An illustration of a cross section of the fluid at the first instant of such a self-organized "Bénard's roll". After X.-F. Shen [2.21]

β is the called saturation coefficient. Now (2.68) should be rewritten as

$$\frac{\mathrm{d}I}{\mathrm{d}t} = (D - A_c) - \beta I^2. \tag{2.69}$$

This is the frequently used light intensity equation for the laser.

For (2.69) there are two stationary solutions:

$$I_1 = 0, \text{ and } I_2 = (D - A_c)/\beta. \tag{2.70}$$

Let the whole system be in equilibrium or near-equilibrium. When the atom population number at the excited level is much less than that at the ground level, $D < A_c$. In such a case, I_2 would be less then zero. This solution for I_2 is meaningless, because light intensity is never negative. Therefore, (2.69) has only one solution with no lasing. Based on linear stability analysis, this solution of disorder is stable.

If the intensity of the photon pump increases, D will increase with the atom population number at the excited level. When the intensity of the photon pump is strong enough for a reversed atom population, i.e. the gain coefficient is greater than the wear coefficient ($D > A_c$), the nonzero stationary solution I_2 becomes positive and meaningful.

Based on linear stability analysis, the solution $I_1 = 0$ becomes unstable, while the solution $I_2 = (D - A_c)/\beta$ becomes a stable solution. The result is illustrated in Fig. 2.7. When $D < A_c$, there is a solution of $I_1 = 0$. When $D > A_c$, there are two solutions: a stable solution (solid line) of $I_2 = (D - A_c)/\beta$ and an unstable solution (dashed line) of $I_1 = 0$.

The laser example also shows that all these photons are emitted with the same frequency, the same direction and the same phase. This is a further example of self-organization.

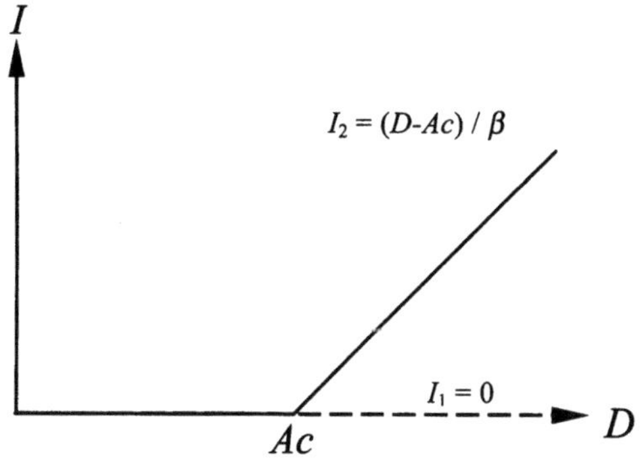

Fig. 2.7. Relation between the light intensity I and the gain coefficient D. The solid line represents the stable solution, while the dashed line represents the unstable solution. After X.-F. Shen [2.21]

2.8.3 Chemical Oscillation

Chemical oscillation shows another kind of dissipative structures, which illustrates how the breaking of time–translation symmetry leads to oscillatory behavior. A special phenomenon of chemical oscillation was observed in a chemical reaction system by Belousov in 1958 and earlier by Bray in 1921. Later Zhabotinsky reported the ordered picture produced by these reactions. Their reports initially met with skepticism, but their observations were later well confirmed. It is difficult to find an explanation in classical thermodynamics, but it is easy to get explanations in modern thermodynamics. In examples 1 and 2 mentioned above, two different approaches have been used to find theoretical explanations for dissipative structures. The key factor in chemical oscillation is the self-catalysis effect of some intermediates in chemical reactions. The theoretical explanation on the basis of reaction coupling for the chemical oscillation, Belousov–Zhabotinsky reactions (or B–Z reactions, for short), will be dealt with in the last chapter of the book.

3. High-Pressure Synthesis and Low-Pressure Vapor Growth of Diamond Seeds

3.1 Early History of Diamond Synthesis

Since ancient times, there has already been a desire for "a long life" and "riches". In the ancient East the former was more attractive. For instance, it was said that The First Emperor of the Qin Dynasty of China sent about three thousand boys and girls to the east ocean for seeking medicines for "immortal long life". Later, the "Daoist" Ge Hong tried to "prepare fair pills" for "immortal long life". In the ancient West the latter might be more attractive. It was believed that there was a kind of "philosophical stone", which had the special capacity of changing anything into gold. That was "alchemy". Although immortality pill makers and alchemists could never achieve their illusions, some early aspects in science had been developed.

For instance, the modern word "chemistry" was originally from the ancient word "alchemy". Due to continuous exploration and tests, it was gradually understood that there are different elements in nature and that it is impossible to extract gold by chemical methods from minerals, which does not contain the element gold. After the element theory was accepted, "alchemy" gradually disappeared. Due to the development of science it was a surprise to find that diamond, graphite and coal are allotropes. Diamond is very precious, and dazzlingly brilliant, while graphite and coal are back and dirt cheap. Both diamond and graphite are carbon. Therefore, synthetic diamond became a new goal in the pursuit of wealth.

As another example, as early as the 16th century (about 1594) a famous Chinese ancient scientist Shi-Zheng Li (1518–1593, Ming Dynasty) in his book *Encyclopedia of Herbs and Medicines* wrote: "For preparing cinnabar 2 *Jin* solid sulfur was put in a new pan, and then mercury 1 *Jin* was added. The mixture was stir-heated till the bright particles of free mercury disappeared. The pulverized powder was then kept in a jar. The jar was covered by a stone-plate fixed with iron-lines, and daubed by salt mud. After heating by fire and cooling down, cinnabar powder can be found on the sidewall inside the jar, and cinnabar crystals can be obtained near the cover. — cited from Yan Hu's book '*Secrets of Elixir*'." [3.1–3.3] The cinnabar is HgS (mercury sulfide). "*Jin*" was a Chinese weight unit [1 *Jin* equals about 0.5 kg]. That described the preparation procedure of cinnabar from sulfur and mercury by the chemical vapor deposition (CVD) process. That might be the first written record of the ancient CVD process in publication. Such a technique was often used for "making fair pills" in ancient times in China, although the modern term of "chemical vapor deposition (CVD)" was first proposed by John M. Blocher, Jr. et al. in the 1960s [3.4]. Blocher, as chairman, organized international CVD conferences many times, and promoted academic exchanges on CVD, so he was honored as "Sir CVD" in CVD forums. Blocher had mentioned the above ancient history of CVD at the 7th European CVD Conference in Perpiginon, France, 1989. Since the 1950s and 1960s, modern CVD technology had been widely used for cutting tool coatings and the fabrication of integrated circuits. It has become a very important method of manufacturing thin films and semiconductor materials. The introductions to CVD here is due to the fact that the technology has been used for the manufacture of diamond films, the thermodynamics of which will be discussed in detail in this book.

It was said that natural diamond had been found probably several thousand years ago. Diamond has always seemed to be a kind of rare precious treasure or a symbol of wealth. Diamond is the hardest material. Its English name was originally from the Greek word $\alpha\delta\alpha\mu\alpha\zeta$, which means unsubjugable, so the character of diamond can be seen. Moreover, diamond has many other outstanding characteristics. For instance, the heat conductivity of diamond is better than that of silver and copper. Diamond is economically very valuable. In the 17th or 18th centuries, investigation started to explore the composition of diamond. For instance, in 1675 Isaac Newton (1641–1727) on the basis of comparison between diamond and some combustible gases pointed out that diamond must also be combustible. After that, Robert Boyl (1627–1691) confirmed that diamond really changed in the flame. In 1694 the Italian Academy of Science organized a performance to show diamond burning out at the focus of a lens. In this way Newton's judgment had been demonstrated. In 1772 A.L. Lavoisier (1743–1794) studied the gas formed by combustion of diamond. It was found that the gas made lime-water turbid. Therefore, it was known that carbon dioxide was produced. In 1797 S. Tennant burnt diamond in a closed metallic container filled with oxygen. The carbon content of the produced gas was just the same as the weight of the burnt diamond. In this way, it was demonstrated that diamond is a crystal form of carbon. Later, in 1799 H. Guilton demonstrated that graphite is another crystal form of carbon, so graphite and diamond are allotropes. Coal is also an allotrope of carbon, but it is a noncrystalline form of carbon with more impurities. The price difference between diamond and graphite is so large, that the synthesis of diamond from graphite (or coal) is very attractive [3.5].

Because diamond has the highest density among all the allotropes of carbon and because of the close relationship between the original ore of natural diamond and volcanic eruption, it was supposed that the formation of synthetic diamond should be under high temperature and high pressure. Air should also be isolated from the produced diamond to prevent it from burning.

Since the 19th century and the beginning of the 20th century the success of synthetic diamond or diamond-like crystals had been declared several times. The most attractive one among them was J.B. Hanney's announcement [3.6]. During 1880–1882 he declared that if a carbon-containing substance together with metallic lithium were heated inside a sealed iron tube, after cooling synthetic diamond crystals were obtained. However, later on many people repeated his experiments, and none of them met with success. Therefore, many people had questioned the reality of his statement, but he had given no satisfactory answer. In 1943 some exhibits labeled "Hanney's diamond" were found among collections of the British Museum. Thus, F.A. Bannister of the Mineral Department of the British Museum and K. Lonsdale of the Royal Institute cooperated to check these samples by X-ray analysis [3.7]. The conclusion was that one of Hanney's samples is definitely not a diamond, while some of the others are pure diamond and some are diamonds with impurities. Such results could not confirm the reality of Hanney's work, because "Whether Hanney did in fact succeed in producing diamonds must in the last resort be dicided by the success or otherwise with which his experiments can be repeated." [3.8] In 1893 H. Moissan declared that he melted carbon-containing iron in a crucible at 3000 °C, and then the crucible was cooled in water [3.9]. The solidified iron was etched by acid, and a kind of crystal insoluble in acid was obtained. Its hardness was harder then corundum. It was also said that its density also approached that of diamond. However, it was demonstrated by other researchers that what Moissan obtained was not diamond. Later, C. Persons had also declared that he obtained synthetic diamond [3.10]. In 1928 a paper entitled *The Problem of the Artificial Production of Diamond* was published. After a careful analysis of the existing evidence, the writer stated: "The conclusion seems inevitable thst diamonds have not yet been produced in the laboratory, and that investigators have been misled into regarding as diamond various transparent, single-refracting minerals which happen to be very resistant to chemical reagents" [3.7].

3.2 Thermodynamic Predictions

In the history of diamond research and exploration the development of thermodynamics must be mentioned. In Chap. 1 we saw that classical equilibrium thermodynamics had been well developed in the 19th century and at the beginning of the 20th century. Significant success had been achieved not only in the study of heat engines but also in industrial manufacturing and chemical research. Theoretical calculations and predictions derived from classical equilibrium thermodynamic can be quantitative in many cases. Powerful authority had been set up due to the strict deduction of classical thermodynamics. Research on the synthesis of diamond was also built on the basis of thermodynamics. In the 1920s the equilibrium phase diagram of carbon had already been understood to a certain degree. That is, at ordinary low pressure diamond is a metastable phase, while graphite is a stable phase. Therefore, in the regime of equilibrium thermodynamics the spontaneous transformation from graphite to diamond at low pressure is impossible.

When the author visited Stanford University in 1996, M. Boudart told a story that happened in France at the beginning of the 20th century. There was a gentleman with high social position, who hoped to get synthetic diamond. Everyday before he left his workplace, a sealed iron tube with graphite inside it was put into a furnace. In the morning of the next day he opened the tube after cooling to see what happened in it. Each day there was no diamond, but one day he found something different from that found on other days. He carefully looked at it. It was really diamonds. He declared that he had obtained synthetic diamond. However, thermodynamics had already developed at that time, so public opinion disagreed with him. The gentleman had to check his experiment. It was found that his assistant knowing that his boss was interested in getting diamond, put some real diamonds in the opened tube. Similar things had happened several times in the past. It made lasting impressions on people that diamond seemed impossible to synthesize at low pressure, or that low-pressure diamond synthesis violated the thermodynamics.

In 1938 F.D. Rossini and R.S. Jessup calculated the equilibrium line between the diamond and graphite phases below 1200 K according to data for the combustion heat, specific heat, compressibility, and heat expandability of both diamond and graphite [3.11]. On this basis R. Berman and F. Simon made some revisions, took the change of the Gibbs free energy ΔG as a criterion of phase equilibrium, and extended the phase line between diamond and graphite to higher than 1200 K [3.12]. This is the Berman–Simon line which is well known in the forum of diamond synthesis. Its mathematical expression is

$$\Delta G_1\ (\mathrm{kJ\ mol^{-1}}) = \{p\ (\mathrm{Pa}) - [2.73 \times 10^6 T\ (\mathrm{K}) + 7.23 \times 10^8]\}\Delta V\ (\mathrm{m^3\ mol^{-1}}). \quad (3.1)$$

To understanding the predictability and guidance of thermodynamics, the following transformation reaction from graphite, denoted by C (gra), into diamond, denoted by C (dia), could be taken as an example to illustrate a simple calculation and discussion [3.13, 3.14]:

$$\mathrm{C(gra)} = \mathrm{C(dia)}. \quad (3.2)$$

From thermodynamic data books, the heat of formation of C (gra) and C (dia) at standard temperature (298.15 K) and standard pressure (1 atmosphere, p^0 = 101.325 kPa) are 0 and 1.8962 kJ mol^{-1}, respectively. The entropies of C (gra) and C (dia) at standard temperature and standard pressure are 5.6940 and 2.4389 kJ K^{-1} mol^{-1}, respectively. By calculation, it can be shown that

$$\Delta H^0_{298} = 1.8962 - 0 = 1.8962\ \mathrm{kJ\ mol^{-1}} \quad (3.3)$$

$$\Delta S^0_{298} = 2.4389 - 5.6940 = -3.2551\ \mathrm{J\ K^{-1} mol^{-1}} \quad (3.4)$$

$$\Delta G^0_{298} = \Delta H^0_{298} - T\Delta S^0_{298} \tag{3.5}$$
$$= 1.8962 - (298) \times (-0.0032551) = 2.866 \text{ kJ mol}^{-1},$$

that is, the formation Gibbs free energies of C (gra) and C (dia) at standard temperature and standard pressure ΔG^0_{298} are 0 and 2.87 kJ mol^{-1}, respectively. It is also known that the densities of C (gra) and C (dia) at standard temperature and standard pressure are 2.261×10^3 and 3.513×10^3 kg m^{-3}, respectively. From (1.22) in Chap. 1 it is known that

$$\left(\frac{\partial G}{\partial p}\right)_T = V. \tag{3.6}$$

The corresponding equation can be obtained for a reaction as

$$\left(\frac{\partial \Delta G}{\partial p}\right)_T = \Delta V. \tag{3.7}$$

Therefore, it might be possible to change the pressure to get a negative value of the Gibbs free energy of the transformation from graphite to diamond:

$$\Delta G(p) - \Delta G\left(p^0\right) = \int_{p^0}^{p} \Delta V \mathrm{d}p = \Delta V\left(p - p^0\right) \tag{3.8}$$

$$\Delta G(p) = \Delta G\left(p^0\right) + \Delta V\left(p - p^0\right) = 2.87 \text{ kJ mol}^{-1} \tag{3.9}$$
$$+ \left[\left(\frac{12.011}{3.513} - \frac{12.011}{2.260}\right) \times 10^{-6}\text{m}^3\text{mol}^{-1}\right]$$
$$\times (p - 101.325 \text{ kPa}).$$

If $\Delta G(p)$ ¡ 0 was required, then the pressure should be p ¿ 1.52×10^6 kPa, i.e., about 15 000 times that of the atmosphere. It was predicted that graphite might be transformed into diamond at very high pressure.

3.3 Success of the High-Pressure Process

A limitation of thermodynamics is that some predictions about the possibility of a process could be made, but there are still some uncertainties about the realities of the process. Sometimes, based on thermodynamics, a reaction is considered to be possible, but it does not happen practically. A hydrogen–oxygen gas mixture with a molar ratio of 2:1 at room temperature can be taken as such an example. However, if a spike or a catalyst of platinum black is introduced, the gas mixture will immediately explode. Therefore, for the realization of the transformation from graphite to diamond, the temperature should rise to a certain high degree. That required high temperature and high pressure technology. From the p-T equilibrium phase diagram of carbon it could be seen that if the temperature for running the transformation from graphite to diamond is at more than one thousand degree, then the pressure should be several tens or hundreds of thousand times that of the atmosphere. It is a truism that our attitude toward the physical world is conditioned by the scale of magnitudes provided in our terrestrial environment. We tend to classify pressure and temperature as high or low by comparing them with 10^2 kPa and a room temperature of 298 K despite the fact that almost all matter in the universe exists under very different conditions. For instance, at the center of the Earth by no means a large astronomical body, the pressure is about 4×10^8 kPa, and at the center of Sun the pressure is about 10^{12} kPa. It was a very difficult technique during the first half of the 20th century to achieve such temperature and pressure conditions required for the transformation from graphite to diamond. The American scientist P.W. Bridgeman of Harvard University was a pioneer in studying high-pressure physics, and worked fruitfully. Due to his contribution to high-pressure physics he received the Nobel Prize for Physics in 1946.

The high pressure concerned here was about ten thousand atmosphere or more. Ordinary containers were not able to suffer such a high pressure, but a large three-dimensional structure could be constructed. The simplest high-pressure apparatus is that of a tetrahedral anvil, which provides considerable support to the four central pistons by each hydraulic press. The high-pressure chamber is a very small tetrahedral volume. If a cubic high-pressure volume was desired, six hydraulic presses would need to be in arrangement. However, such a revision changes only the shape of the high-pressure volume but not the pressures. Due to the unceasing efforts of Bridgeman and his coworkers at Harvard, pressures of up to 400 kbar (40 MPa) were achieved. The attainment of such high pressures is made possible by the construction of pressure vessels of alloys such as Carboloy (tungsten carbide cemented with cobalt).

Once the necessary pressure is achieved by designed structural apparatus, then the next difficulty at such high pressures is the high temperature which is required for diamond synthesis. This depends on the durability of materials, which need to have properties quite unlike those with which we are familiar. One more important difficulty is the leakage of the high-pressure vessel, which is constructed by several carbide pistons from each side. There are gaps between the pistons. Pyrophyllite (wonderstone) is a hydrated aluminum silicate ($Al_2O_3\cdot$ 4 $SiO_2\cdot$ H_2O), which is used as a gasketing material, because it transmits the applied pressure without too great a frictional loss and is stable under most operate conditions. All these technological difficulties have gradually been solved one by one.

Based on the theoretical prediction of thermodynamics and the technical support of high-pressure physics, on 16th December 1954 F.P. Bundy, H.T. Hall and their coworkers at the American General Electric Company met with success on high pressure processes by which the first synthetic diamond crystals from graphite or other carbon materials had been obtained. In 1955 Bundy, Hall et al. announced the achievement of high pressure diamond synthesis in the journal *Nature* [3.15]. Metals were added as catalysts into the reaction mixture. That was very important for the success of the process, so the process was called a high-temperature and high-pressure catalytic process or simply a high-pressure catalytic process. Figure 3.1a shows a freshly grown single crystal of diamond, half-buried in the center of its catalyst metal growth medium, now frozen. It was prepared at 6 GPa and 1700–1800 K by using nickel as the catalyst. Figure 3.1b shows a photograph of high-pressure diamond crystals which is commercially available. A high-pressure apparatus that provides considerable support to the inner piston is the tetrahedral anvil, as shown in Fig. 3.2. As the pistons advance, some pyrophyllite is extruded between the anvils, and this effect helps to protect the carbide faces at the region of greatest pressure. This apparatus, designed by Tracy Hall, was used for the first commercial production of synthetic diamonds.

Now it is estimated that if there were no catalysts, the transformation from graphite to diamond would be achieved at about 15 GPa and 3000–4000 K. Up to now, no available container could withstand these conditions. If metallic catalysts (nickel, molybdenum, cobalt and so on) are adopted, synthetic diamond can be obtained at 5.5–6 GPa and 1600–1800 K in about 1 minute. Later on after continuous improvements the high pressure catalytic process had become the main process of diamond syntheses in industry. Now, this kind of high-pressure apparatus with the tetrahedral structure is rarely used; the most general form has two hydraulic presses from opposite directions, as shown in Fig. 3.3. Figure 3.4 shows an example of such packages of raw materials. The production of synthetic diamonds by the high pressure catalyst process worldwide may be about 600 million carats (120 thousand tons) per year. South African De Beer Co. and American General Electric Company are the biggest enterprises in diamond. Synthetic diamonds are widely used in different industrial applications. After 15 years efforts in 1970 the American General Electric Company announced the success of synthetic diamond of jewel grade. In 1971 a nuclear temperature gradient process was reported, and it was said that several transparent diamonds with 0.30, 0.31 and 0.39 carat, respectively, were obtained. However, the production cost was too high, so it was still not able to compete with natural diamond. In 1986 and 1987 some bigger crystals of diamond were prepared by the high-pressure process in the former Soviet Union and South Africa. The largest single crystal was bigger than ten millimeters but had bad transparency or bad color. The prices of these big crystals were not cheap, and they did not replace natural diamond in jewelry.

Synthetic diamonds are mainly used for mechanical or exploration industries. The high-pressure apparatus and processes mentioned above are static. High pressure can remain stable within a certain interval of time. During that interval of time the phase transformation from graphite to diamond can take place. Another way to achieve high temperature and high pressure is a dynamic and instantaneous method. The highest laboratory pressures (about

Fig. 3.1. (**a**) A single crystal of diamond and (**b**) commercial production of diamond crystals (40 ×) prepared by the high-pressure catalyst process. (a) from W.J. Moore [3.16] with permission from Pearson Educatuin Limited. Thanks to Xiao-Hu Fang for photograph (b) provided with copyright permission

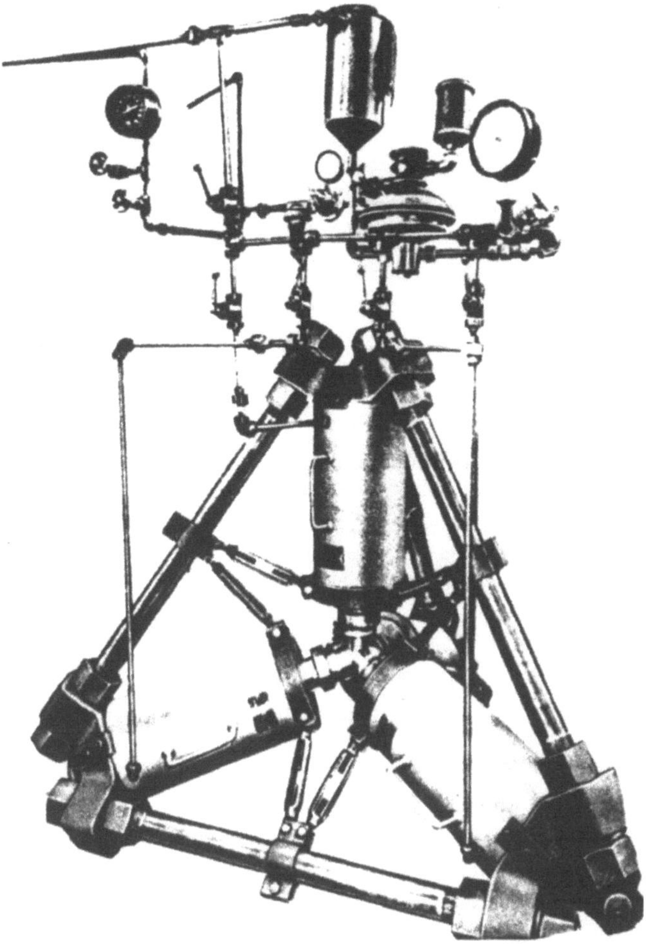

Fig. 3.2. A high-pressure apparatus with a tetrahedral structure. From W.J. Moore [3.16] with permission from Pearson Educatuin Limited

20–30 GPa) have been achieved by dynamic methods in which a shock wave produced by compressed gas or explosives is caused to travel through the specimen. The instantaneous high-temperature and high-pressure conditions are achieved by the shock wave of an explosion, so the process is called the explosion process. In 1961, B.J. Alder and R.M. Christian found evidence for diamond formation in graphite that was shock loaded at high temperature. Figure 3.5 shows the phase diagram of carbon summarized at that time. Because the instantaneous high pressure in the explosion process lasted only for a few microseconds, Alder humorously remarked: "We were millionaires for a microsecond." Figure 3.6 shows the flying plate apparatus for the explosion process. First the top electric detonator 1 ignites the explosive bar 2. The

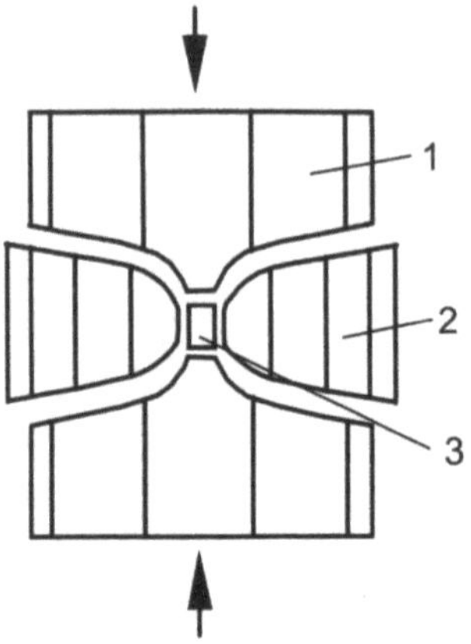

Fig. 3.3. An apparatus with two hydraulic presses from opposite directions. (1) piston, (2) anvil, (3) chamber. After X.-H. Fang [3.17]

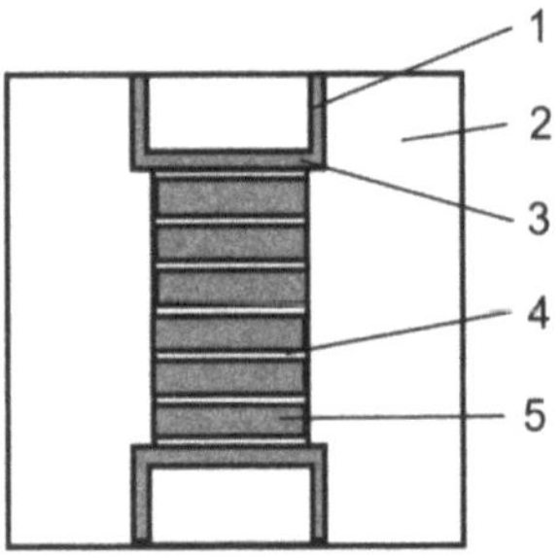

Fig. 3.4. A package of raw materials for the high-pressure catalyst process. (1) steel ring, (2) pyrophyllite, (3) metallic cover, (4) catalyst, (5) graphite. After X.-H. Fang [3.17]

iron bar 3 through an angle of α flies to the explosive flat 4. The explosion of the flat 4 has the iron plate 5 fly through an angle of β to the TNT main package of explosive 6. The explosive force of the main package has the flying plate 7 destroy the supports 8. The simultaneous lash on the graphite plate produces an instantaneous high-temperature and high-pressure condition lasting a few microseconds, so that the transformation from graphite to diamond is achieved. However, the particles of diamond prepared by the explosion process are usually more tiny, so this process is rarely used in industrial production.

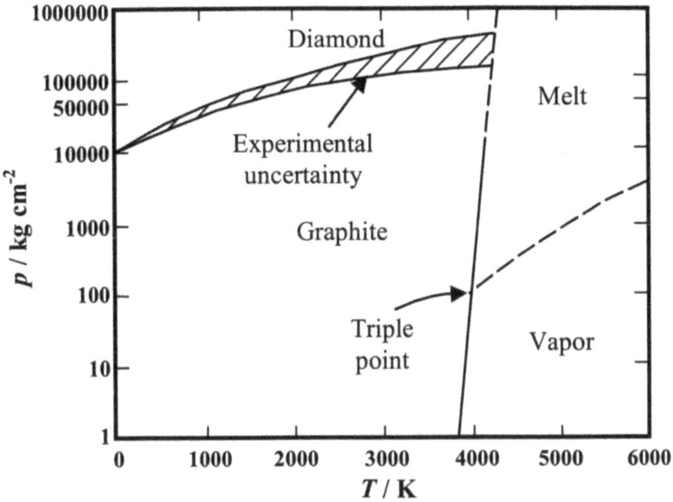

Fig. 3.5. Equilibrium phase diagram of carbon with logarithmic pressure summarized in the 1960s. From W.J. Moore [3.16] with permission from Pearson Educatuin Limited

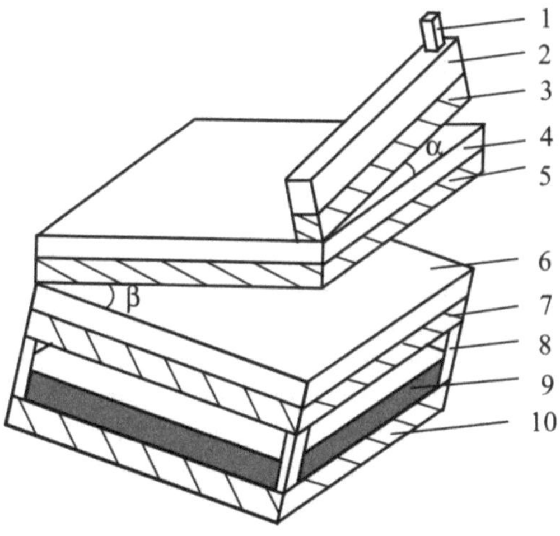

Fig. 3.6. An illustration of the explosion process and construction of the flying plate. (1) electric detonator, (2) explosive bar, (3) iron bar, (4) explosive flat, (5) iron plate, (6) TNT main package, (7) flying plate, (8) supports, (9) graphite plate, (10) iron plate. After X.-H. Fang [3.17]

3.4 Early Exploration of Low-Pressure Vapor Growth of Diamond Seeds

As mentioned above, at the end of the 18th century it was already known that diamond is an allotrope of carbon, so that diamond synthesis from graphite or even from coals had been explored in different ways. In the 19th century and at the beginning of the 20th century, all these trials were basically running at simple and crude workshops. The pressures used were much lower than the predicted high pressure, which was calculated or estimated by equilibrium thermodynamics for diamond being a stable phase. Therefore, all these trials failed. The hasty declarations of the success of the production of synthetic diamond on the basis of erroneous experimental results during that time had finally damaged the reputation of the persons concerned. That also made most scientists believe that diamond syntheses at low pressure were completely impossible. In particular, after the announcement of the American General Electric Company in December of 1954, the success of the high-pressure process became a firm demonstration of the theoretical power of classical thermodynamics. The development of matters is often complex and tortuous. There was some undeserved praise for classical thermodynamics following the success of diamond synthesis. There was a prevalent argument that diamond synthesis was only possible at very high pressures, and impossible to meet with success at low pressures. This argument had also been expressed in some textbooks. For instance, "In the last century, many experiments were running for the transformation from graphite into diamond. However, all these experiments failed in the end. In fact, thermodynamics had already demonstrated that the transformation from graphite to diamond was only possible at a pressure of over 15 000 times that of the atmosphere, so that all trials at lower pressure than this must be useless and failed." [3.18] In sentences such as these, "thermodynamics" only represented classical equilibrium thermodynamics. Strictly speaking, the statement should be revised to "Based on equilibrium thermodynamics the spontaneous transformation from graphite to diamond will be impossible, if the pressure is lower than 15000 times that of the atmosphere." The premise "Based on equilibrium thermodynamics" and key word "spontaneous" are very important, and should never be omitted. Something, which is impossible in equilibrium thermodynamics, may be possible in nonequilibrium thermodynamics. Something, which is impossible by a spontaneous reaction, may be possible by a nonspontaneous reaction with inputting external energy. Even according to equilibrium thermodynamics, synthetic diamond at low pressure may be impossible to get from graphite but may be possible to get by pyrolysis of some other gaseous hydrocarbons with higher chemical potentials. That should never be excluded from discussions. In the following paragraph the general conditions for crystal growth from the vapor phase will be discussed.

3.5 Thermodynamic Conditions for Crystal Growth from the Vapor Phase

According to a large number of experimental observations, a solid (usually a crystal) can grow from the gas (or vapor) phase. Besides the fact that the growth temperature should be below its melting point (the melting point of diamond is as high as 3550 °C, and low-pressure diamond growth from the vapor phase is usually at about 1000 °C.), the following two conditions should be satisfied:

1. The vapor phase should be in a supersaturated state.
2. If there is no nucleus, the supersaturation degree should be able to cover the nucleus formation barrier.

More details are given below.

3.5.1 Supersaturation Degree

When the system is in equilibrium between two phases, the crystal or solid is not able to grow; that is, the vapor phase must be in a supersaturated state during crystal growth. Figure 3.7 shows a p-T (pressure-temperature) phase diagram of a single component system. If the vapor phase of the system is located on the gas–solid line (such as at point D), crystal growth may not take place. If it comes into the solid phase region (such as at point E) crystal growth from the vapor phase will be possible. Points D and E represent the states (p^e, T) and (p, T), respectively. Here, $p > p^e$. Usually, $p/p^e = \alpha$ is defined as the saturation ratio, and $\Lambda = \alpha - 1 = (p/p^e) - 1$ is defined as the supersaturation degree. p is the practical pressure, when the vapor phase is at temperature T, and p^e is the equilibrium vapor pressure at the same temperature.

For a gas–solid phase transformation of 1 mole of the substance from $E \rightarrow D$ at a constant temperature, the change of the Gibbs free energy will be

$$\begin{aligned}\Delta G &= \int_p^{p^e} V \mathrm{d}p = \int_p^{p^e} \frac{RT}{p} \mathrm{d}p = -RT \ln \frac{p}{p^e} \\ &= -RT \ln \alpha = -RT \ln (1 + \Lambda) \,.\end{aligned} \tag{3.10}$$

Therefore, if the supersaturation degree is greater than zero, $\Delta G < 0$, crystal growth from the vapor phase will take place spontaneously.

If the average volume of a single atom is v_m, then the change of the Gibbs free energy for the gas–solid phase transformation of a unit volume of the gas phase will be

$$\Delta g_v = -\frac{k_\mathrm{B} T}{v_\mathrm{m}} \ln \alpha = -\frac{k_\mathrm{B} T}{v_\mathrm{m}} \ln (1 + \Lambda) \tag{3.11}$$

where, k_B is the Boltzmann constant.

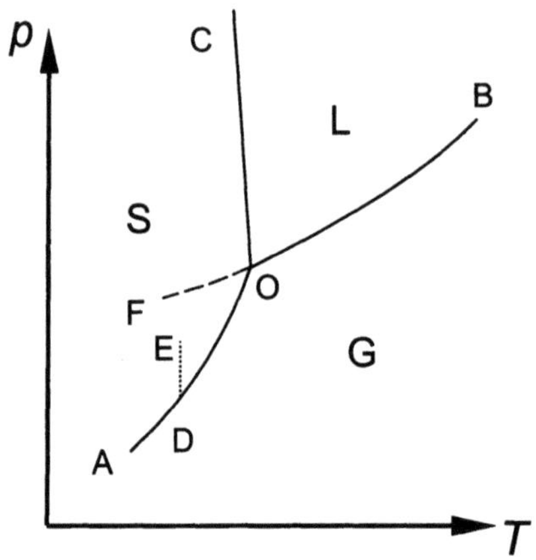

Fig. 3.7. p-T phase diagram of a single-component system

3.5.2 Nucleation Barrier

Nucleation can be divided into two kinds: homogeneous nucleation and heterogeneous nucleation [3.19, 3.20]. Under certain conditions, nucleus formation directly in the vapor phase is called homogeneous nucleation, while nucleus formation in the system on external particles (such as dust, surfaces of solid particles or a crystal nucleus) is called heterogeneous nucleation.

Homogeneous nucleation The following analysis is for the nucleus formation process from the vapor phase. Molecules always move perpetually in the vapor phase. However, their velocities are different from each other. Due to the fluctuation of energy sometimes molecules can be connected to each other to form small groups with different sizes. These groups can grow to form a nucleus by absorbing other new molecules, they can also separate into individual molecules. These groups may be called crystal embryos.

If the system is in a supersaturation state, after the formation of crystal embryos some of the vapor molecules become molecules inside crystal embryos. Meanwhile due to the formation of the crystal embryo the free energy decreases, i.e., ΔG_V is negative. The decrease of the volume free energy is the driving force for crystallization. On the other hand, after the formation of the crystal embryo a vapor–embryo interface forms. The molecular energies are different for molecules at the interface and inner molecules of the crystal embryo. The inner molecules are surrounded by other molecules. The attractive forces from different directions are equal for each molecule inside the crystal embryo, so these forces counteract each other. However, these are different for molecules at the interface. The molecular density is high inside the crystal embryo, so the attracting force between the molecules is strong. On the other hand, the molecular density is small in the vapor phase, so the attracting force between molecules in the vapor phase is weak. If a molecule moves from the inner part of the crystal embryo to the surface and increases the surface area of the crystal embryo, work is needed to overcome the attracting force. The loss of work equals the energy gain of the surface. The free energy needed at constant temperature and pressure for increasing a unit area of surface is called the surface energy, denoted by ε. Its dimension is J m^{-2}. Due to the surface energy, when inner molecules of the crystal embryo becomes molecules at the surface of the crystal embryo, the free energy of the crystal embryo surface will increase, i.e., ΔG_S is positive. The increase of the surface free energy is the resistance force for crystallization. In this way, the total change of free energy during the formation of a crystal embryo is

$$\Delta G = \Delta G_V + \Delta G_S. \tag{3.12}$$

Supposing that the crystal embryo is of a spherical shape of radius r, then

$$\Delta G = \frac{4}{3}\pi r^3 \Delta g_v + 4\pi r^2 \varepsilon \tag{3.13}$$

where, Δg_v is the Gibbs free energy change during the formation of per unit volume of the crystal embryo. According to (3.13), the decrease of the volume free energy is directly proportional to r^3, while the increase of the surface free energy is directly proportional to r^2. Figure 3.8 shows the curves of the changes of different free energy terms with the change of crystal embryo radius r. The curve of the total free energy change ΔG is accompanied by two curves ΔG_V and ΔG_S. From the figure it can be seen that there is a maximum value of ΔG^* on the curve of ΔG. The critical Gibbs free energy ΔG^* corresponds to the critical radius r^* of the crystal embryo. From the viewpoint of thermodynamics, if $r < r^*$, the crystal embryo is still not able to grow spontaneously. The reason is that if it grows, the free energy of the system will increase. On the contrary, if the crystal embryo shrinks and disappears, the free energy of the system will decrease. Therefore, when $r > r^*$, the crystal embryo may grow spontaneously, and the total free energy of the system decreases with the increase of r. The crystal embryo, which can grow spontaneously, is called the crystal nucleus. If the radius of the crystal embryo is $r = r^*$, then the possibility for growing just equals the possibility for shrinking. In this case the crystal embryo is in its critical state for transfer from crystal embryo to nucleus. Such a nucleus is called a critical nucleus or a critical crystal embryo.

The critical radius r^* corresponds to the maximum value of the free energy change ΔG^*, so the condition for the existence of the critical nucleus is

$$\frac{\mathrm{d}\Delta G}{\mathrm{d}r} = 0. \tag{3.14}$$

Differentiation of (3.13) gives $4\pi r^{*2}\Delta g_v + 8\pi r^* \varepsilon = 0$, and

$$r^* = -\frac{2\varepsilon}{\Delta g_v}. \tag{3.15}$$

Substitution in (3.13) gives

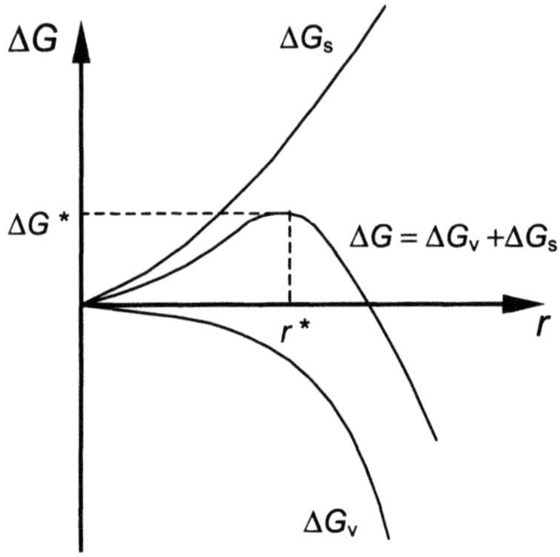

Fig. 3.8. Relationship between ΔG and the radius of the crystal nucleus

$$\begin{aligned} \Delta G^* &= \frac{4}{3}\pi\left(-\frac{2\varepsilon}{\Delta g_v}\right)^3 \Delta g_v + 4\pi\left(-\frac{2\varepsilon}{\Delta g_v}\right)^2 \varepsilon \\ &= -\frac{32}{3}\frac{\pi\varepsilon^3}{\Delta g_v^2} + \frac{16\pi\varepsilon^3}{\Delta g_v^2} = \frac{16}{3}\frac{\pi\varepsilon^3}{\Delta g_v^2} = \frac{4}{3}\pi r^{*2}\varepsilon . \end{aligned} \tag{3.16}$$

It can be seen that ΔG^* just equals 1/3 of the critical nucleus surface energy. This means that during the formation of the critical nucleus, the decrease of the volume free energy can only compensate for 2/3 of the surface free energy. The other 1/3 of the surface free energy must be provided by the exterior. This part of the energy is called the nucleation energy or nucleation barrier. The nucleation energy is provided by the energy fluctuation in the vapor phase, that is, it is provided by random heat motion of the surrounding molecules. From the microscopic viewpoint, due to the random heat motion of molecules the energy at different places is not uniform. It always fluctuates and deviates above or below the average energy. This is called energy fluctuation. If inside the vapor phase the energy fluctuation is equal to or higher than the magnitude of the critical nucleation energy, then a nucleus forms there.

According to (3.15) and (3.11), r^* is inversely proportional to Δg_v, and Δg_v is also related to the supersaturation degree of the vapor phase. The greater the supersaturation degree of the vapor phase, the lower the values of r^*, and thus the easier it is to form nuclei. If there are a lot of crystal nuclei, then a lot of tiny crystals will form. Therefore, if bigger crystal grains are needed, the growth condition, in which only a few nuclei exist, should be controlled. In this way conditions can be controlled to meet necessities.

Heterogeneous nucleation The above discussion of nucleation is only for that happens inside the clean vapor phase, and directly from the vapor phase. If inside the vapor phase there are some solid materials (such as a substrate), then the nucleus will form first on the solid surface. That is called heterogeneous nucleation.

Figure 3.9 shows a nucleus β with a spherical cap of radius r on the substrate surface S from the α vapor phase. θ is the contact angle between the nucleus and the substrate surfaces. The free energy change of the formation of a new nucleus can be divided into two parts, i.e., the change of the volume free energy and the change of the surface free energy.

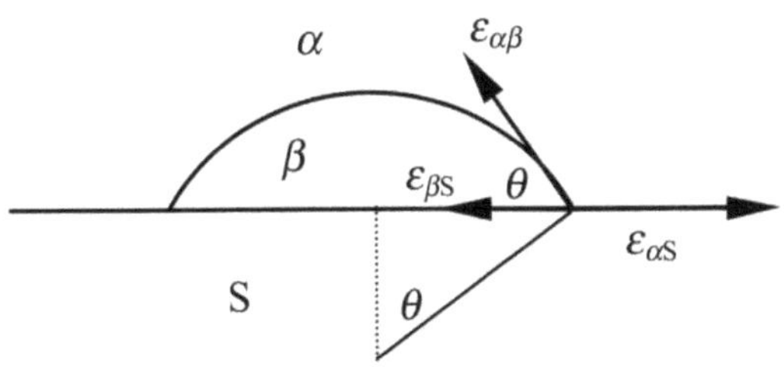

Fig. 3.9. Nucleation on a substrate surface

After the formation of the spherical cap the change of the volume free energy is $\Delta G = V\Delta g_v$. Here the volume V of the spherical cap is

$$V = \int_\theta^0 \pi\left(r\sin\theta\right)^2 r\mathrm{d}\left(\cos\theta\right) = \frac{\pi r^3}{3}\left(2 - 3\cos\theta + \cos^3\theta\right) \tag{3.17}$$

$$\Delta G_V = \frac{\pi r^3}{3}\left(2 - 3\cos\theta + \cos^3\theta\right)\Delta g_v. \tag{3.18}$$

The surface free energy of the spherical cap consists of two parts. The first part is the surface energy of the spherical cap contacted with the α phase:

$$\Delta G_{S_1} = \varepsilon_{\alpha\beta}\int_0^\theta 2\pi\left(r\cos\theta\right) r\mathrm{d}\theta = 2\pi r^2\left(1 - \cos\theta\right)\varepsilon_{\alpha\beta}. \tag{3.19}$$

The second part is the interface energy between the β phase and the substrate S:

$$\Delta G_{\mathrm{S}_2} = -\pi\left(r\sin\theta\right)^2\left(\varepsilon_{\alpha\mathrm{S}} - \varepsilon_{\beta\mathrm{S}}\right) \tag{3.20}$$

$$\begin{aligned}\Delta G_{\mathrm{S}} &= \Delta G_{\mathrm{S}_1} + \Delta G_{\mathrm{S}_2} \\ &= 2\pi r^2\left(1 - \cos\theta\right)\varepsilon_{\alpha\beta} - \pi\left(r\sin\theta\right)^2\left(\varepsilon_{\alpha\mathrm{S}} - \varepsilon_{\beta S}\right),\end{aligned} \tag{3.21}$$

where, $\varepsilon_{\alpha\beta}$, $\varepsilon_{\alpha S}$ and $\varepsilon_{\beta S}$ are the interface energies between α-β, α-S (substrate), and β-S, respectively. It may be supposed that the value of the interface energy equals the value of the interface tension. At the connecting point among the three phases α, β and S, all these interface tensions must satisfy the mechanical equilibrium condition, so

$$\varepsilon_{\alpha S} = \varepsilon_{\beta S} + \varepsilon_{\alpha\beta} \cos\theta. \tag{3.22}$$

After the formation of the spherical cap the free energy change of the whole system is

$$\begin{aligned} \Delta G_{\text{non}} &= \frac{\pi r^3}{3}\left(2 - 3\cos\theta + \cos^3\theta\right)\Delta g_v + 2\pi r^2 (1-\cos\theta)\,\varepsilon_{\alpha\beta} \\ &\quad + \pi (r\sin\theta)^2 \left(\varepsilon_{\alpha S} - \varepsilon_{\beta S}\right) \\ &= \left(4\pi r^2 \varepsilon_{\alpha\beta} + \frac{4\pi r^3}{3}\Delta g_v\right) f(\theta) \end{aligned} \tag{3.23}$$

$$f(\theta) = \frac{(2+\cos\theta)(1-\cos\theta)^2}{4}. \tag{3.24}$$

According to $\frac{d\Delta G}{dr} = 0$, the critical curvature radius of the spherical cap nucleus is

$$r^* = -\frac{2\varepsilon_{\alpha\beta}}{\Delta g_v}. \tag{3.25}$$

The critical energy of nucleation is

$$\Delta G_{\text{non}}^* = \frac{16\pi\varepsilon_{\alpha\beta}^3}{3\Delta g_v^2} \cdot f(\theta) = \Delta G \cdot f(\theta). \tag{3.26}$$

ΔG_{non}^* represents the critical energy of heterogeneous nucleation, while ΔG^* represents the critical energy of homogeneous nucleation.

The change range of the contact angle θ is $180° \geq \theta \geq 0$, so the change range of $f(\theta)$ is $1 \geq f(\theta) \geq 0$. Therefore, (3.26) shows that nucleation on the substrate is much easier than the homogeneous nucleation in free space. According to (3.22), the magnitude of θ is related to the interface energies between substrate, vapor and nucleus. Figure 3.10 shows the relationship between $f(\theta)$ and θ. It can be seen that when $\theta = 0°$, $f(\theta) = 0$ and $\Delta G_{\text{non}}^* = 0$. This means that there is no necessity of three-dimensional nucleation, and that there is no barrier for the transformation from vapor to crystal. Such a situation can be found if the material and the crystal structure of the growing material is exactly the same as that of substrate. For instance, diamond grows from the vapor phase on diamond. The crystal seed or nucleus is exactly the same as the depositing thin film material. There is no barrier to nucleation, that is to say, if there are only diamond crystal seeds without graphite seeds in the pyrolysis furnace during the decomposition of methane, the diamond growth will be favorable. During the derivation of the foregoing formulae, for simplicity a spherical nucleus (in homogeneous nucleation) or a spherical cap nucleus (in heterogeneous nucleation) is supposed. These results are still suitable in principle for different geometric crystals grown from the vapor phase.

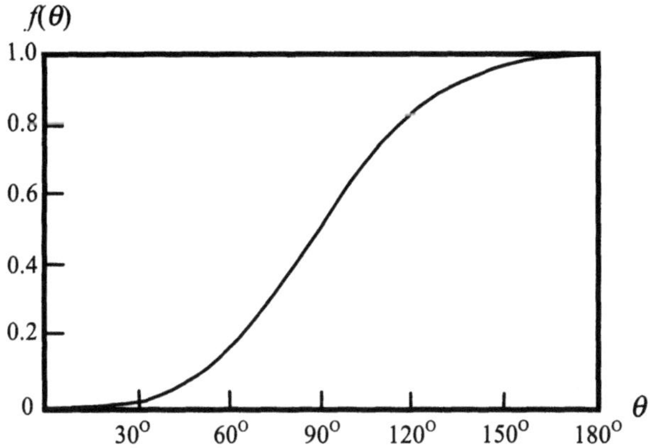

Fig. 3.10. Relationship between $f(\theta)$ and the contact angle θ

3.6 Diamond Seed Growth from the Vapor Phase

While these high-pressure experiments were going on, the low-pressure experiments were also progressing nicely. At the end of the 1950s low-pressure experiments were mainly based on the pyrolysis of methane, acetylene, other hydrocarbons, and even carbon halides to get carbon films. These carbon films were usually amorphous, graphite, or a mixture constructed by sp^2 and sp^3 carbon atoms (perhaps a mixture of nanometer diamond and nanometer graphite, or a tiny crystal mixture of diamond and graphite). In the last chapter it was mentioned that there is a very small difference in the free energy between graphite and diamond, i.e., 0.3 eV $atom^{-1}$ or 2.9 kJ mol^{-1} at room temperature. If the temperature is high enough during deposition, carbon atoms will have high enough activities to tend to form crystals, so the initial step will be nucleation. During the formation of a diamond nucleus or graphite nucleus the nucleation barrier must be overcome. If the height of the barrier is close to the free energy difference between graphite and diamond, then a new plan could be put forward: A small amount of diamond powder as seeds are put at the pyrolysis region in the furnace, as shown in Fig. 3.11. In this way carbon atoms can grow a diamond structure on the surfaces of the diamond powder. The growth of diamond crystals on diamond surfaces from the vapor phase does not need to overcome the nucleation energy, while the formation of a new graphite nucleus on a diamond surface must overcome the nucleation barrier. The nucleation of graphite on a diamond surface may be more difficult than the growth of diamond on diamond surfaces, so in such a process it is possible to grow diamond without the nucleation of graphite for a short time. According to the equilibrium phase diagram of carbon, at ordinary low pressures graphite is more stable than diamond, so in pyrolysis processes there is still the possibility of forming graphite nuclei. Once the graphite nuclei are formed, carbon atoms coming from pyrolysis will tend to grow a graphite solid phase on the graphite nuclei, or to form a solid mixture consisting of sp^2 and sp^3 carbon atoms (perhaps amorphous). It was very difficult to prepare synthetic diamond in this process. However, it was a kind of test for diamond synthesis, and from a theoretical point of view there was nothing violating classical equilibrium thermodynamics.

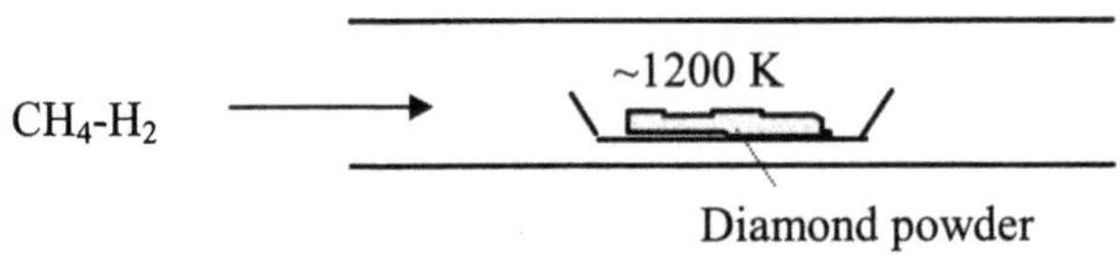

Fig. 3.11. Setup of a simple pyrolysis process for diamond seed growth from the vapor phase

An obvious amount of diamond growth from the vapor phase on the diamond powder could not be obtained after one run. Soon after inputting methane (hydrogen as a diluent) inside the pyrolysis furnace the original white or colorless diamond powder changed into gray or black. Then the methane was shut off and the hydrogen flow kept going on. Thus the hydrogen etches the graphite layer on the surface of the diamond powder, and then the diamond powder becomes white or colorless. In diamond crystals carbon atoms connect to each other by sp^3 σ bonds. This four-valence saturated structure of diamond is difficult to be gasified by hydrogen. The original diamond powder and the as-grown diamond layer are basically preserved. In this way diamond can be grown layer by layer, until the weight increase of the diamond powder is measurable after several cycles. Therefore, diamond growth from the vapor phase at low pressures had been demonstrated.

It is also possible to accelerate the graphite etching rate in the hydrogen atmosphere by increasing the furnace temperature or increasing the hydrogen pressure in the reactor. However, the diamond growth rate is so slow (less than 0.01 $\mu m\ h^{-1}$) that no practical application has been found. Due to these doubts about low-pressure processes, research on low-pressure diamond synthesis at this stage had not attracted much attention.

At about the same time the research work of W.G. Eversole at the American Union Carbon Co. and the research work of B.V. Deryagin, B.V. Spitsyn, D.V. Fedoseev et al. at the Physical Chemistry Institute of the former Soviet began [3.21–3.24]. After a patent application in about 1959 Eversole ended his work on it, while Deryagin, Spitsyn, Fedoseev continued their research, and made much progress. Later on the American J.C. Angus also started working on this project, and attended international conferences or symposiums on diamond held in the former Soviet Union [3.25]. Generally, the West did not pay much attention to low-pressure diamond synthesis at that time. Angus wrote in his paper in 1968 that "In the light of this history it is remarkable that the question has received so little serious attention." and that "... it is surprising that this remarkable work has received so little attention." [3.25] In spite of this, the general attention was still not paid to it, mainly because in this process graphite must be periodically removed from the seed crystals and diamond growth will cease; the rate of diamond growth was so low that after about ten deposition–cleaning cycles in one day the weight of diamond increased only by about 5 mg without quality identification.

The exploration of low-pressure diamond synthesis at this stage had met some positive results. However, remarkable work on activated low pressure diamond syntheses was started in about 1970 by Deryagin, Spitsyn, and Fedoseev et al. Activated low-pressure diamond growth from the vapor (or gas) phase had opened up a new era in the history of synthetic diamond. The growth rate of diamond in activated low-pressure processes is several orders of magnitude faster than the simple pyrolysis process in the 1950s and 1960s. Moreover, the new process was continuous, while the old diamond seed growth at ordinary low pressure was discontinuous as mentioned above. The more important difference is that the transformation of graphite to diamond can be achieved only by the new activated process. From the viewpoint of thermodynamics, the old process could still be explained on the basis of classical equilibrium thermodynamics, but the new activated process could be satisfactorily and even quantitatively explained only by modern thermodynamics. In fact, since the middle of the 1980s the term "low-pressure diamond growth from the vapor phase" in the literature has generally only represented activated low-pressure diamond growth from the vapor phase, which was initiated by Deryagin, Spitsyn, Fedoseev et al., and Setaka, Matsumoto, et al. (see the next chapter). Low-pressure diamond synthesis with activation and without activation are two different technologies and two different development stages in nature, and should never be confused. In this book later on, if no other notation is given, the term "low-pressure diamond growth from the vapor phase" only means activated low-pressure diamond growth from the vapor phase.

4. Activated Low-Pressure Diamond Growth from the Vapor Phase

4.1 Success of Activated Low-Pressure Vapor Growth Process

In the early stage of low-pressure diamond synthesis there are two periodic circular steps:

1. The methane pyrolysis and diamond growth step.
2. The hydrogen etching graphite step.

Due to the sustained research effort of B.V. Deryagin, B.V. Spitsyn, and D.V. Fedoseev et al., it was found that the introduction of atomic hydrogen with superequilibrium concentration obviously accelerated the etching rate of graphite in the second step, while there was nearly no influence on the diamond etching rate [4.1–4.11]. Thus, a radically new idea was brought to mind. If superequilibrium atomic hydrogen (SAH, for short) was introduced during the pyrolysis of methane, carbon atoms produced by the pyrolysis of methane might grow on the surfaces of diamond, and the undesired graphite might simultaneously be etched away. After experiments it was confirmed that this process was not only reliable but also produced very nice diamond crystals on a nondiamond substrate. Photographs of such beautiful diamond crystals were published in in 1976, as shown in Fig. 4.1. Meanwhile, the diamond growth rate had been increased by several orders of magnitude, and was about several micrometers per hour. The practical timetable for the success of the activated low pressure process was shifted to an earlier date, because the chemical transport reaction (CTR) process had been successfully used for transporting carbon atoms via gas phase from graphite to diamond at low pressure in about 1970. A photograph of the epitaxial diamond layer grown on diamond seeds at low pressures similar to that of Fig. 4.2b was published in 1971. In comparison with the discontinuous circular simple pyrolysis process mentioned in the last chapter the obvious outstanding result of the new process is the growth of very beautiful diamond crystals on the nondiamond substrate. Another outstanding result is that diamond growth took place with simultaneous graphite etching, and diamond growth was possible from graphite as the unique carbon source. In this way low-pressure diamond synthesis had come into its new era.

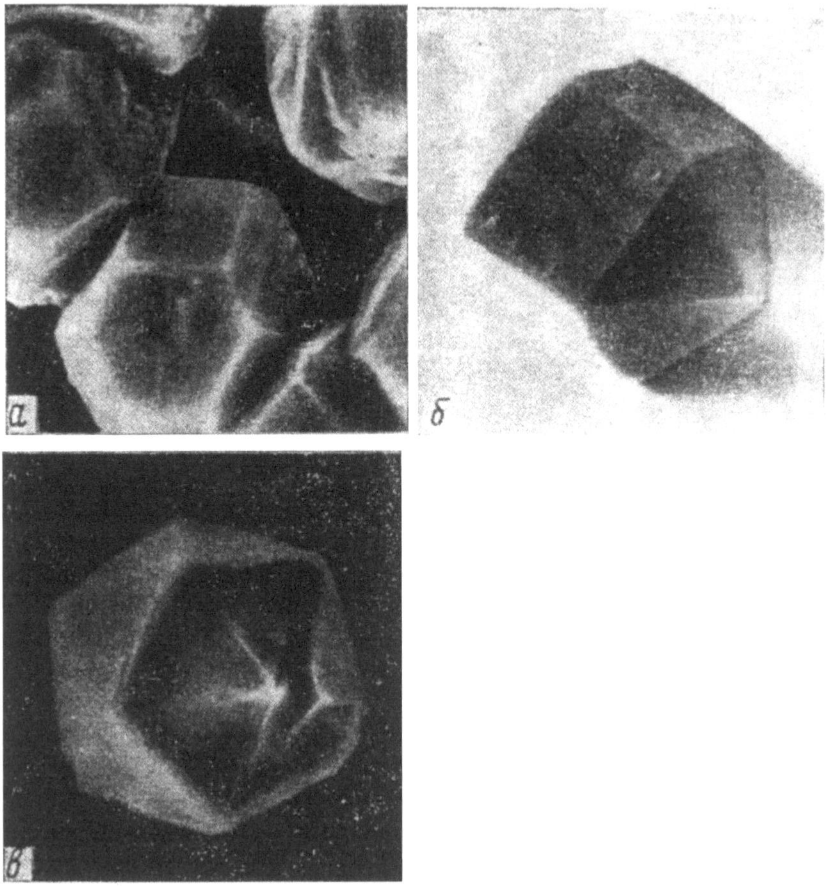

Fig. 4.1. Beautiful diamond crystals grown from the vapor phase on a nondiamond substrate by Deryagin, Spitsyn et al. in 1976. From B.V. Deryagin, B.V. Spitsyn et al. [4.7] with permission from Nauka Publishers, Moscow

The spurious or mistaken claims of the success of diamond synthesis repeated in the last century, together with the lack of high-pressure equipment, made it seem impossible to synthesize diamond at low pressures. Another impression gained from classical equilibrium thermodynamics was that the transformation from graphite to diamond could be achieved only under very high pressure. The prediction of high-pressure diamond synthesis derived from classical equilibrium thermodynamics had also been regarded as an absolute thermodynamic conclusion. In particular, after the success of the high

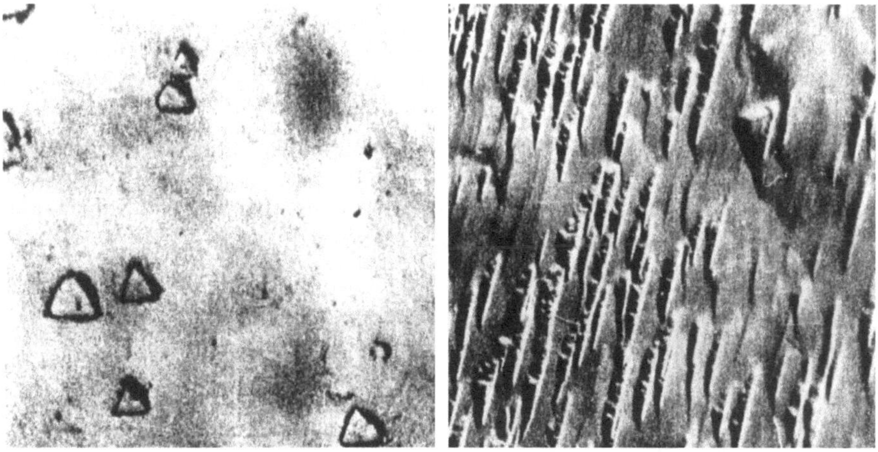

Fig. 4.2. Morphology of epitaxial diamond films on natural diamond substrate, (**a**) 3.7 μm thick grown on a (111) natural diamond substrate (1200 ×), (**b**) 2 μm thick grown on a polished natural diamond surface with crystallographic orientation about 5° off the (100) plane (800 ×). Reprinted from B.V. Spitsyn [4.10] with permission from Elsevier Science. A similar to (b) photograph of epitaxial diamond film on natural diamond substrate had been published in 1971 [4.3]

pressure process, a very general viewpoint in the 1960s and 1970s was that there was no possibility of synthesizing diamond at low pressures. Therefore, until the middle of the 1980s most scientists did not accept the achievement of activated low pressure diamond synthesis in the former Soviet Union. It was even sometimes jokingly called "alchemy" when that creative result was talked about, that is, activated low-pressure diamond growth from the vapor phase was still considered to be absolutely impossible.

In about 1980 N. Setaka, Y. Sato, and S. Matsumoto et al. of the National Institute for Research in Inorganic Materials of Japan confirmed the research of Deryagin, Spitsyn, and Fedoseev et al. on activated low-pressure diamond growth from the vapor phase [4.12–4.17]. Much progress had also been made by Setaka et al. on research on this project. They improved the experimental equipment, designed new processes and published photographs of beautiful diamond crystals (or diamond thin films) in 1982, as shown in Fig. 4.3, grown by hot filament chemical vapor deposition (HFCVD) at about 10 kPa and 1000–1300 K on silicon (Si) or molybdenum (Mo) substrates from 1% CH_4 in a hydrogen gas mixture. The temperature of the hot filament was 2200–2400 K. The main effect of the hot filament was to produce more hydrogen atoms. The concentration of the atomic hydrogen produced is much higher than the equilibrium concentration at the substrate temperature (1200 K), so it is called superequilibrium atomic hydrogen (SAH). The existence of superequilibrium atomic hydrogen plays a key role in activated low-pressure diamond growth. Later on microwave and radio-frequency plasma was also successfully used for generating superequilibrium atomic hydrogen instead of the hot filament, and all these processes were successful in getting low-pressure diamond growth from the vapor phase. It was found that nearly all hydrocarbons (and even graphite only) could be used as precursors for low-pressure diamond growth. Later it was also found that oxygen-containing or halogen-containing hydrocarbons could also be used as precursors. Setaka et al. have published a series of papers since 1982. In this way the influence of the activated low-pressure diamond process grew.

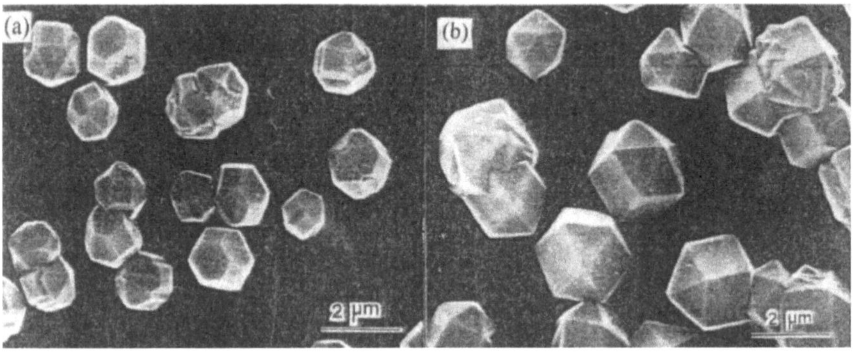

Fig. 4.3. Photographs of diamond crystals grown from the vapor phase at low pressure by the hot filament method were published in 1982. From S. Matsumoto, Y. Sato, M. Tsutsumi and N. Setaka [4.13] with permission from Kluwer Academic Publishers

These experimental reports on activated diamond growth from the vapor phase published in journals by former Soviet and Japanese scientists, and the photographs and data in these reports had already firmly supported their conclusions, but general attention was still not paid to it. In 1984 Rustum Roy of Penn State University visited the National Institute for Research in Inorganic Materials. Setaka told him about the authenticity of activated low-pressure diamond growth from the vapor phase initiated in the former Soviet Union, and showed him very beautiful diamond crystals grown at low pressure from the vapor phase similar to those in Fig. 4.3 by microscopy. The visual evidence was both convincing and stimulating, so on his return Roy started a project on activated CVD diamond with funding from the Office of Naval Research. After a short time he succeeded in producing diamond films. Roy announced his results in February 1986. Since then activated CVD diamond growth has become a major focus of research throughout the world. In 1984 Zeng-Sun Jin of Jilin University of China was a visiting scientist in Tokyo. Jin also visited Setaka's lab. On his return to China Jin met with success in hot filament diamond growth also in 1986 [4.18]. Since 1986, papers on activated CVD diamond had increased rapidly. More and more photographs of beautiful diamond crystals or films grown by activated CVD processes were published, such as those shown in Figs. 4.4 and 4.5. At the low crystallization temperature nuclei consist of isotropic polycrystalline diamond crystals, as shown in Figs. 4.5c and 4.5d. With increase of substrate temperature the situation may change drastically, as shown in different photographs of Figs. 4.4 and 4.5. That will be discussed at the end of the next chapter with theoretical treatment.

The preparation technology was not only hot filament or microwave processes, but also oxygen–acetylene combustion torch processes among others [4.19]. The combustion process was also first started in Japan. The combustion gases are acetylene, and a lot of hydrocarbons. In fact, diamond growth at low pressure from the vapor phase can be achieved by any form of external energy and any carbon-containing sources. In particular, due to the success of the combustion process, it may be possible to grow diamond at home or in the workshop with an oxygen–acetylene torch. It was a hearsay that in 1930s or 1940s, workers found a diamond on the ground of a workshop in which oxygen–acetylene torches were used for cutting steel and iron plates. Perhaps the diamond was produced by the oxygen–acetylene torches. Of course, this is only an interesting guess and hearsay. In fact, up to now the biggest single crystals of diamond grown by the activated low-pressure diamond process are only several hundred micrometers in magnitude, such as shown in Fig. 4.4a. Diamond synthesis of jewelry grade at low pressure lies in the future.

Karl E. Spear in 1987 published a review of the history of the activated low pressure diamond process. He said that "The new diamond technology has a curious history that is a story of scientific caution and communication difficulties. It was perhaps influenced by the mystique that surrounds diamonds — a mystique that has led some scientists to joke of 'alchemy' when discussing their (note: former Soviet scientists) investigations. The activated vapor process was first recognized in the Soviet Union, then in Japan, and finally, in the past few years, in the West." [4.20]

4.2 Setup of Activated Low-Pressure Growth from the Vapor Phase

As mentioned above, in all activated low-pressure CVD processes superequilibrium atomic hydrogen or other activated particles (such as superequilibrium atomic oxygen) must be generated in sufficient concentrations. The hot filament process has many advantages, such as simple equipment, good quality of diamond thin films, good stability and reliability for operation, so the process has been widely used. A lot of experimental data reported were based on this process. The typical diamond growth conditions in this process are: CH_4 1%-H_2 99% gas mixture as carbon sources, 2200–2400 K for hot filament, molybdenum, tungsten or tantalum as materials for the hot filament,

Fig. 4.4. Photographs of diamond crystals and thin films grown by activated low pressure CVD process: (**a**) (top left) a large single crystal, (**b**) (top right) a ⟨100⟩ orientated diamond film, (**c**) (bottom left) a ⟨111⟩ orientated diamond film and (**d**) (bottom right) a step of diamond film. Thanks to Z.-S. Jin of Jilin Univesity for photographs provided with copyright permission

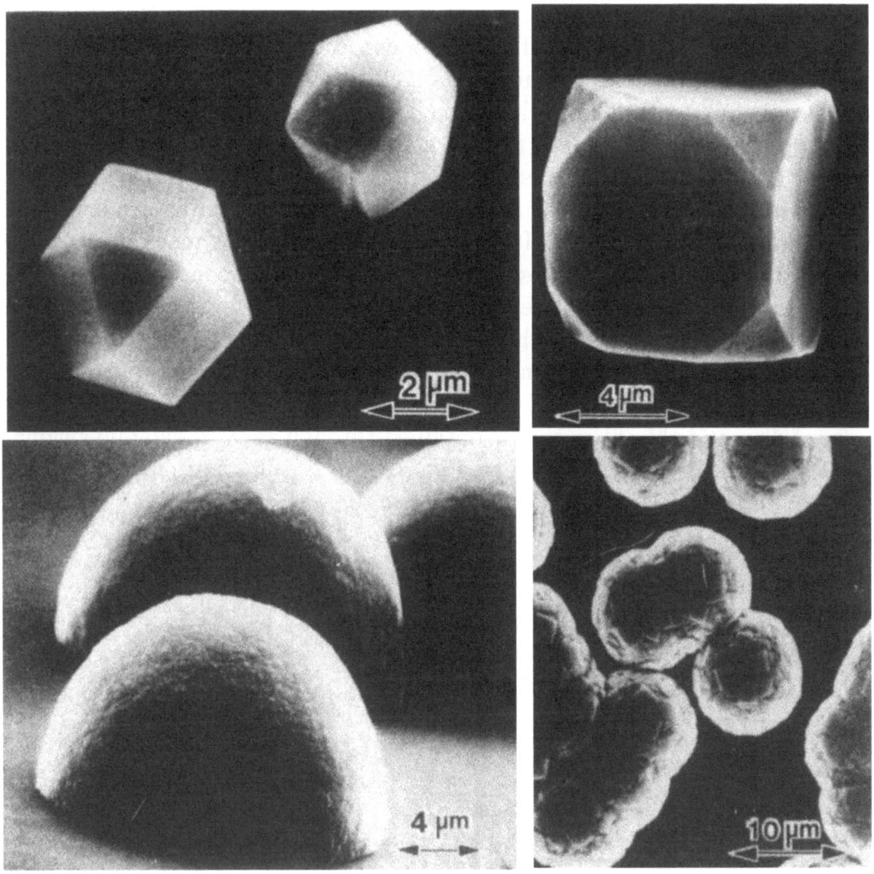

Fig. 4.5. Beautiful photographs of diamond grown on silicon wafers by activated low pressure CVD process: (**a**) and (**b**) (top, left and right) beautiful cubo-octahedral single crystals with very high quality; (**c**) and (**d**) (bottom, left and right) "cauliflower-like" polycrystalline diamond particles. From N. Setaka [4.16] with permission from the Materials Research Society

1073–1273 K for the substrate temperature, and 10 kPa (0.1 atmosphere) of the total pressure. The hot filament temperature and the substrate temperature are easy to measure, so in this book thermodynamic calculations and discussions are mainly about the hot filament process. The results of thermodynamics can also be used for microwave or radio-frequency plasma processes. One shortcoming of the hot filament process is that it is difficult to introduce a little more oxygen, because the hot filament is made of tungsten, tantalum and other refractory metallic materials. If there is too much oxygen introduced, the hot filament will burn out rapidly. In microwave or other radio-frequency, DC or AC generated plasma processes there is no such limitation to the oxygen content in raw materials. Moreover, the activation temperature of neutral particles (¿ 3000 K) in plasma is higher than the activation temperature (i.e., hot filament temperature, usually 2200–2400 K) in the hot filament process. The higher temperature is a little favorable to generate atomic hydrogen. The working pressure in the microwave process (about 2–5 kPa) is lower than that in the hot filament process. The diamond growth rate in the hot filament process and the microwave process is usually about 10 μm h^{-1} or more. The plasma torch process depends on an arc discharge to generate atomic hydrogen. Its activation temperature is much higher than that in the hot filament process, and the corresponding concentration of superequilibrium atomic hydrogen is also higher. That is favorable to diamond growth, so in the plasma torch process the highest growth rate of diamond thin films is up to about 1 mm h^{-1}. The equipment for the combustion process may be the simplest, and its operation is also very convenient. A large area of diamond thin films can be obtained through an X-Y two-dimensional movement on the substrate surface. Figures 4.6–4.12 illustrate the equipment used in different processes.

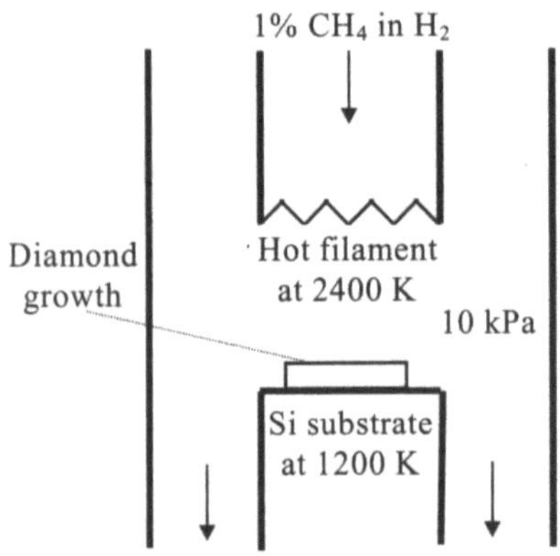

Fig. 4.6. Hot filament process

From these illustrations it can be seen that there is no necessity to have diamond nuclei during activated low-pressure diamond crystal growth on a nondiamond substrate from the vapor phase. Graphite etching may take place with simultaneous diamond growth. If a high temperature gradient chemical transport reaction (HGCTR) process or a closed system microwave plasma process are used, then diamond can be obtained from graphite as a unique carbon source, as shown in Figs. 4.11 and 4.12, respectively. Obviously, as mentioned in the last chapter in the low-pressure simple methane pyrolysis process diamond is a metastable phase, while in activated low-pressure processes discussed here diamond becomes a stable phase. The stability change of phases is the key difference between these two processes. Up to now there

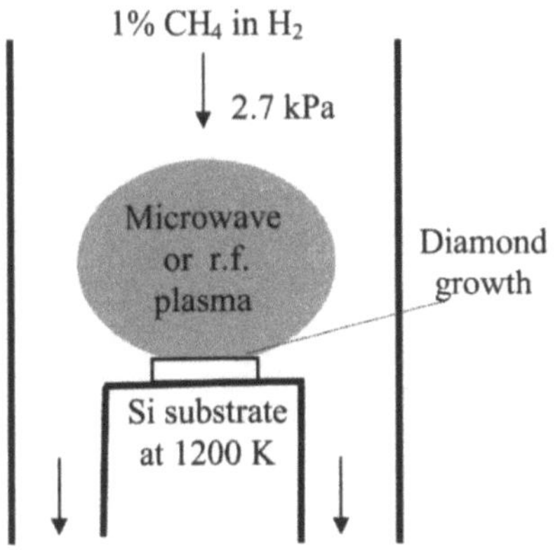

Fig. 4.7. Microwave or r.f. plasma process

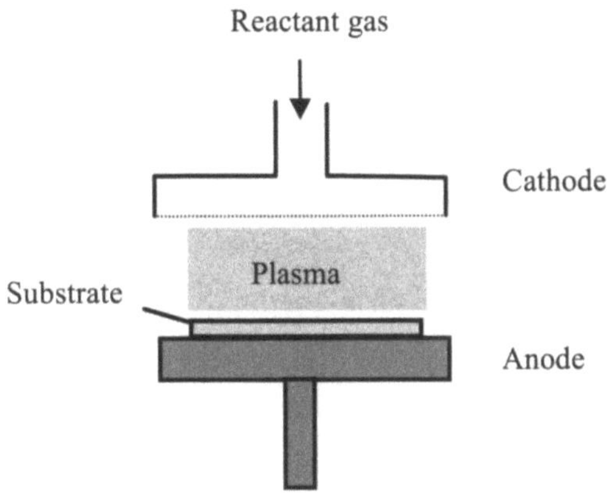

Fig. 4.8. DC plasma process

is still confusion between these two processes in the literature. Both of theses processes are sometimes still called metastable diamond growth processes or diamond growth under metastable conditions. This is an unsuitable or at least not a strict statement. Because the low-pressure simple pyrolysis process for diamond growth has now stepped down from the stage of history, so in the literature and in this book below the term "low pressure diamond process" basically means only activated low-pressure diamond growth from the vapor phase. The discussions in this book are based on thermodynamics, so it is very important to clearly distinguish these two different processes. Therefore, for the latter process the word "activated" will still often be added to emphasize the distinction both in process and in concept.

Figure 4.13 illustrated the apparatus used in the Institute of Physical Chemistry of the former Soviet Union at the early time (maybe during 1965–1967). Hydrogen and methane were separately introduced into the system.

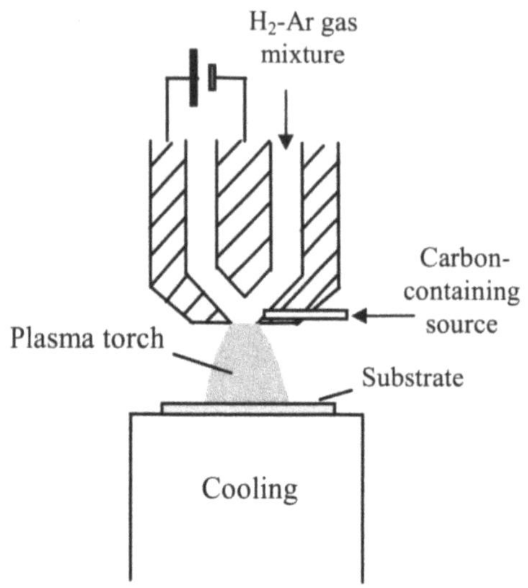

Fig. 4.9. Plasma torch process

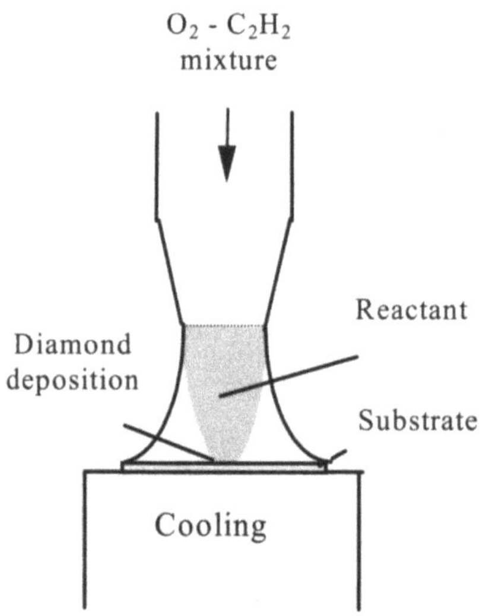

Fig. 4.10. Combustion process

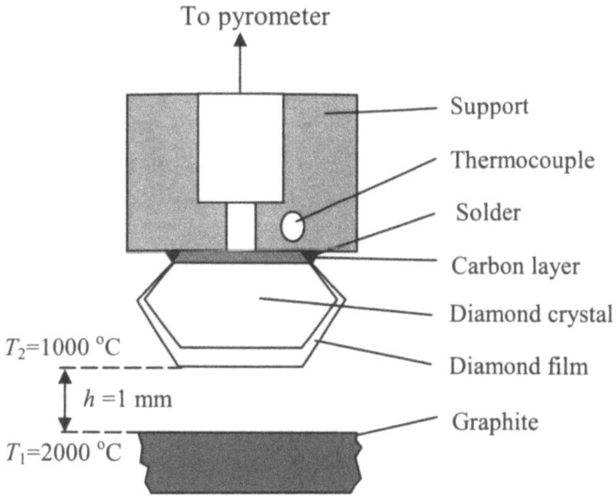

Fig. 4.11. High temperature gradient chemical transport (HGCTR) process (at 1.6 kPa in H_2). Diamond is obtained from graphite. From B.V. Spitsyn [4.9] with permission from The Electrochemical Soc., Inc.

Superequilibrium atomic hydrogen was produced by the tungsten heater at the very high temperature. Although such a apparatus was used for polycrystalline diamond film growth on the diamond seeds at that time, the importance of superequilibrium atomic hydrogen was very clearly shown.

4.3 Preferential Etching Kinetic Model of SAH

According to experimental observation, after the introduction of superequilibrium atomic hydrogen the most outstanding performance is the diamond growth taking place with simultaneous graphite etching [4.9,4.17,4.21]. There was difficult to explain this phenomenon by classical thermodynamics, so Spitsyn proposed a kinetic model on the basis of the preferential etching of graphite by superequilibrium atomic hydrogen [4.10]. The key viewpoints of this model are as follows.

1. If there is only methane in the reactant gas, the methane pyrolysis rates for producing graphite or diamond can be adjusted by selecting the kinetic conditions so that the deposition or growth rate of diamond may be a little higher than the deposition rate of graphite, as shown in Fig. 4.14a.

2. If there exists only hydrogen with atomic hydrogen, the etching rate of graphite by atomic hydrogen is much higher than the etching rate of diamond, as shown in Fig. 4.14b.

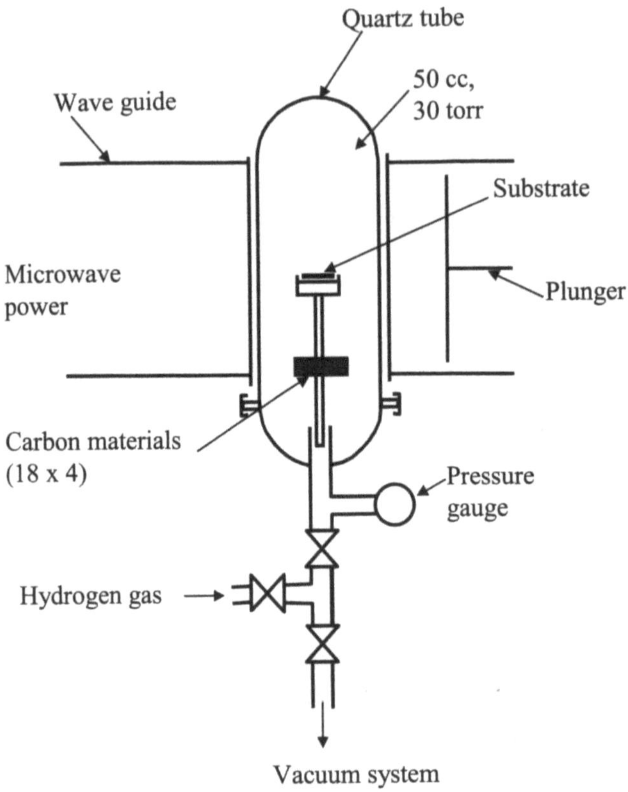

Fig. 4.12. Closed microwave plasma process. Diamond is obtained from graphite. From N. Setaka [4.17] with permission from The Electrochemical Soc., Inc.

3. If both methane and hydrogen with a superequilibrium concentration of atomic hydrogen are used, then due to the combination of both the methane pyrolysis effect and the atomic hydrogen etching effect the situation that the diamond growth rate is positive, while the graphite growth rate is negative (i.e., etching), may be presented as shown in Fig. 4.14c.

Figure 4.14 illustrates the key points of the model. Here, r and e represent the growth rate and the etching rate, respectively. The subscripts d and g represent diamond and graphite, respectively.

4.4 Kinetic Control Model

At that time there was another kinetic control model, in which the so-called "metastable growth" of diamond is emphasized [4.22–4.24]. The model suggested that during low-pressure diamond growth from the vapor phase the

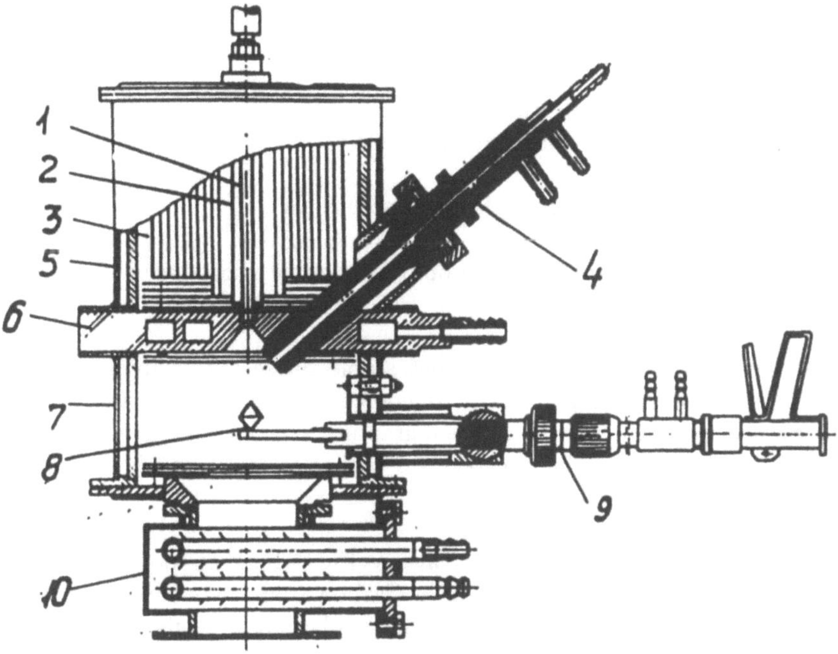

Fig. 4.13. Superequilibrium atomic hydrogen introduced for activated low-pressure diamond growth at early time in the Institute of Physical Chemistry of the former Soviet Union. (1) tungsten tube for hydrogen supply, (2) tungsten heater, (3) Mo screens, (4) methane supply, (5) bell-jar, (6) water-cooled copper diaphragm, (7) reaction chamber, (8) diamond seed crystal, (9) positioner, (10) water-cooled exhaust. From B.V. Spitsyn [4.9] with permission from The Electrochemical Soc., Inc.

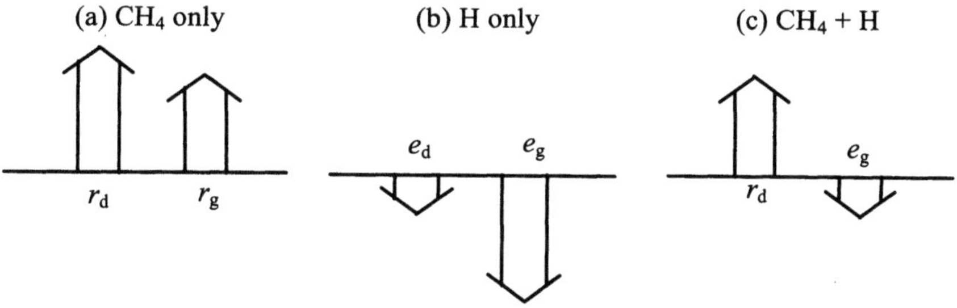

Fig. 4.14. An illustration of preferential etching kinetic model of superequilibrium atomic hydrogen

process be controlled by kinetic effects but not by thermodynamic effects. The key point of view may be described as follows.

1. Low-pressure diamond growth from the vapor phase is controlled by kinetic effects but not by thermodynamic effects.

2. Hydrocarbons are of higher chemical potentials.

3. During the pyrolysis of hydrocarbons it is possible to produce both graphite and diamond.

4. At low pressures diamond is a metastable phase and graphite is a stable phase. However, the free energy difference between graphite and diamond is very small, and the barrier between graphite and diamond is high enough. Therefore, if the kinetic parameters are well selected, it will be possible to get diamond but not graphite.

For illustration, the model may be drawn as Fig. 4.15. From points 2, 3 and 4, the details of the changes are as follows. Carbon atoms in hydrocarbons are of higher chemical potential energy. During pyrolysis of gaseous hydrocarbons, carbon atoms can drop into both the graphite and diamond phases of lower chemical potential energies. Although the energy level of the graphite phase is a little lower, it is still possible to make carbon atoms drop mainly into the diamond metastable phase and to growth diamond by selecting the kinetic parameters. This explanation in the early stage might be suitable for the low-pressure simple pyrolysis process, but for the activated low-pressure CVD diamond process there is no way of providing a clear and logical explanation, especially for diamond growth with the simultaneous graphite etching.

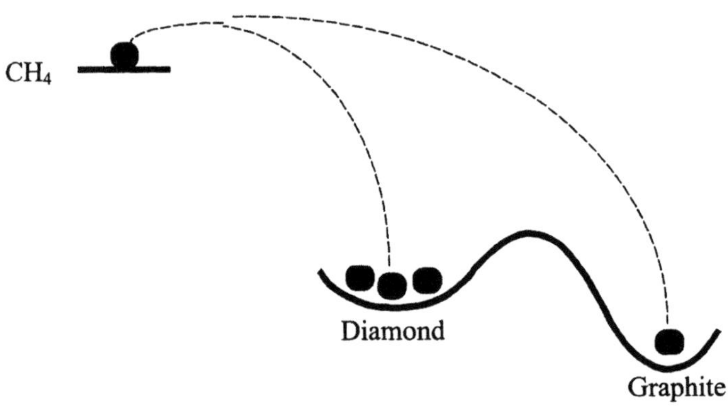

Fig. 4.15. An illustration of kinetic controlled model

The main problem in the kinetic control model was that the kinetics were considered to contradict thermodynamics for activated low-pressure diamond growth from the vapor phase. It is well known that kinetics from the microscopic viewpoint and thermodynamics from the macroscopic viewpoint always compensate each other. There are only the following three possible situations in their relationships.

1. If a change has already kinetically happen in practice, it must also be thermodynamically possible.

2. If a change has been correctly predicted to be thermodynamically impossible, then it should never take place kinetically.

3. If a change is thermodynamically possible, it may be impossible to carry out in practice due to its kinetic rate being too slow.

In the case of activated low-pressure diamond growth from the vapor phase which has already taken place in practice, it will not be thermodynamically impossible, or out of thermodynamic control.

The similar statement can be found in our previous paper (1994) that "There had been an incorrect point of view that the growth of activated CVD diamond is controlled only by kinetics, not thermodynamics. It is incorrect, because kinetics and thermodynamics are compensated each others from different sides, but not against each other. As a basic point of view in physics and chemistry should be that any practical kinetically possible process must be thermodynamically possible without any exception, but, inversely, thermodynamically possible process may be kinetically impossible, or its kinetic rate approximately equals to zero. A famous example is ($2H_2$ $+O_2$) mixture. The formation of water from the mixture is absolutely thermodynamically possible even at room temperature, but in fact the kinetic rate of the water formation reaction nearly equals zero, i.e., the reaction is kinetically impossible at room temperature without a catalyst (such as platinum black) or an initiation (such as a spark)." [4.25]

4.5 Some Thermodynamic Theoretical Models of the 1980s

As mentioned above the former Soviet pioneering researchers mainly took the preferential etching model of superequilibrium atomic hydrogen as their theoretical base of activated low-pressure diamond growth. This is basically a nice kinetic model, but it was difficult to make further quantitative calculations. However, an incorrect viewpoint of the kinetic control model had influenced many researchers in the activated CVD diamond forum. They believed that activated low-pressure diamond growth was controlled only by its kinetics but not by thermodynamics. In the middle of the 1980s, activated CVD diamond growth had become a major focus of research throughout the world. A lot of researchers brought knowledge from different specialties, such as physics, chemistry, mechanics, electronics, materials, and so on. However, the basic thermodynamic question of why diamond can grow under activated low pressure conditions had not yet been satisfactorily answered. In fact, that also presented a challenge to classical thermodynamics.

In the 1980s, the equilibrium phase diagram of carbon was already been known, as shown in Fig. 4.16. There was a key puzzle for theoretical researchers. It was well confirmed in classical equilibrium thermodynamics that at low pressures diamond is a metastable phase, while graphite is a stable phase. Why in the new diamond process can "metastable" diamond growth and "stable" graphite etching take place simultaneously at low pressures? Which is more stable, diamond or graphite at such low pressures? Therefore, activated low-pressure diamond growth from the vapor phase was regarded as a "thermodynamic paradox" perhaps "violating the second law of thermodynamics". In particular, in a chemical transport reaction, as shown in Figs. 4.11 and 4.12, it was confirmed that synthetic diamond crystals can be obtained from graphite as the unique carbon source inside a closed system. Graphite transforms into diamond via the gas phase under such activated low pressure conditions. To sum up, during most of the 1980s there was little research from the theoretical viewpoint into this thermodynamic puzzle. In 1989 three theoretical models were proposed. Some scientists had already realized that the activated low-pressure diamond growth system was a nonequilibrium system. Thus different kinds of revisions or assumptions had been made to classical equilibrium thermodynamics, and that resulted in different explanations. However, most scientists are used to use equilibrium thermodynamics, so to cast off such long-term dominant influences of classical equilibrium thermodynamics should spend some time in a zigzag way.

4.5.1 Quasiequilibrium Model

M. Sommer, K. Mui and F.W. Smith proposed a thermodynamic quasiequilibrium model in 1989 [4.26]. There is a quasiequilibrium method, which had been proposed and used in research work on oxides by J.C. Batty and R.R. Stickney in 1969 [4.27]. This method includes many kinetic steps and introduced many kinetic parameters. According to the quasiequilibrium method and kinetic theory, conclusions similar to that from equilibrium thermodynamics may be derived. The key assumption of the qausiequilibrium model for activated low-pressure diamond synthesis was that thermodynamic equilibrium existed between the solid carbon surface and the vapor species desorbed from it. In the treatment of graphite or diamond growth from the vapor phase, various vapor species collided with graphite or diamond surfaces and were absorbed on these solid surfaces. Then the equilibrium adsorbed species were

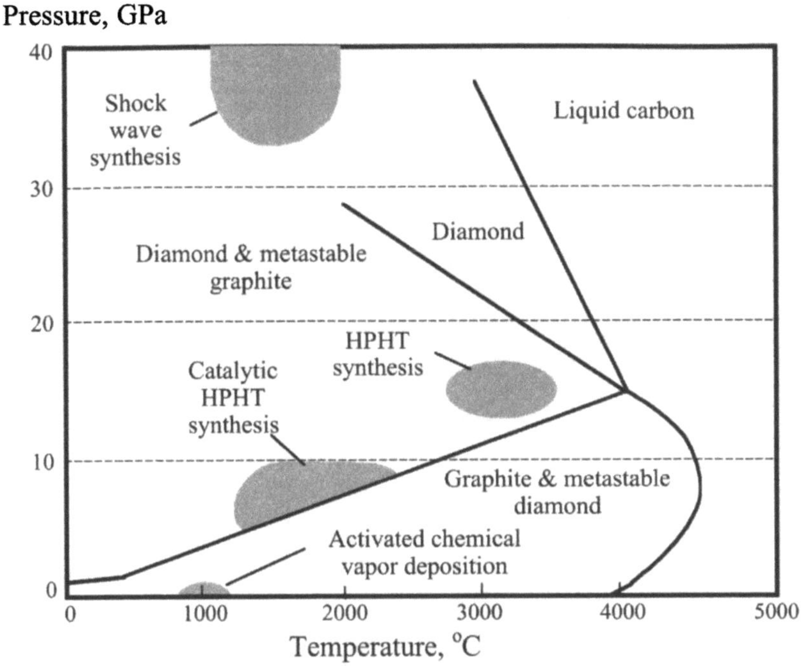

Fig. 4.16. Equilibrium phase diagram of carbon in the late 1980s

evaporated or desorbed on graphite or diamond surfaces. Several additional kinetic (nonthermodynamic) parameters of sticking (or equilibration) coefficients and desorption (or accommodation) coefficients had to be introduced. A quasiequilibrium phase diagram of the carbon-hydrogen binary system was obtained, as shown in Fig. 4.17. In this phase diagram it can be found that the gas–solid phase line of graphite was located on the left low concentration side, while the gas–solid phase line of diamond was located on the right side. Therefore, the solid phase region of graphite was bigger than that of diamond, and there was a single graphite solid phase region between these two gas–solid phase lines. The whole phase diagram was calculated at low pressures. The conclusion agreed with the well-known fact that graphite is a stable phase and diamond is a metastable phase at ordinary low pressure. According to this phase diagram graphite is easier to grow or deposit than diamond. All of these conclusions agreed with the conclusions coming from classical equilibrium thermodynamics. In fact, this phase diagram was basically an equilibrium phase diagram of the carbon–hydrogen binary system but calculated by a quasiequilibrium method.

According to experimental observation in the literature, graphite is more easily etched at activated low pressure during diamond growth from the vapor phase, so Sommer et al. chose the desorption coefficient of hydrogen

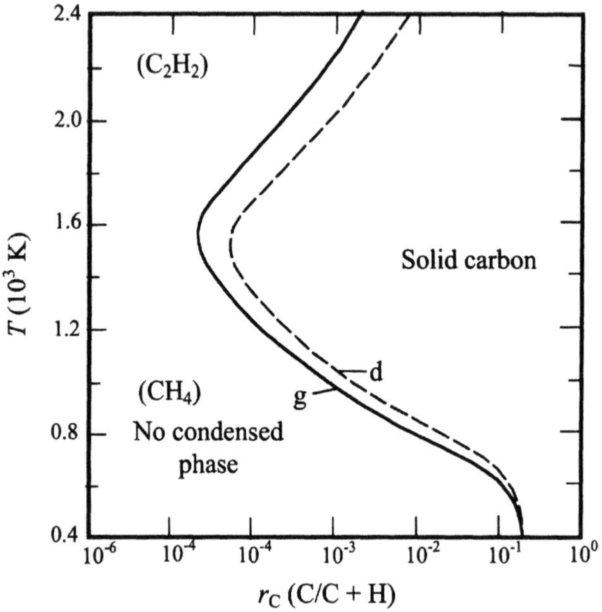

Fig. 4.17. Phase diagrams of the carbon–hydrogen binary system calculated by the quasiequilibrium method at a total pressure of 4.8 kPa. Assuming that the H_2 desorption coefficients on both graphite and diamond surfaces equal 100%. Reprinted from M. Sommer, K. Mui and F.W. Smith [4.26] with permission from Elsevier Science

molecules on the graphite surface into the gas phase as the key kinetic parameter for this phenomenon. It was assumed that the desorption coefficient of hydrogen molecules, which were absorbed and underwent a reaction equilibrium on the graphite surface, did not equal 100% as used in Fig. 4.17 but equals 20%, while the desorption coefficient of hydrogen molecules on the diamond surface equals to 100% and remained unchanged both in Figs. 4.17 and 4.18 for activated low-pressure diamond growth from the vapor phase. Although no detailed theoretical basis of this assumption had been provided, this treatment was very useful. After this modification the graphite phase line shifted to the right side of the diamond phase line, and thus a single diamond solid phase region emerged between the gas–solid phase line of diamond and the gas–solid phase line of graphite in the modified quasiequilibrium phase diagrams. Figure 4.18 is the phase diagram of the carbon–hydrogen binary system calculated by the modified quasiequilibrium method. Experimental data for synthetic diamond grown in the hot filament process and reported by Mastumoto were just dropped inside the diamond phase region of the phase diagram. That was the success of the model. However, there were at least the following obvious insufficiencies.

1. There was no clear theoretical foundation for the assumption of a 20% desorption coefficient of hydrogen molecules on the graphite solid surface but 100% desorption coefficient of hydrogen molecules on the diamond surface. Moreover, other kinetic parameters were also estimated.

2. There was no close relationship with the important role of superequilibrium atomic hydrogen. That would involve a series of problems.

3. If the substrate temperature were equal to the hot filament temperature in the hot filament process the graphite phase line of the original quasiequilibrium phase diagram should meet the graphite line of the modified quasiequilibrium phase diagram at the same point; however, this did not happen in Figs. 4.17 and 4.18.

Due to the strictness of thermodynamics, more serious problems could be deduced from these insufficiencies.

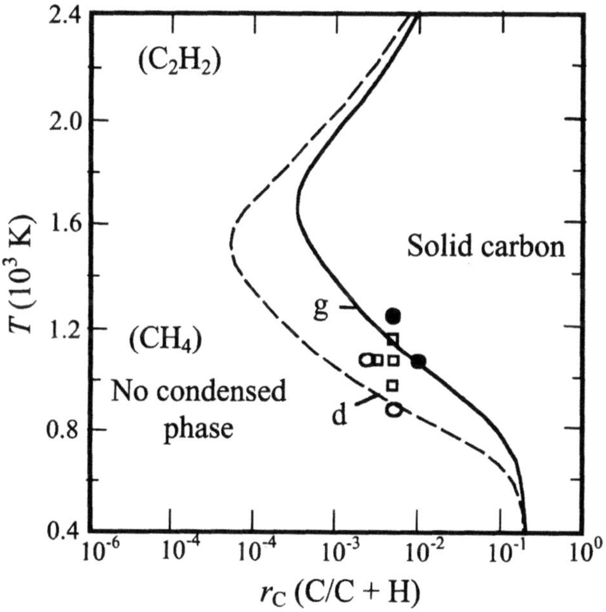

Fig. 4.18. Phase diagrams of the carbon–hydrogen binary system calculated by the revised quasiequilibrium method at a total pressure of 4.8 kPa. Assuming that the H_2 desorption coefficient on graphite is 20% but on diamond is still equal to 100%. Reprinted from M. Sommer, K. Mui and F.W. Smith [4.26] with permission from Elsevier Science

4.5.2 Surface Reaction Thermodynamic Model

In 1989, Yarbrough proposed another surface reaction thermodynamic model for the activated low-pressure diamond growth process [4.28, 4.29]. The key points of view were as follows.

1. Activated low-pressure diamond growth from the vapor phase was a surface reaction process, so the nature of the reaction might be controlled by the structure of its surface(s) during growth. That is, the process depended on the thermodynamic change of the surface reaction.

2. The expression of the basic surface reaction was

$$C_2 \text{ [graphite, (0001)]} + H^o = C_2H \text{ [diamond, (111)]}. \tag{4.1}$$

The structures of the graphite (0001) surface and the diamond (111) surface in hydrogen are shown in Fig. 4.19. ΔH_f represented the heat of formation, and H^o represented atomic hydrogen.

3. Due to hydrogenation of the diamond surface, the hydrogenated diamond surface C_2H [diamond, (111)] was more stable than the graphite surface C_2 [graphite, (0001)], so the above reaction should go to the right-hand side.

According to this theoretical model Yarbrough had estimated the stabilities of the hydrogenated diamond surface and the graphite surface. The conclusion was that the relative stabilities of the hydrogenated diamond surface and graphite surface changed with the concentration of hydrogen atoms. Figure 4.20 showed that when the concentration of hydrogen atoms was high enough, the diamond phase would be more stable than the graphite phase, while the graphite phase is more stable than the diamond phase at ordinary low-pressure conditions. The advantage of this model was the introduction of the well-known important role of superequilibrium atomic hydrogen into its theoretical treatment. However, there were also some basic shortcomings in this model.

1. As pointed out by Yarbrough himself, the large uncertainties involved, such as the arbitrary nature of some of the estimates, the assignment of bulk entropies to surfaces and the various steric factors, meant that the theoretical results were not convincing. Thermodynamics is very strict. In particular, the result is very sensitive to the thermodynamic data used. Arbitrary data may result in arbitrary conclusions.

2. Yarbrough's surface reaction equation was not a complete reaction equation. If the reaction could take place as in the expression, then a free-standing so-called "hydrogenated diamond surface" without any bulk to support it

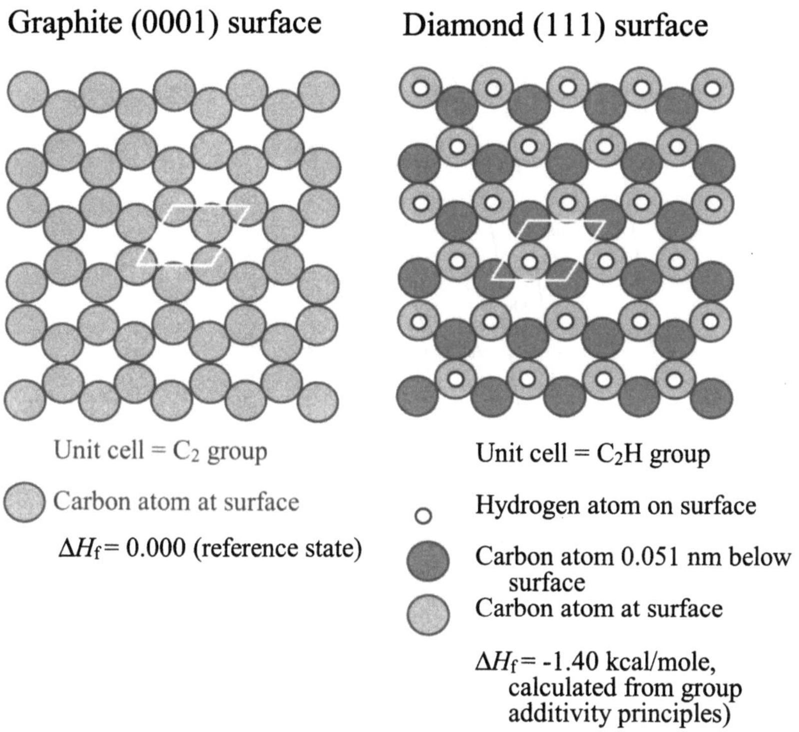

Fig. 4.19. Graphite (0001) surface and diamond (111) surface in hydrogen. For graphite (0001) surface $\Delta H_f = 0.000$ (reference state). For diamond (111) suface $\Delta H_f = -1.400$ kcal/mole (calculated from group additivity principles). Reprinted from [4.28, 4.29] with permission from TMS (The Minerals, Metals & Materials Society)

would be produced. The surface reaction must be connected with another well-known surface abstract reaction of atomic hydrogen on diamond surfaces. If such a revision were made, the total reaction equation would be nearly the same as our chemical pump reaction equation except for a small difference of a coefficient between them. This will be discussed later.

In the note of Fig. 4.20, Yarbrough wrote that "Also shown is the expected (equilibrium) partial pressure of atomic hydrogen at the lowest substrate temperature (600 °C) for which calculations were made. For comparison the expected partial pressure of atomic hydrogen at 10 torr and 1000 °C is about 1.66×10^{-7} atm or $\log[H^\circ] = -6.78$. At 2000 °C and 10 torr, typical operating conditions in the hot wire synthesis of diamond, the expected partial pressure of atomic hydrogen (at the filament) is 2.09×10^{-3} atm or $\log\,[H^\circ] = -2.68$. Clearly if a significant fraction of the atomic hydrogen

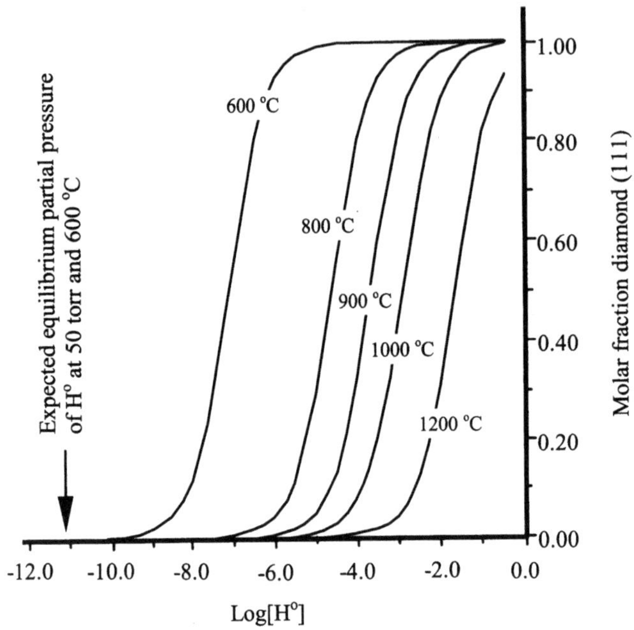

Fig. 4.20. Variation of mole fraction diamond (111) versus atomic hydrogen partial pressure for five different surface temperatures. Reprinted from [4.28, 4.29] with permission from TMS (The Minerals, Metals & Materials Society)

formed in a plasma or at a hot wire persists to form a steady state partial pressure significantly above its equilibrium value for a given surface temperature, diamond not graphite is the indicated product. At very high surface temperature, extremely high partial pressure of atomic hydrogen would be required for steady state diamond growth."

4.5.3 Defect-Induced Stabilization Model

In 1989 Y. Bar-Yam and T.D. Moustakas also proposed a defect-induced stabilization model for activated low-pressure diamond film deposition in the journal *Nature* [4.30]. Bar-Yam et al. said: "Under such growth conditions diamond should be unstable relative to graphite, yet diamond is formed in practice." So the basic point of view of their model was as follows.

1. During activated low-pressure CVD diamond growth high vacancy concentrations were present near the growth face. To explain the reason why diamond but not graphite could grow, a comparison of stabilities between diamond with high vacancy concentrations and graphite with high vacancy concentrations should be made.

2. It was assumed that the formation free energy for diamond and graphite with a concentration n of vacancies was $F = F_{\mathrm{b}} + nF_{\mathrm{d}}$. Here, F_{b} is the binding energy of the materials with negligible defect concentrations and F_{d} is the free energy of formation of a defect or vacancy. Preferential diamond growth would occur if the formation energy of vacancies was smaller in diamond and

$$n > \frac{F_{\mathrm{b}}\,(\mathrm{dia}) - F_{\mathrm{b}}\,(\mathrm{gra})}{rF_{\mathrm{d}}\,(\mathrm{gra}) - F_{\mathrm{d}}\,(\mathrm{dia})} \tag{4.2}$$

where r is the ratio of vacancies in graphite and diamond. Thus, a significant energy difference in the energies of defect formation might change the balance between the close binding energies of graphite and diamond.

The vacancy formation energy in diamond and in graphite was calculated by them as 7.2 eV and 7.6 eV, respectively. It is known that the difference in the cohesive energies of diamond and graphite is about 0.03 eV. According to (4.2), assuming $r \approx 1.8\%$, vacancies at the growth surface would stabilize diamond with respect to graphite, so in the activated low pressure CVD process it was favorable to grow diamond with a lot of defects instead of growing graphite with a lot of defects.

Based on this point of view the energy levels are illustrated in Fig. 4.21. Obviously, this theoretical model could not explain why perfect graphite crystals were not grown with the lowest energy level. In fact, as mentioned above, nice crystals can be obtained by the activated low-pressure CVD diamond process, and even diamond epitaxial layers on a diamond substrate can also be prepared. All of these were difficult to explain from the defect-induced stabilization model.

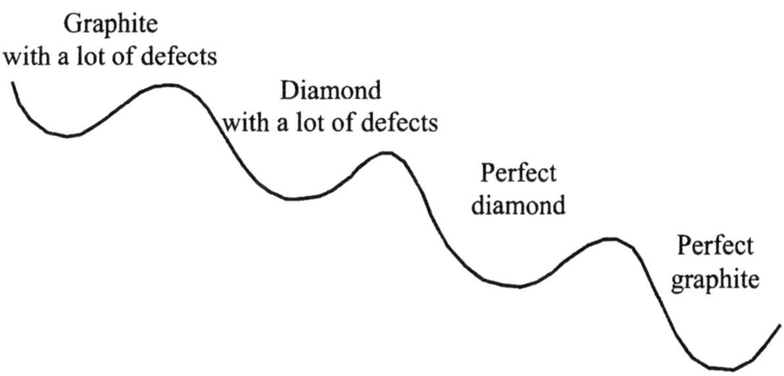

Fig. 4.21. An illustration of the changes of energy level introduced by defects

In the February 1990 issue of the journal *Science*, Yarbrough and Messier, in their paper *Current Issue and Problems in the Chemical Vapor Deposition of Diamond* after a brief review of the theoretical models and thermodynamic explanations made the comment that "All of these formulations suffer from being inconsistent with at least some experimental results."from [4.31] That was an objective comment on these models, including the model proposed by Yarbrough himself, and reflected the theoretical situation in activated CVD diamond research. The reaction mechanisms for the growth of diamond from hydrocarbon vapors were also of intense interest at that time. However, since methane (or methyl) and acetylene were the principal species in the reaction chemistry, the special reaction sequences will not be discussed in detail in this book. In spite of that there were different shortcoming in thermodynamic models or explanations proposed in 1989 or earlier for activated CVD diamond growth, and even some of them were incorrect, but from these theoretical treatments much useful information and ideas were still provided; for instance, the form of phase diagram adopted in Sommer's model; and the effect of superequilibrium atomic hydrogen had been directly under consideration and had been connected with the relative stabilities between diamond (surface) and graphite (surface) in Yarbrough's model, and so on. All of these were the background for the appearance of a new theoretical model, which was proposed in April 1990 as the chemical pump model or a reaction coupling model and will be discussed in the next chapter [4.32].

5. Reaction Coupling Model

5.1 Chemical Pump Model

In 1990 Ji-Tao Wang and Jan-Otto Carlsson proposed the chemical pump model [5.1–5.6]. The key point of view of this model was the importance of inputting an external energy for steady diamond growth at low pressures. During activated diamond growth from the vapor phase at low pressure the external energy was carried into the system in the form of superequilibrium atomic hydrogen (SAH), which was atomic hydrogen in high concentration and hence with higher energy. Moreover, SAH inputted a kind of chemical energy selectively into a definite solid phase, so that the relative stabilities between solid phases (diamond and graphite) were changed. During the association of superequilibrium atomic hydrogen the energy was released in chemical form into the graphite phase, so that the chemical potential of carbon atoms of the graphite phase increased. This phenomenon was called a "chemical pump" effect. The chemical energy released during the association of superequilibrium atomic hydrogen changed graphite from a stable phase into a metastable phase, while diamond became a stable phase. That was the reason why diamond can grow steadily at activated low pressures.

The important role of superequilibrium atomic hydrogen has been generally recognized. SAH should be connected with the stability change between the graphite and diamond phases. The stability change between the graphite and diamond phases was caused by inputting the external energy. Hot filament, plasma or other energy sources generated molecular hydrogen to produce a sufficient amount of superequilibrium atomic hydrogen or other free radicals. The change of the relative stability between graphite and diamond was due to the interaction between activated particles (such as superequilibrium atomic hydrogen) and surfaces of these two solid phases. The external energy introduced by activated particles changed the graphite phase from the original stable state into a more unstable state (graphite is etched), and increased the relative stability of the diamond phase. The atomic hydrogen increases the chemical potential of the carbon atoms in the graphite phase under these conditions. This is very similar to the effect of a chemical pump. The chemical pump through its action on the surfaces of graphite and diamond changed the relative stabilities of solid phases.

5.1.1 Mechanism of the Chemical Pump

The hybridization of the carbon atoms in graphite and diamond is different (sp^2 and sp^3, respectively). Clusters on the graphite and diamond surfaces are different during the growth process. The clusters on the (0001) graphite surface in a hydrogen environment can be described as aromatic polycyclohydrocarbons, whereas the clusters on a (111) diamond surface in the same environment are aliphatic. For brevity, only the simplest six-member ring clusters are discussed. The simplest six-member aromatic hydrocarbon is benzene, and cyclohexane is the simplest six-member aliphatic hydrocarbon. A single six-member cluster on a graphite surface in hydrogen can then be treated as a benzene molecule, which is physically adsorbed on the graphite surface, see Fig. 5.1a. However, a single six-member cluster on a (111) diamond surface in hydrogen can be considered as a cyclohexane molecule, which has adsorbed chemically on the surface, reacted with the diamond hydrogenated surface and released three hydrogen molecules, see Fig. 5.1b. The graphite clusters are unsaturated. On acceptance of superequilibrium atomic hydrogen (SAH) more saturated hydrocarbons (cyclohexene, cyclohexane, decahydronaphthalene, etc.) are formed. The atomic hydrogen provides sufficient energy for decomposition of the clusters into smaller fragments (CH_3, CH_4, C_2H_2, C_2H_4, C_2H_6, etc.). This results in etching of the graphite clusters as well as in etching of the graphite itself.

The enthalpy change for the reactions between graphite and molecular hydrogen and between graphite and atomic hydrogen, which yield different hydrocarbons, illustrates the specific influence of atomic hydrogen on the etching of graphite. Without atomic hydrogen the enthalpy change curve passes a maximum, which can be regarded as an activation barrier for graphite etching in molecular hydrogen (Fig. 5.2). However, for atomic hydrogen a continuous decrease in enthalpy is obtained on the path to CH_4. For atomic hydrogen the calculations were performed for 25% atomic hydrogen, i.e., one hydrogen atom takes part in the formation of one CH_4 molecule. The diamond clusters are saturated, which means that a reaction path similar to that of graphite etching could not be found. No etching (or only weak etching) of diamond by atomic hydrogen could be found.

The favorable influence of atomic hydrogen on diamond growth has been discussed in many papers in the literature. Here, only a few additional comments of the relevance to this study are given. A hydrogenated (111) diamond surface contains a lot of tertiary carbon atoms (the tertiary carbon atom is defined as a carbon atom with three C—C σ single bonds and one C—H σ single bond). On impingement of atomic hydrogen the tertiary carbon atoms become free radicals and molecular hydrogen is formed. The heats of formation of a free tertiary carbon radical, molecular hydrogen and atomic hydrogen are 33 kJ mol^{-1} (based on t-butyl), 0 and 218 kJ mol^{-1} at 298.15 K, respectively. Thus free tertiary carbon radicals are easily formed. These radicals stabilize the (111) surface somewhat (energy decrease). With an increased stability

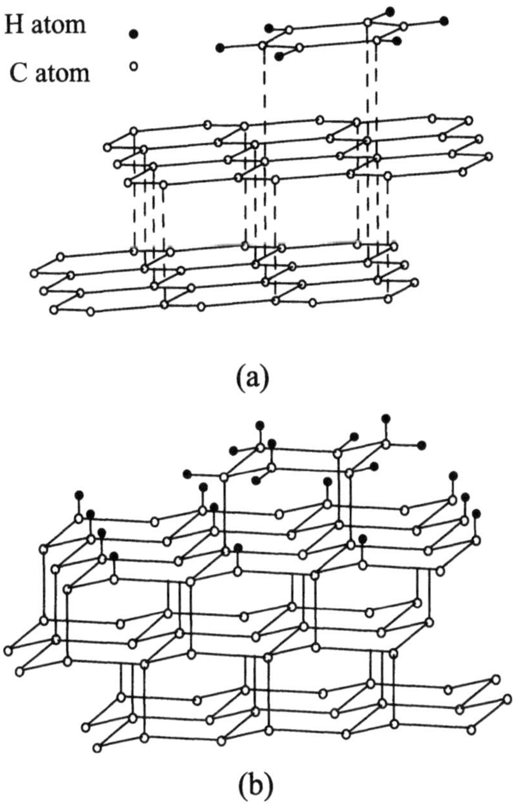

Fig. 5.1. The cluster on the graphite surface is benzene-like, whereas the cluster on the diamond surface is cyclohexane-like. (**a**) Cluster on a (0001) graphite surface and (**b**) cluster on a (111) diamond surface in hydrogen

of this surface the formation of gaseous hydrocarbons (etching) is reduced. The tertiary carbon radicals on the surface may act as sinks for hydrocarbon radicals supplied from the vapor and thus support diamond growth.

For a (0001) graphite surface the situation is completely different. There are no tertiary, secondary (the secondary carbon atom is defined as a carbon atom with two C—C σ single bonds and two C—H σ single bonds) or primary carbon atoms (the primary carbon atom is defined as a carbon atom with one C—C σ single bond and three C—H σ single bonds) but a big π bond exists. As discussed above a six-member cluster on a graphite surface in an atomic hydrogen environment is benzene-like. The simplest radical of benzene is phenyl. The heat of formation of a free phenyl radical is 335 kJ mol^{-1} at 298.15 K, which is much higher than the heat of formation of a tertiary carbon radical on a (111) diamond surface. Therefore, the existence of atomic hydrogen does not promote growth on the graphite surface.

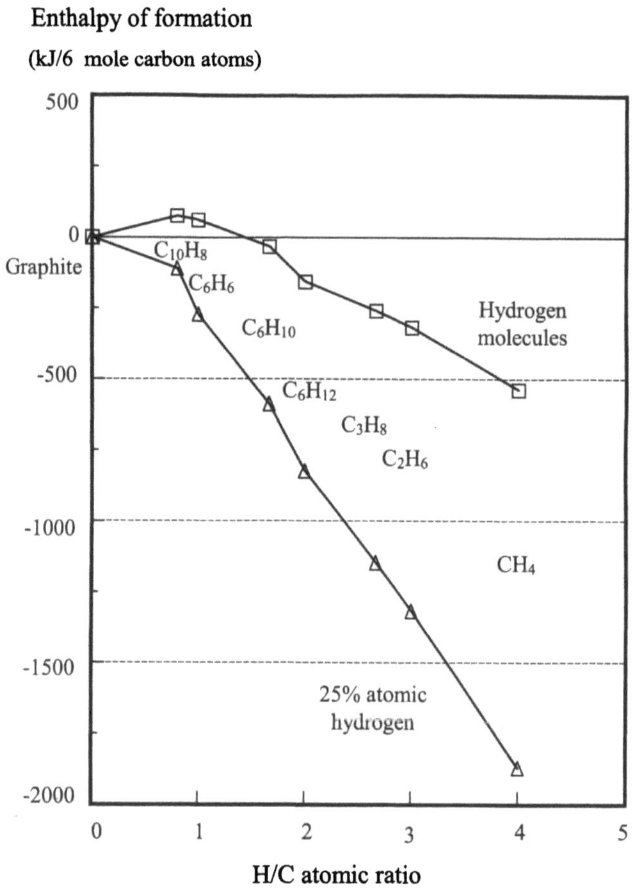

Fig. 5.2. Enthalpy change at 1000 K for the reaction between graphite and molecular hydrogen and the reaction between graphite and atomic hydrogen (25% atomic hydrogen)

In summary, the effect of atomic hydrogen in concentration higher than the equilibrium concentration (superequilibrium atomic hydrogen) can be described in the following way.

1. A (0001) graphite surface is easily etched. The superequilibrium atomic hydrogen increases the energy of the surface and transforms it into an unstable state, i.e., a strong chemical pump effect on the graphite surface is obtained. Growth on the graphite surface requires free surface radicals. However, the energy needed to form such radicals, for instance a phenyl radical, is high. This makes growth difficult.

2. A (111) diamond surface with its tertiary carbon atoms is very different from a (0001) graphite surface. Weak etching on the (111) diamond surface

in atomic hydrogen occurs, i.e., a weak pumping effect on the (111) surface is obtained. Free surface radicals are more easily formed on diamond than on graphite, promoting growth on diamond (a weak negative pump effect). Therefore, the etching and growth promoting effects of superequilibrium atomic hydrogen on the (111) diamond surface nearly offset each other (i.e., the chemical pump effect approaches zero).

The overall effect of superequilibrium atomic hydrogen from an energy point of view is to increase the energy of graphite and to change the energy of diamond slightly. The pump medium "superequilibrium atomic hydrogen" pumps, with its chemical energy, the normally stable graphite phase to a metastable state.

5.1.2 Chemical Pump Reaction

In the previous section, the chemical pump effect was discussed in detail from the viewpoint of the microscopic mechanism and process. However, from the viewpoint of thermodynamics the energy levels of the initial state and the final state are more interesting. The energy levels of the initial state and the final state can be used as a criterion of the reaction direction and reaction degree. For brevity, only a carbon–hydrogen binary system (such as a gas mixture of CH_4 with hydrogen used as reactant) in the hot filament diamond process will be discussed in this chapter. Suppose that the hot filament temperature $T_{\text{activated}}$ is 2200–2400 K, the total pressure p is less than 101.325 kPa, and T represents the substrate temperature.

To understand whether graphite [denoted by C (gra)] or diamond [denoted by C (dia)] is more favorably produced in terms of energy, reaction (5.1) should be written out first. It is well known that under equilibrium low-pressure conditions graphite is a stable phase, while diamond is a metastable phase, so the Gibbs free energy change of the reaction (5.1), denoted by subscript 1, is greater than zero, i.e., $\Delta G_1 > 0$. Then

$$\mathrm{C(gra)} = \mathrm{C\,(dia)}\,. \qquad (5.1)$$

In Chap. 1 it was shown that the direction criterion of a spontaneous reaction at constant temperature and pressure is the sign of the Gibbs free energy change of the reaction ΔG. If ΔG of the reaction is smaller than zero, the spontaneous direction of the reaction will be toward the right-hand side, otherwise, if ΔG is bigger than zero, the spontaneous direction of the reaction will be toward the left-hand side. Because $\Delta G_1 > 0$ here, it represents the spontaneous direction of reaction (5.1) being toward the left side. That is, reaction (5.1) is nonspontaneous.

However, in the same system there is another reaction (5.2), the association of superequilibrium atomic hydrogen, taking place simultaneously:

$$H = 0.5H_2. \tag{5.2}$$

For the carbon–hydrogen system in the activated low-pressure CVD diamond growth process, the important role of superequilibrium atomic hydrogen is generally recognized. Of course, the association of superequilibrium atomic hydrogen, i.e., reaction (5.2) denoted by subscript 2, can take place spontaneously, because $\Delta G_2 \ll 0$. This reaction releases a lot of energy and produces hydrogen molecules.

The sum of reaction (5.1) and (5.2) with a generalized parameter χ (called the pump parameter) is
(5.3) = (5.1) + χ (5.2),

$$C(gra) + \chi H = 0.5\chi H_2 + C(dia). \tag{5.3}$$

The whole reaction, i.e., reaction (5.3) denoted by subscript 3, takes place at constant total pressure and at constant substrate temperature. The hot filament temperature is much higher than the substrate temperature, so the hot filament generates atomic hydrogen. If χ is not too small, then according to $\Delta G_3 = \Delta G_1 + \chi\Delta G_2$, ΔG_3 must be smaller than zero. Due to $\Delta G_3 < 0$, reaction (5.3) going toward the right side completely agrees with basic thermodynamic laws. That is to say, according to basic thermodynamic laws, at low pressure graphite can react with superequilibrium atomic hydrogen and produce molecular hydrogen and diamond. The reaction (5.3) here is called a chemical pump reaction.

5.2 Calculation of the Chemical Pump Effect

There are still some difficulties in quantitatively calculating the chemical pump reaction (5.3). Rewrite the reaction equation in the form:

$$C(gra) + \chi\,(H - 0.5H_2) = C(dia) \tag{5.4}$$

$$\Delta G_4 = \Delta G_3 = \Delta G_1 + \chi\Delta G_2.$$

If χ is not too small, the Gibbs free energy change of the whole reaction will still be smaller than zero, i.e., $\Delta G_4 < 0$. The direction of reaction (5.4) should be toward the right side. This reaction can be used for diamond growth. The left side of reaction (5.4) can be treated as a whole, and thermodynamic data of [C(gra) + χ(H − 0.5 H_2)] can be used instead of thermodynamic data of the original graphite during thermodynamic calculations. In this way, the chemical pump effect has already been included in the thermodynamic calculations. The routing Gibbs free energy minimization program for thermodynamic calculations, such as EKVICALC [5.7] or THERMOCALC, can still be used. For calculation of the carbon–hydrogen binary system all possible gaseous species, such as H, H_2, CH_4, C_2H_2, CH_2, CH_3, C_2H_4, C_2H_6, and so on, should be considered. All these thermodynamic data can be found in standard thermodynamic data books and in the literature [5.8, 5.9].

Figure 5.3 shows the calculated stability diagram for the C-H system with different values of the chemical pump parameter χ. When $\chi = 0$, that means no chemical pump effect, i.e., the diagram is just the same as the ordinary phase diagram of the carbon–hydrogen system. If $\chi = 0.1$, 0.2, 0.3, and 0.4, then diagrams with different degrees of the chemical pump effect are obtained. The calculations were performed for conditions where experimental data are available (total pressure 4.8 kPa, filament temperature 2200 K). As can be seen the pump effect drastically changes the position of the stability region of graphite. If the substrate temperature equals the filament temperature (2200 K) there is no pump effect, if the substrate temperature is higher than 2200 K no superequilibrium atomic hydrogen exists, and hence the graphite stability region is unchanged. If the substrate temperature is below 1400 K, then the equilibrium concentration of atomic hydrogen at the filament is much bigger than that at the substrate temperature, i.e., $[\mathrm{H}]_{\mathrm{filament}} \gg [\mathrm{H}]_{\mathrm{substrate}}$ and therefore the pump is effective. In the temperature range 1400–2200 K, which is outside the normal diamond growth temperature, the borderlines are sketched. Only for a minor pump effect $\chi \approx 0.05$) is diamond as stable as graphite.

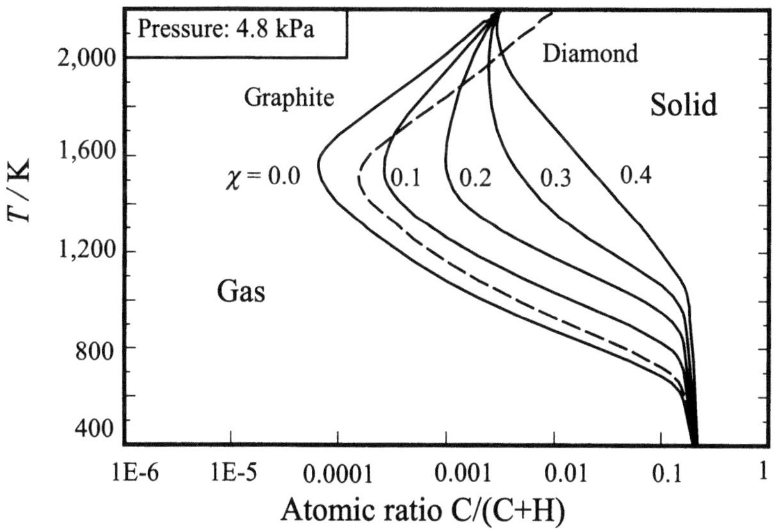

Fig. 5.3. Influence of χ value on thermodynamic stability in the C-H system (total pressure, 4.8 kPa; filament temperature, 2200 K)

Figure 5.4 includes some data points from the experimental investigation by Matsumoto et al. [5.10] As discussed in the original paper of Matsumoto et al., the temperature reported was too low since the effect of substrate heating by the hot filament could not be taken into account because of the temperature measurement technique. With the experimental conditions used it is reasonable that the hot filament above the substrate surface will increase the substrate temperature by 100 K, which, for the comparison here, has been added to the reported temperature. For $\chi = 0.2$ good agreement between the calculated diamond stability region and Matsumoto's experimental observations is obtained. Finally, Fig. 5.5 shows the stability diagram of the carbon–hydrogen system at a lower total pressure (1.3 kPa). For a lower total pressure the diamond stability region is shifted towards a lower carbon concentration.

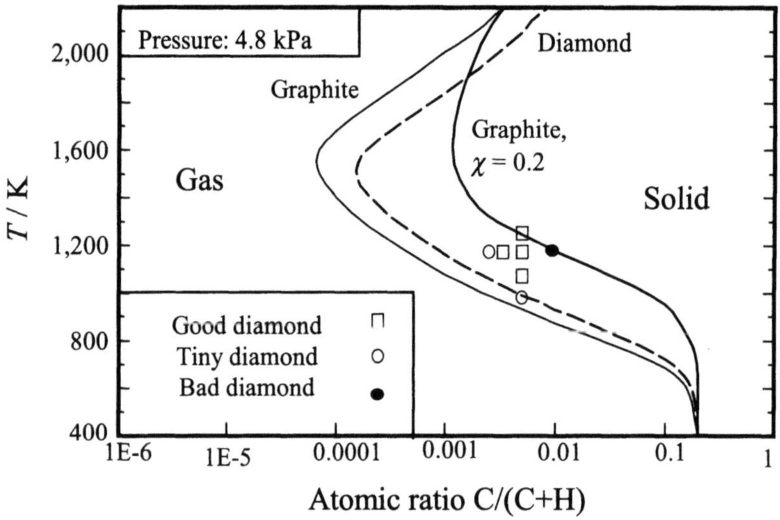

Fig. 5.4. Comparison between experimental data and calculated stability diagram (total pressure, 4.8 kPa; filament temperature, 2200K; $\chi = 0.2$)

During the impingement of activated particles with higher energy on the solid surface, the thermodynamic stabilities of different solid phases may be changed significantly for the allotrope system with difference structures. For instance, for a carbon system at 1200 K, atomic hydrogen very effectively etches graphite. Atomic hydrogen of superequilibrium concentration creates a chemical pump effect on the graphite surface, and increases the energy level of graphite. In the activated low-pressure diamond growth process atomic hydrogen of high concentration quickly makes diamond more stable than graphite. That is, superequilibrium atomic hydrogen pumps the energy level of graphite up to a new level, so that diamond becomes a stable phase. To sum up, in the activated low-pressure diamond growth process, there are two very important basic points. First, activated particles (such as atomic hydrogen) must be of superequilibrium concentration. Second, there must be special

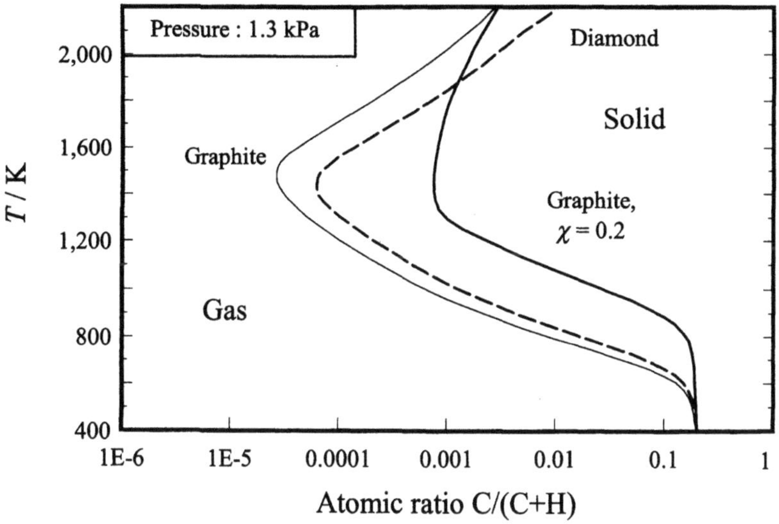

Fig. 5.5. Calculated stability diagram for the carbon–hydrogen system at lower total pressure; and the diamond stability region is shifted towards a lower carbon concentration (total pressure, 1.3 kPa; filament temperature, 2200K; $\chi = 0.2$)

surface structures of different phases. For instance, the structures of diamond and graphite are quite different. That will be favorable to a chemical pump reaction taking place, to change the original stabilities of different phases, and to get unexpected results. That is the most important contribution with wide implications provided by the success of activated low-pressure diamond growth from the vapor phase.

To sum up, the chemical pump model not only solved the puzzle or so called "thermodynamic paradox" between experimental observations during activated low-pressure diamond growth from the vapor phase and certain conclusions of classical thermodynamics, but also combined the two main kinds of models based on kinetics and thermodynamics, see Fig. 5.6. A common point of view of kinetic models is that diamond is a metastable phase while graphite is a stable phase at low pressures, and due to kinetic conditions (such as the fact that the diamond etching rate is much smaller than the graphite etching rate in the existence of SAH) favorable to practical diamond growth the diamond growth rate can much higher than the practical growth rate of graphite and even graphite can be etched away by SAH, see Fig. 5.6a. The common point of view of all thermodynamic models are that under special conditions (different conditions were emphasized in different models, such as the emphasis on absorption and desorption, surface reaction or defects) the stabilities of diamond and graphite have already changed, and in practice diamond becomes a stable phase while graphite becomes a metastable phase, so only diamond can grow, see Fig. 5.6b. Both of these common points of view can be unified in the chemical pump model, see Fig. 5.7.

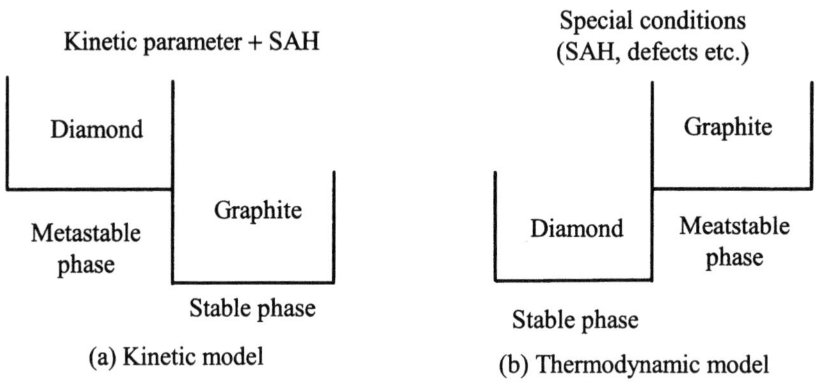

Fig. 5.6. Two kinds of theoretical model

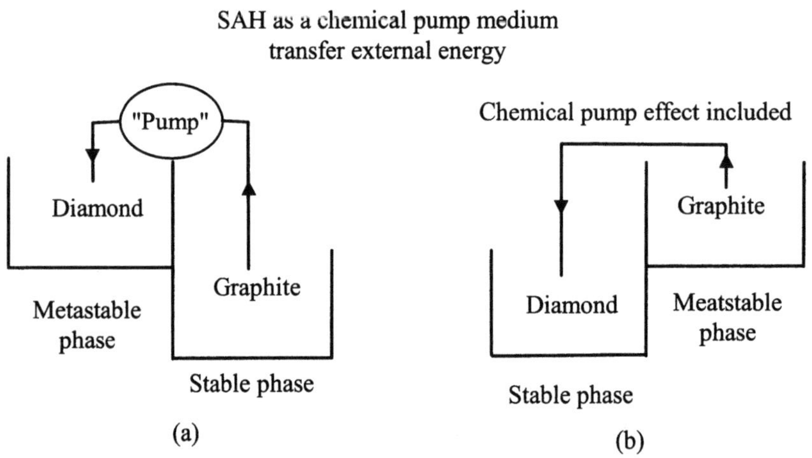

Fig. 5.7. Two kinds of illustrations for the chemical pump model

Based on the chemical pump model diamond at low pressures is a metastable phase of higher energy, while graphite is a stable phase of lower energy, but due to the chemical pump carbon atoms can be transferred from a lower energy graphite phase into a higher energy diamond phase, see Fig. 5.7a. In this way, the model explains diamond growth from the vapor phase taking place together with simultaneous graphite etching. On the other hand, the chemical pump model can also emphasize that due to the chemical pump effect graphite has already become a metastable phase, while diamond has become a stable phase, see Fig. 5.7b. Therefore, graphite can be transformed into diamond via the gas phase. The situation is very similar to an ordinary water pump pumping water from a lower pool to a higher pool. Two explanations can be given. One can say that due to the pump, water can be transferred from the lower pool to the higher pool. One can also say that after the addition of the water-head of the pump the energy level of the lower pool is already higher than the energy level of the higher pool, so that water in the lower pool can flow into the higher pool.

5.3 Reaction Coupling Model

As mentioned above, a chemical pump model was proposed in 1990. The model casts off arbitrary assumptions in the explanation of activated low-pressure diamond growth from the vapor phase, and makes important creative progress in the thermodynamic calculation method. However, before the fall of 1994, there were rather obvious problems in its calculated results. According to the calculated results in 1990, if the substrate temperature was equal to the hot filament temperature (i.e., there existed no superequilibrium concentration of atomic hydrogen) the gas–solid phase line of activated graphite did not meet the original equilibrium phase line of graphite at the same point. That was the reason why we had to write in our paper of 1990 that "In the temperature range 1400–2200 K, which is outside the normal diamond growth temperature, the borderlines are sketched" [5.1] and in our paper of 1994 that "Due to the instability of solid diamond at high temperature the over 1500 K parts of dashed lines in Fig. 2 and Fig. 3 also have no practical meanings." [5.6] That was due to the hidden troubles and insufficiencies of the chemical pump reaction (5.1)–(5.4), in the following two respects.

1. The expression of (5.2), i.e., $H = 0.5\ H_2$ was not strict enough. The expression could not distinguish the following situations at constant temperature and pressure:
(a) When atomic hydrogen is of superequilibrium concentration, then
H (superequilibrium) $= 0.5\ H_2$, $\Delta G_2 < 0$.
(b) When atomic hydrogen is of equilibrium concentration,
H (equilibrium) $= 0.5\ H_2$, $\Delta G_2 = 0$.
(c) When atomic hydrogen is of underequilibrium concentration,
H (underequilibrium) $= 0.5\ H_2$, $\Delta G_2 > 0$.
Obviously, these three situations are very important for a theoretical model for diamond growth. Without the notation (superequilibrium), (equilibrium), or (underequilibrium), the expression itself is not able to determine the value

and even the sign of ΔG_2. That was not permitted for any strict thermodynamic model.

2. If some kind of notation (superequilibrium), (equilibrium) or (underequilibrium) is added into the equation to distinguish them, then the thermodynamic data of superequilibrium atomic hydrogen should not be used in the same way as the standard thermodynamic data of equilibrium atomic hydrogen for the calculation as done in the previous section in this chapter. Correspondingly, the value and the meaning of activated graphite (gra*) and χ will be different from the previous values. A careful derivation for thermodynamic data of superequilibrium atomic hydrogen and activated graphite should be made first.

Of course, this problem would not have happened in ordinary equilibrium thermodynamic calculations. In other words, in ordinary equilibrium thermodynamic calculations the atomic hydrogen is always assumed to be in equilibrium, as the situation of 1(b). It is well known that standard thermodynamic data of carbon, hydrogen, and simple hydrocarbons can be found in data books. For any reaction taking place at constant temperature and pressure A + B = C + D, $\Delta G = G_{\mathrm{D}} + G_{\mathrm{C}} - G_{\mathrm{A}} - G_{\mathrm{B}}$. If the G value of the Gibbs free energy of any component could be changed arbitrarily, then any value and even the sign of ΔG might be obtained. For instance, if a bigger value of G_{D} is chosen, then perhaps $\Delta G > 0$, and that would mean the reaction not being spontaneous. However, if a smaller value of G_{D} is chosen, then maybe $\Delta G < 0$, and that would mean the reaction being spontaneous. Such an arbitrary selection, evaluation, or change of thermodynamic data will make thermodynamic calculation meaningless.

For the solution of the first problem, H* was used instead of the previous notation of H for representing atomic hydrogen of superequilibrium concentration. Therefore, the previous reaction (5.1)–(5.4) was rewritten in the stricter form of expressions together with the notation of temperature, pressure, etc., that is,

$$\mathrm{C(gra)} = \mathrm{C(dia)}, \ (T, p \leq 10^5 \ \mathrm{Pa}) \qquad (5.5)$$
$$\Delta G_1 > 0,$$

$$\mathrm{H}* = 0.5\mathrm{H}_2, \ (T_{\mathrm{activated}} \gg T, p \leq 10^5 \mathrm{Pa}) \qquad (5.6)$$
$$\Delta G_2 \ll 0.$$

Here, H* represents atomic hydrogen of superequilibrium concentration.
(5.7) = (5.5) + χ (5.6) gives

$$\mathrm{C(gra)} + \chi\mathrm{H}* = 0.5\chi\mathrm{H}_2 + \mathrm{C(dia)}, \ (T, p \leq 10^5 \mathrm{Pa}) \qquad (5.7)$$
$$\Delta G_3 = \Delta G_1 + \chi \Delta G_2$$

and (5.7) can also be rewritten as

$$\begin{aligned} \mathrm{C(gra*)} &= \mathrm{C(dia)}, \quad (T, p \leq 10^5\mathrm{Pa}) \\ \Delta G_{3'} &= \Delta G_3 . \end{aligned} \tag{5.8}$$

Here,

$$\mathrm{C(gra*)} = \mathrm{C(gra)} + \chi\,(\mathrm{H*} - 0.5\mathrm{H_2})\,. \tag{5.9}$$

If the reaction rate ratio $\chi = r_2/r_1 >\mid \Delta G_1/\Delta G_2 \mid$ is not too small, (i.e., superequilibrium atomic hydrogen is of sufficient concentration), then we must have $\Delta G_3 < 0$, that is, reaction (5.7) or reaction (5.8) should be toward the right side. gra* represents activated graphite which means graphite activated by superequilibrium atomic hydrogen. It should be noted that the notation of superequilibrium atomic hydrogen has been changed formally from its previous form of H in reaction (5.1)–(5.4) into the notation of H* in reaction (5.5)–(5.8). Naturally, thermodynamic data of atomic hydrogen of superequilibrium concentration should not directly use the standard thermodynamic data or functions of atomic hydrogen of equilibrium concentration. Correspondingly, the values and definitions of activated graphite gra* and χ in reaction (5.9) are quite different from the previous ones. In comparison with criterion (2.18) of reaction coupling at constant temperature and pressure in Chap. 2, i.e., $[(\Delta G_1)_{T,p} > 0,\ (\Delta G_2)_{T,p} < 0$ and $\{(\Delta G_1)_{T,p} + \chi(\Delta G_2)_{T,p}\} \leq 0]$, it can clearly be found that the physical meaning of the coupling parameter χ is the reaction rate ratio, i.e., $\chi = r_2/r_1$. In the above formulae, p is the total pressure, T is the substrate temperature, $T_{\mathrm{activated}}$ is the activated temperature, which in the hot filament process equals the temperature of the hot filament. In microwave, radio-frequency, DC or AC plasma processes, the activated temperature equals the temperature of neutral particles in the plasma (3000–3500 K).

For distinguishing the important change of notation of superequilibrium atomic hydrogen from H to H*, since the fall of 1994 we usually rename this theoretical model the nonequilibrium thermodynamic reaction coupling model or the reaction coupling model for short [5.11–5.24]. From this name it can be found that the concept of the chemical pump is the same as the concept of reaction coupling or coupled reaction proposed by D. Burk in 1931 [5.25] and by Th. De Donder in about 1936 [5.26]. After knowing the earlier work of reaction coupling, the renaming from chemical pump model to reaction coupling model is suitable to show respect for the previous concept and also to have the stricter academic meaning. However, the names of the chemical pump and chemical pump model are advantageous for vividness and ease of understanding. Moreover, this concept of "pump" is already very popular in biochemistry for the description of many activated diffusion processes in which a substance can be "diffused" or "pumped" from a low concentration part to a high concentration part. For instance, Na^+, K^+-ATP pump, etc., are well known terms in biochemistry. Therefore, the term the chemical pump can be used as well.

5.4 Thermodynamic Data for Activated Graphite

For solving the second problem mentioned in Sect. 5.3, the derivation of thermodynamic data for superequilibrium atomic hydrogen and the corresponding activated graphite must be discussed. First, a clear definition of superequilibrium atomic hydrogen should be given: Superequilibrium atomic hydrogen is defined here as atomic hydrogen of equilibrium hydrogen atom concentration at activated temperature $T_{\text{activated}}$ (i.e., the hot filament temperature T_{filament} in the hot filament process), so for the substrate temperature T atomic hydrogen is of superequilibrium concentration. The derivation of thermodynamic data for superequilibrium atomic hydrogen and activated graphite will be discussed as follows. In brief, the derivations here are limited only to an activated (hot filament) temperature equal to 2400 K, but in two different ways.

1. Gibbs free energy method

(1) If the substrate temperature equals the activated (hot filament) temperature, the whole system may be treated as an equilibrium system, so at 2400 K the free energy $G_{\text{gra}*}^{2400}$ (representing the Gibbs free energy of the activated graphite at 2400 K) should equal the free energy of equilibrium graphite G_{gra}^{2400}:

$$G_{\text{gra}*}^{2400} = \left[G_{\text{gra}}^{2400} + \chi\left(G_{\text{H}*}^{2400} - 0.5G_{\text{H}_2}^{2400}\right) = G_{\text{gra}}^{2400}\right] \tag{5.10}$$

that is,

$$G_{\text{H}*}^{2400} = 0.5G_{\text{H}_2}^{2400}. \tag{5.11}$$

From thermodynamic data book (D.R. Stull et al. [5.8]), the standard Gibbs free energy of molecular hydrogen at 2400 K is

$$G_{\text{H}_2}^{2400} = -400.290 \text{ kJ mol}^{-1}. \tag{5.12}$$

Therefore, the Gibbs free energy of superequilibrium atomic hydrogen at 2400 K is

$$G_{\text{H}*}^{2400} = -200.145. \text{ kJ mol}^{-1}. \tag{5.13}$$

For comparison, the standard Gibbs free energy of atomic hydrogen

$$G_{\text{H}}^{2400} = -117.429 \text{ kJ mol}^{-1} \tag{5.14}$$

which was used in our modeling work before the fall of 1994, is cited here from thermodynamic data books. In fact, the difference between these free energies is due to the difference of the partial pressure of superequilibrium atomic hydrogen from the standard pressure.

(2) From thermodynamics, $G = H - TS$. For an idea gas, the values of the free energy G and the entropy S are closely related to the partial pressure (or concentration), but the value of the enthalpy H is not. Approximately, $H_{\text{H}*} = H_{\text{H}}$, so

$$S_{H*} = S_H - (G_{H*} - G_H)/T \tag{5.15}$$

$$\begin{aligned} S_{H*}^{2400} &= \left[S_H^{2400} + \left(\frac{200145 - 117429}{2400} \right) \text{ J mol}^{-1}\text{K}^{-1} \right] \\ &= \left[S_H^{2400} + 34.465 \text{ J mol}^{-1}\text{K}^{-1} \right]. \end{aligned} \tag{5.16}$$

The heat capacity of superequilibrium atomic hydrogen at constant pressure C_p is taken to be the same as that of atomic hydrogen in the equilibrium state. Therefore, the equality is valid for the whole range of interest, i.e.,

$$S_{H*}^{2400} = S_{H*}^{298} + \int_{298}^{2400} \frac{C_p}{T} \mathrm{d}T \tag{5.17}$$

and

$$S_H^{2400} = S_H^{298} + \int_{298}^{2400} \frac{C_p}{T} \mathrm{d}T \tag{5.18}$$

so

$$\begin{aligned} S_{H*}^{298} &= \left[S_H^{298} + 34.465 \right] \text{ J mol}^{-1}\text{K}^{-1} \\ &= [114.609 + 34.465] \text{ J mol}^{-1}\text{K}^{-1} = 149.073 \text{ J mol}^{-1}\text{K}^{-1}. \end{aligned} \tag{5.19}$$

(3) According to these superequilibrium atomic hydrogen data and expression (5.9) for activated graphite, thermodynamic data for activated graphite with different χ values can be obtained, as listed in Table 5.1.

Notice that in (5.12) and (5.14) standard thermodynamic data, i.e., data at standard pressure (101.325 kPa or 1 atmosphere) are used. If thermodynamic data at different pressures are substituted into (5.12) and (5.14), then the corresponding thermodynamic data for activated graphite at different pressures can be obtained. However, in fact, after the calculation it can be found that the data listed in Table 5.1 can be used without obvious error in the pressure range of interest in this book (0.010–101.325 kPa).

2. Equilibrium constant method

For superequilibrium atomic hydrogen and activated graphite data a more careful derivation from equilibrium constants can be introduced. In the derivation in the last paragraphs an approximation of the partial pressure of molecular hydrogen equal to the total pressure had been used. In fact, if the total pressure is lower and the hot filament temperature is higher, then an obvious deviation may be introduced. To remove this deviation, a derivation can be made by an equilibrium constant method. For example,

(1) The basic data of atomic hydrogen and molecular hydrogen at standard temperature and pressure can be found from thermodynamic data books. According to the relationship G–T, i.e., (1.47),

$$G = H_0 - AT \ln T - \frac{1}{2}BT^2 - \frac{1}{2}CT^{-1} - \frac{1}{6}DT^3 + IT$$

Table 5.1. Thermodynamic Data of Activated Graphite (gra*) with Different χ Values

$H_{gra*} = 217.9861\ \chi$ (kJ mol^{-1})			
$S_{gra*} = 5.6861 + 83.7815\ \chi$ (J mol^{-1} K^{-1})			
C_p (J mol^{-1} K^{-1}) $= A + B10^{-3}T + C10^{5}T^{-2} + D10^{-6}T^{2}$			
	298–1000 K	1000–1550 K	1550–2400 K
A	$0.1597 + 5.7635\chi$	$25.5701 + 5.7635\chi$	$25.5701 + 5.9518\chi$
B	$38.8990 + 1.2400\chi$	$-0.2232 + 1.2400\chi$	$-0.2232 - 1.6351\chi$
C	$-1.5154 + 0.2988\chi$	$-39.1476 + 0.2988\chi$	$-39.1476 + 20.6857\chi$
D	$-17.4121 - 1.3416\chi$	$0.0978 - 1.3416\chi$	$0.0978 + 0.1076\chi$

Hot filament: 2400 K, Ref. Temp.: 298 K, Pressure range: 0.010–100 kPa
Definitions of B, C and D here are a little different from those in text

in Chap. 1, the corresponding G^0 values and the ΔG^0 value of the equilibrium reaction H = 0.5 H_2 at standard pressure and activated temperature can be obtained. Based on $\Delta G^0_{\mathrm{m}} = -RT \ln K^{\circ}_p$ or $\Delta G^0 = -nRT \ln K^{\circ}_p$, i.e., (1.85), nondimensional thermodynamic reaction equilibrium constants can be obtained. If all the reactants and products are treated as ideal gases, then

$$K^{\circ}_p = \left(\frac{p^e_{\mathrm{H}_2}}{p^0}\right)^{0.5} \left(\frac{p^e_{\mathrm{H}}}{p^0}\right)^{-1} \tag{5.20}$$

where p^e represents the equilibrium partial pressure. K°_p is a function of temperature and independent of pressure. When the reaction occurs at 2400 K, K°_p (2400 K) = 63.130. Therefore, the calculated equilibrium partial pressure of atomic hydrogen and molecular hydrogen are p^e_{H} (2400 K) = 0.015715, 0.0048853, 0.0014635 and 0.00007661 × 101.325 kPa, and $p^e_{\mathrm{H}_2}$(2400 K) = 0.98428, 0.095115, 0.0085365 and 0.00002339 × 101.325 kPa at activated temperature (2400 K) and the total pressures p = 1, 0.1, 0.01 and 0.0001 × 101.325 kPa, respectively.
(2) From the definition of superequilibrium atomic hydrogen (H*) above, it can be seen that $p_{\mathrm{H}*} = p^e_{\mathrm{H}_2}$(2400 K). The relationship between G and the pressures, i.e., $G_{\mathrm{m}}(p,T) = G^0_{\mathrm{m}}(p^0,T) + RT \ln\left(\frac{p}{p^0}\right)$ was obtained in Chap. 1. According to the procedure mentioned in Chap. 1, and the values of $p_{\mathrm{H}*}$ and $p^e_{\mathrm{H}_2}$(2400 K) the following equation can be obtained:

$$G^{298}_{\mathrm{H}*} = G^{0,298}_{\mathrm{H}} + R \times 298.15 \times \ln\left(\frac{p^e_{\mathrm{H}*}}{p^0}\right) \tag{5.21}$$

and

$$G^{e,\ 298}_{\mathrm{H}_2} = G^{0,\ 298}_{\mathrm{H}_2} + R \times 298.15 \times \ln\left(\frac{p^e_{\mathrm{H}*}}{p^0}\right). \tag{5.22}$$

By substitution of $S_{H*}^{298} = (H_{H*}^{298} - G_{H*}^{298})/(298.15)$, and $S_{H_2}^{e,\,298} = (H_{H_2}^{298} - G_{H_2}^{e,\,298})/(298.15)$, the corresponding entropies of superequilibrium atomic hydrogen and molecular hydrogen can be calculated: $S_{H*}^{298} = 149.14$, 158.84, 168.88 and 193.40 J mol^{-1} K^{-1}, and $S_{H_2}^{e,\,298} = 130.71$, 150.14, 170.19 and 219.24 J mol^{-1} K^{-1} at the filament temperature (2400 K) and the total pressure $p = 1$, 0.1, 0.01 or 0.0001 × 101.325 kPa, respectively. Here, the ideal gas assumption is also adopted. The value of the enthalpy H is independent of the partial pressure (or concentration), so $H_{H*}^{298} = H_{H}^{298}$.
(3) It is known that the standard entropy of graphite at 298 K is $S_{gra}^{298} = 5.6860$ J mol^{-1} K^{-1}. According to (5.9), the definition of activated graphite is C (gra*) = [C (gra) + χ (H*– 0.5 H_2)], and the entropy of activated graphite is $S_{gra*} = [S_{gra} + \chi\,(S_{H*} - 0.5\ S_{H2})]$. By substitution, $S_{gra*} = 29.146$, 29.144, 29.146 and 29.144 J mol^{-1} K^{-1} are obtained for the total pressure 1, 0.1, 0.01 and 0.0001 × 101.325 kPa, respectively. Here, $\chi = 0.28$ is taken. Obviously, in the pressure range of interest (0.0001–1 × 101.325 kPa), the entropy of activated graphite may be treated as a constant value independent of pressure. Based on $G = H - TS$, and (1.47), $G = H_0 - AT\ln T - \frac{1}{2}BT^2 - \frac{1}{2}CT^{-1} - \frac{1}{6}DT^3 + IT$, values of the Gibbs free energy at different temperatures may be obtained.
(4) The enthalpy and the heat capacity of superequilibrium atomic hydrogen have already been treated by the ideal gas approximation, i.e., $H_{H*} = H_H$ and $C_{pH*} = C_{pH}$. Finally, the calculated results are listed in Table 5.2.

Table 5.2. Thermodynamic data for activated graphite (gra*) (coupling parameter $\chi = 0.28$)

$H_{gra*} = 61.036$ (kJ mol^{-1})		
$S_{gra*} = 29.145$ (J mol^{-1} K^{-1})		
C_p (J mol^{-1} K^{-1}) $= A + B10^{-3}T + C10^{5}T^{-2} + D10^{-6}T^{2}$		
298–1000 K	1000–1550 K	1550–2400 K
$A = 1.7735$	$A = 27.184$	$A = 27.2366$
$B = 39.236$	$B = 0.1240$	$B = -0.6810$
$C = -1.4317$	$C = -39.064$	$C = -33.356$
$D = -17.7877$	$D = -0.2778$	$D = 0.1279$

Hot filament: 2400 K, Ref. Temp.: 298 K, Pressure range: 0.010–100 kPa
Definitions of B, C and D here are a little different from those in text

Although this derivation is stricter than the former, substitution of $\chi = 0.28$ in Table 5.1 shows that the two tables are the same, in practice. After these two derivations, the meaning of the thermodynamic data for superequilibrium atomic hydrogen and activated graphite can be understood more clearly. The key point is that the thermodynamic data for superequilibrium atomic hydrogen must be calculated on the basis of the practical nonequilibrium concentration in its corresponding stationary states; that is the important key point of thermodynamic calculations for nonequilibrium systems.

It should be noted that in both of these derivations the concentration of superequilibrium atomic hydrogen is defined as the equilibrium concentration of atomic hydrogen at the activated temperature. If the practical concentration of superequilibrium atomic hydrogen and the χ value can be experimentally measured, it may be found that the practical concentration of superequilibrium atomic hydrogen will be lower and the practical χ value will be higher. However, they will compensate each other, and will prevent too much deviation in the calculated results of nonequilibrium (stationary) phase diagrams.

5.5 New Concept of Nonequilibrium (Stationary) Phase Diagrams

In 1876 Gibbs (1839–1903) deduced the phase rule from thermodynamics, but the application of the phase rule in physics, chemistry, etc., was mainly due to the contribution of H.W.B. Roozeboom's work which started in 1884. Phase diagrams are concerned with the influences of temperature, concentration and pressure on the phase changes of a system. All of these changes may be illustrated by a diagram (or diagrams). Such a diagram is called a phase diagram. After about one hundred years of development, classical equilibrium thermodynamics and the related theory of equilibrium phase diagrams was more or less complete degree. Due to the strictness of their theoretical base and their simple character, equilibrium phase diagrams have played an important role in science and industrial production. For instance, phase diagrams are very useful in research and development on salty lakes. Moreover, there is a special academic discipline of metallography in metallic smelting. Metallography is the essential theoretical basis of metallurgy.

In classical thermodynamics phase diagrams were always regarded as equilibrium diagrams [5.27]. There were only two basic kinds of phase diagrams, i.e., phase diagrams of stable equilibrium systems and phase diagrams of metastable equilibrium systems. Both of them are phase diagrams of equilibrium systems, and are called equilibrium phase diagrams for short. For instance, in a one-component p-T phase diagram a dashed line for supercooling liquid is sometime drawn below the triple point. This is a phase line of a metastable phase diagram. The essential prerequisite for such a dashed phase line is the nonexistence of the stable solid phase. Strictly speaking, when the metastable supercooling liquid phase line has been drawn, the gas–solid phase line and the solid–liquid phase line should no longer be drawn. However, for convenience of applications all of these lines are drawn together in the same phase diagrams. For clarity stable phase lines are drawn by solid lines, while metastable phase lines are drawn by dashed lines. If the phase diagram is only a metastable phase diagram, it is not necessary to do so. Figure 5.8 is a phase diagram for a Fe-C system [5.28]. The dashed lines in Fig. 5.8 construct a typical example of metastable iron–cementite (Fe_3C, iron carbide) phase diagrams. The phase diagram of the metastable iron–cementite equilibrium system is very useful for the heat treatment of steel. The stable equilibrium iron–carbon system should involve equilibrium between a phase of iron (carbon solution in iron) and most of the stable phase of carbon — graphite, see the solid line part of Fig. 5.8. For discussion about certain questions it is more convenient to overlap the stable and metastable phase diagrams with each other, as in Fig. 5.8. It should be emphasized that the metastable phase diagrams are still equilibrium phase diagrams. All of the phases in the metastable equilibrium phase diagram are still in equilibrium between them. However, the metastable equilibrium phase diagram is only a special case of the equilibrium phase diagram, in which at least one thermodynamic stable phase does not appear for kinetic reasons, such as the lack of a crystal nucleus of the stable phase, or due to some kind of hindrance of the formation of the stable phase.

Phase diagrams of nonequilibrium systems may seem to be unimaginable in classical thermodynamics. Because most scientists (including the author) received a university education in classical thermodynamics, many parts of this book still start from the basis of classical thermodynamics. There was no special name (or they were called just stability diagrams) for these phase diagrams calculated on the basis of the chemical pump model before the end of 1994, although there were some problems in the calculation at that time. From the systematization of modern thermodynamics in Chap. 2, there arose a new field of nonequilibrium nondissipative thermodynamics [$\mathrm{d}_i S_1 < 0$, $\mathrm{d}_i S_2 > 0$ and $\mathrm{d}_i S = (\mathrm{d}_i S_1 + \mathrm{d}_i S_2) = 0$]. Because this belongs to the equality part of the second law of thermodynamics, phase diagrams in the new field can still be quantitatively calculated. On the other hand, the new field also belongs to nonequilibrium thermodynamics (the classification is no longer based on the inequality of the second law), so the calculated phase diagrams are nonequilibrium phase diagrams.

Nonequilibrium phase diagrams usually change with time, and should be illustrated on video rather then paper. Therefore, in this book nonequilibrium stationary phase diagrams or so-called phase diagrams of stationary nonequilibrium states will be mainly discussed. If a system is in its stationary states, then there must be some kind of equilibrium between these phases. Otherwise, the system will not be in a stationary state independent of time. The activated low-pressure CVD diamond growth system is a typical system of stationary states, in which the growth rate of diamond crystals or films is rather slow (in the hot filament process about 10 μm h^{-1}) and the

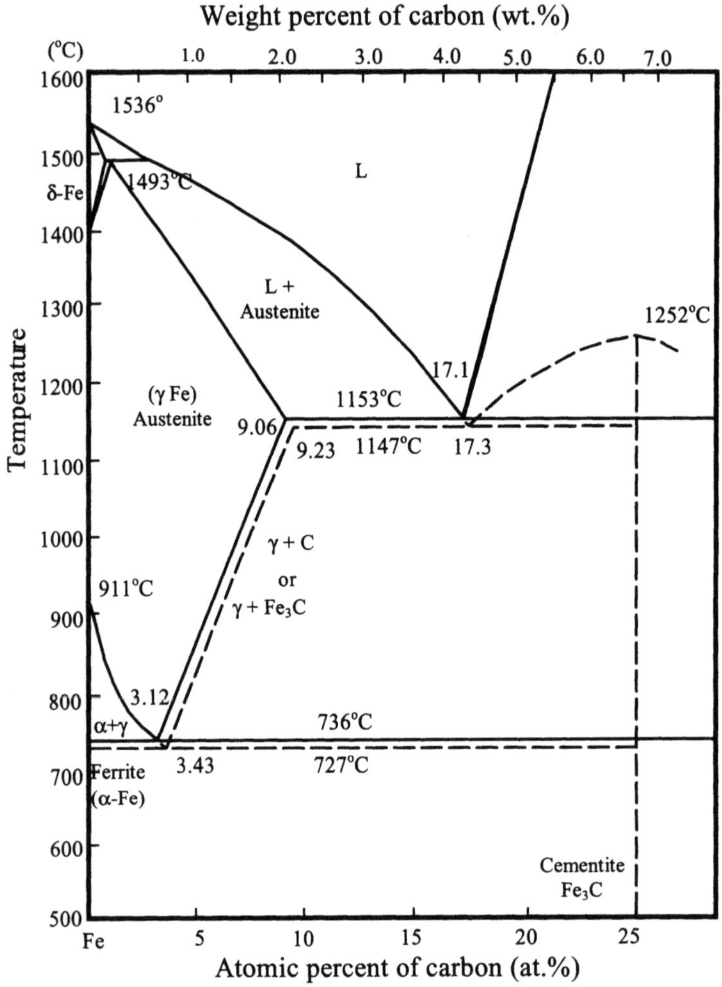

Fig. 5.8. Equilibrium phase diagram of Fe-C system. The dashed lines construct an iron–cementite (Fe_3C, iron carbide) metastable equilibrium phase diagram, while the solid lines construct an iron–graphite stable equilibrium phase diagram

CVD diamond growth system can remain steady for several hours, several days or even longer, so it provides an excellent practical example for research on nonequilibrium stationary phase diagrams. In the following discussions it can be seen that there are still two subclasses of stable phase diagrams and metastable phase diagrams in nonequilibrium stationary phase diagrams. It should be keep in mind that the difference between the equilibrium phase diagram and the nonequilibrium (stationary) phase diagram is different from the difference between the stable phase diagram and the metastable phase diagram. No confusion should be made between these differences. In particular, we should be careful to distinguish a nonequilibrium stationary phase diagram from a metastable equilibrium phase diagram. In any corresponding system of nonequilibrium (stationary) phase diagram there is at least one macroscopic process, while in the corresponding system of metastable equilibrium phase diagrams there is no macroscopic process and there is a lack of a stable phase or there is a hindrance for a chemical reaction.

The terms "nonequilibrium phase diagram" or "nonequilibrium stationary phase diagram" [or "nonequilibrium (stationary) phase diagram", in general] is a radically new concept. Maybe in polymer solutions there are some kinds of nonequilibrium phase diagrams — shearing induced phase diagrams [5.29]. However, they are quite different in meaning.

5.6 Nonequilibrium Nondissipation Principle

It is well known that the calculation principle for equilibrium phase diagrams at constant temperature and pressure is the minimization principle of the Gibbs free energy of the system. All the stable phase diagrams and the metastable phase diagrams calculated by the minimization principle of the Gibbs free energy of isothermal and isobaric systems are equilibrium phase diagrams. The calculation principle for nonequilibrium stationary phase diagrams should be based on the nonequilibrium nondissipation principle (also called the nonequilibrium non-entropy-production principle), i.e., [$\mathrm{d}_iS_1 < 0$, $\mathrm{d}_iS_2 > 0$ and $\mathrm{d}_iS = (\mathrm{d}_iS_1 + \mathrm{d}_iS_2) = 0$], which can be regarded as a special case of the entropy production minimization principle of the system proposed by Prigogine in 1945 for a system in its stationary states, as mentioned in Chap. 2.

The minimization principle of the entropy production for a nonequilibrium system of stationary states is an important step in linear nonequilibrium thermodynamics. According to this principle, in the stationary state the entropy production has its minimum value compatible with some auxiliary conditions to be specified in each case, that is, if a system is in a linear nonequilibrium state close to equilibrium and under an external constrained limitation (or control condition), then the system in its stationary states corresponding to the external constrained limitation (or control condition) is at a minimum of entropy production. Due to the external constrained limitation (or control condition) the system cannot develop into an equilibrium state, but a stationary state independent of time. In Chap. 2 it was discussed and demonstrated that the stationary state of the thermodiffusion phenomenon is of minimum of entropy production. Prigogine had summarized it into a more general form — the minimization principle of entropy production. Prigogine also clearly pointed out that "the stationary conditions are equivalent to the conditions of minimum production of entropy." Therefore, for activated low pressure diamond growth the system having reached a stationary state after development must be at a minimum of entropy production. For phase diagram calculations, [$\mathrm{d}_i S_1 < 0$, $\mathrm{d}_i S_2 > 0$ and $\mathrm{d}_i S = (\mathrm{d}_i S_1 + \mathrm{d}_i S_2) = 0$], i.e., the entropy production of the system equals to zero, is a limitation condition — we called the nonequilibrium nondissipation condition or the nonequilibrium nondissipation principle.

Before going on to the practical method, the equality between the minimization of entropy production and the minimization of the Gibbs free energy dissipation at constant temperature and pressure should be discussed first.

In nonequilibrium thermodynamics, the entropy production per unit time σ (also called the strength of the entropy source or the rate of entropy production) can be expressed in the following general form, see also (2.16):

$$\sigma = \sum_k J_k X_k, \tag{5.23}$$

where J_k and X_k are the thermodynamic flux (or rate) and thermodynamic force (or affinity) of the kth irreversible process, respectively. For chemical reactions

$$\sigma = \sum_\rho r_\rho \left(\frac{A_\rho}{T}\right), \tag{5.24}$$

where r_ρ is the rate of the ρ th reaction and T is the temperature. If the system is made up of M components, then the affinity of the ρth reaction

$$A_\rho = -\sum_{i=1}^{M} \nu_{i\rho}\mu_i, \tag{5.25}$$

where $\nu_{i\rho}$ is the stoichiometric coefficient of the ith component in the ρth reaction. Substitution of (5.25) in (5.24) gives

$$\sigma = -\frac{1}{T}\sum_\rho \sum_{i=1}^{M} r_\rho \nu_{i\rho}\mu_i. \tag{5.26}$$

Due to the reactions the change rate of the ith component $\frac{dn_i}{dt}$ equals

$$\frac{\mathrm{d}n_i}{\mathrm{d}t} = \sum_\rho \mu_{i\rho} r_\rho \tag{5.27}$$

so

$$\sigma = -\frac{1}{T}\sum_{i=1}^{M} \mu_i \frac{\mathrm{d}n_i}{\mathrm{d}t}. \tag{5.28}$$

On the other hand, the Gibbs free energy of the system G is

$$G = \sum_{i=1}^{M} n_i \mu_i. \tag{5.29}$$

Differentiation of G with respect to t gives

$$\frac{\mathrm{d}G}{\mathrm{d}t}=\sum_{i=1}^{M}\left(\mu_i\frac{\mathrm{d}n_i}{\mathrm{d}t}\right)+\sum_{i=1}^{M}\left(n_i\frac{\mathrm{d}\mu_i}{\mathrm{d}t}\right). \tag{5.30}$$

$\frac{d\mu_i}{dt}$ in the second right-hand term will be discussed first. At constant temperature and pressure $\frac{d\mu_i}{dt}=0$ for any solid component. Therefore, only gaseous components should be considered. If gaseous components are treated as idea gases, then

$$\mu_i=\mu_i^0+RT\ln\left(\frac{n_i}{\sum n_i}p\right), \tag{5.31}$$

where p is the total pressure and μ_i^0 is the standard chemical potential of the ith component.

$$\frac{\mathrm{d}\mu_i}{\mathrm{d}t}=RT\frac{\mathrm{d}\ln\left(\frac{n_i}{\sum n_i}p\right)}{\mathrm{d}t}=RT\frac{\sum n_i}{n_i}\frac{\mathrm{d}\left(\frac{n_i}{\sum n_i}\right)}{\mathrm{d}t}$$
$$=RT\frac{\sum n_i}{n_i}\left[\frac{\frac{\mathrm{d}n_i}{\mathrm{d}t}}{\sum n_i}-\frac{n_i\sum\frac{\mathrm{d}n_i}{\mathrm{d}t}}{(\sum n_i)}\right]=RT\left[\frac{\mathrm{d}n_i}{\mathrm{d}t}-n_i\frac{\frac{\mathrm{d}n_i}{\mathrm{d}t}}{\sum n_i}\right] \tag{5.32}$$

$$\sum_{i=1}^{M}\left(n_i\frac{\mathrm{d}\mu_i}{\mathrm{d}t}\right)=RT\sum_{i=1}^{M}\left[\frac{\mathrm{d}n_i}{\mathrm{d}t}-n_i\frac{\frac{\mathrm{d}n_i}{\mathrm{d}t}}{\sum n_i}\right]=0 \tag{5.33}$$

Substitution of (5.33) in (5.30) leads to

$$\frac{\mathrm{d}G}{\mathrm{d}t}=\sum_{i=1}^{M}\left(\mu_i\frac{\mathrm{d}n_i}{\mathrm{d}t}\right). \tag{5.34}$$

The entropy flow terms had already been excluded during the introduction of the concept of entropy production. Therefore, (5.28) for the entropy production term is suitable for either an isolated system or an open system. At constant temperature and pressure (5.34) can be deduced. During the calculation the influence of mass flow on the free energy had already been excluded. By comparison between (5.34) and (5.28), it can be seen that at constant temperature and pressure

$$\sigma=-\frac{1}{T}\frac{\mathrm{d}G}{\mathrm{d}t}\ \text{or}\ \frac{\mathrm{d}G}{\mathrm{d}t}=-T\sigma. \tag{5.35}$$

From this it can be found that the nonequilibrium nondissipation principle can be written in the nonequilibrium non-entropy-production form in general for any system, i.e., $[\Delta_i S_1<0,\ \Delta_i S_2>0$ and $\Delta_i S=(\Delta_i S_1+\Delta_i S_2)=0]$, and is equivalent to the nonequilibrium non-(Gibbs free energy)-dissipation form for a system at constant temperature and pressure, i.e., $[(\Delta G_1)_{T,p}>0$, $(\Delta G_2)_{T,p}<0$ and $\{(\Delta G_1)_{T,p}+\chi(\Delta G_2)_{T,p}\}=0]$. If $[(\Delta G_1)_{T,p}$ approaches to zero, $[(\Delta G_2)_{T,p}$ must also approach to zero for remaining $\{(\Delta G_1)_{T,p}+$

$\chi(\Delta G_2)_{T,p}\} = 0]$. The whole criterion will approach to $[(\Delta G_1)_{T,p} = 0$, $(\Delta G_2)_{T,p} = 0$ and $\{(\Delta G_1)_{T,p} + \chi(\Delta G_2)_{T,p}\} = 0]$. That is the Gibbs free energy minimization principle for a typical equilibrium state. Therefore, the minimization principle of the Gibbs free energy in ordinary calculations for the equilibrium phase diagrams can be regarded as a special case of the nonequilibrium nondissipation principle. The change from the entropy production minimization principle to the nonequilibrium nondissipation principle and to the Gibbs free energy minimization principle is continuous in the region close to equilibrium (i.e., linear region). Therefore, the present commercially available programs for the Gibbs free energy minimization can still be used with a certain of nonequilibrium constrained conditions for the calculation of nonequilibrium (stationary) phase diagrams of interest in this book.

5.7 Calculation of Nonequilibrium Phase Diagrams

The detailed calculation method for nonequilibrium (stationary) phase diagrams will be described and discussed from two different viewpoints as follows.

1. The whole discussion may be divided into three steps
(1) According to the reaction coupling theorem, a complete expression set of reactions (5.5)–(5.9) for activated low-pressure diamond growth from the vapor phase can be written out. Based on experimental observations it can be shown that the existence of superequilibrium atomic hydrogen has no obvious influence on either diamond growth from the vapor phase or diamond etching in the vapor phase, while it has significant influence on graphite etching in the vapor phase. Obviously, this is due to graphite being an sp^2 unsaturated carbon structure which is very easy to react with superequilibrium hydrogen atoms and accelerate graphite etching. From the viewpoint of reaction coupling, it may seem that due to the energy released by the reaction (5.6) of the association of hydrogen atoms, the chemical potential of graphite has been increased. Therefore, reaction (5.7) can be rewritten as reaction (5.8), i.e., C (gra*) = C (dia), (T, $p \leq 10^5$ Pa). Here, C (gra*) = [C (gra) + χ(H* − 0.5 H_2)].

Note that if the influence of superequilibrium hydrogen atoms is treated in a more general way, then the following reaction equation may be obtained

$$\mathrm{C(gra*)} = \mathrm{C(dia}^S), (T, p \leq 10^5\mathrm{Pa}). \tag{5.36}$$

Here, C (gra*) is called activated graphite, while C (dia^S) is called stabilized diamond:

$$\mathrm{C(gra*)} = [\mathrm{C(gra)} + \chi \times \eta(\mathrm{H*} - 0.5\mathrm{H}_2)] \tag{5.37}$$

$$\mathrm{C(dia}^S) = [\mathrm{C(dia)} - \chi \times (1 - \eta)\,(\mathrm{H*} - 0.5\mathrm{H}_2)]. \tag{5.38}$$

For simplicity, η is basically treated as 1 in this book. That agrees also with a lot of experiments.

(2) As mentioned above, thermodynamic data for superequilibrium atomic hydrogen and activated graphite can be deduced.
(3) Thermodynamic data for activated graphite are used instead of thermodynamic data of ordinary graphite, and then ordinary programs based on the Gibbs free energy minimization principle can still be used for the calculation of nonequilibrium stationary phase diagrams.

The description of the three-step calculation procedure above has been repeated many times in our previous papers except for the lack of the deduction of superequilibrium hydrogen atom data at an earlier time. However, during the phase diagram calculation for low-pressure diamond growth from the vapor phase, different kinds of phase lines may be involved, which may be difficult to understand. The following discussions will focus on different kinds of phase lines.

2. Calculation for different kinds of phase lines

(1) Calculation of phase line between the gas phase and the stable solid phase in equilibrium phase diagrams

The calculation for the equilibrium phase line between the solid phase of graphite [C (gra)] and the gas phase in a low-pressure carbon-hydrogen system belongs to this kind of calculation. The basic reaction equation (or a group of equations) is

$$\mathrm{C(gra)} = \nu_1\mathrm{H_2} + \nu_2\mathrm{CH_4} + \nu_3\mathrm{C_2H_2} + ..., \ \Delta G = 0 \ (\text{constant } T, \ p). \tag{5.39}$$

Note that a hydrogen molecule is also one of the initial reactants, and ν_1 has a negative value. The reaction may be rewritten in the simpler form

$$\mathrm{C(gra)} = \{\nu_i\mathrm{C}_{X_i}\mathrm{H}_{Y_i}\}_{\mathrm{gra}}, \ \Delta G = 0 \ (\text{constant } T, \ p) \tag{5.40}$$

where $\{\nu_i\mathrm{C}_{X_i}\mathrm{H}_{Y_i}\}_{\mathrm{gra}}$ represents hydrocarbons in the gas phase in equilibrium with the solid graphite phase. Molecular hydrogen ($X = 0$, $Y = 2$) and atomic hydrogen ($X = 0$, $Y = 1$) have been included. According to the mass conservation law, $\sum\nu_iX_i = 1$, $\sum\nu_iY_i = 0$. At constant temperature and pressure $\Delta G = 0$ may represent a system with a single process being in equilibrium. The Gibbs free energy of the system will no longer decrease. Therefore, programs based on the minimization principle of the Gibbs free energy can be used for calculation of these stable phase lines in equilibrium phase diagrams.

(2) Calculation of the phase line between the gas phase and the metastable solid phase in equilibrium phase diagrams

The calculation of the equilibrium phase line between the solid phase of diamond [C (dia)] and the gas phase in a low-pressure carbon-hydrogen system belongs to this kind of calculation. The basic reaction is

$$\mathrm{C(dia)} = \{\nu_i\mathrm{C}_{X_i}\mathrm{H}_{Y_i}\}_{\mathrm{dia}}, \ \Delta G = 0 \ (\text{constant } T, \ p) \tag{5.41}$$

where $\{\nu_i C_{X_i} H_{Y_i}\}_{\rm dia}$ may represent hydrocarbons in the gas phase in equilibrium with the solid diamond phase. Molecular hydrogen ($X = 0$, $Y = 2$) and atomic hydrogen ($X = 0$, $Y = 1$) have also been included. According to the mass conservation law, $\sum \nu_i X_i = 1$, $\sum \nu_i Y_i = 0$. For metastable equilibrium phase diagrams ordinary programs based on the minimization principle of the Gibbs free energy can still be used for the calculation. Note that the concentrations of hydrocarbons with respect to the diamond gas–solid phase line are different from that with respect to the graphite gas–solid phase line, i.e., $\{\nu_i C_{X_i} H_{Y_i}\}_{\rm dia} \neq \{\nu_i C_{X_i} H_{Y_i}\}_{\rm gra}$. It should also be pointed out that during the calculation of the metastable gas–solid phase line the stable solid phase must be excluded.

(3) Calculation of both the stable gas–solid phase line and the metastable gas–solid phase line in nonequilibrium (stationary) phase diagrams

From the above discussion, it can be seen that during the calculation of the metastable gas–solid phase line the stable solid phase must be excluded, otherwise, the gas–solid phase line for the metastable solid phase will not be obtained. This problem will not be met in the calculation of the gas–solid phase line for the stable solid phase. However, if the irrelevant metastable solid phase is still excluded in the calculation of the gas–solid phase line for the stable solid phase, there will be no influence on the calculated results. Due to the complexity of nonequilibrium (stationary) phase diagrams, the stable solid phase and the metastable solid phase may be changed in different temperature ranges. Therefore, it will be better to prevent interference during the calculation of any gas–solid phase line by ensuring that any irrelevant solid phase is previously excluded. In practical calculations of nonequilibrium (stationary) phase diagrams the following two gas–solid phase lines should be calculated, that is,

$$\mathrm{C(gra*)} = \{\nu_i C_{X_i} H_{Y_i}\}_{\rm gra*}\,, \quad \Delta G = 0 \ (\text{constant } T,\ p) \tag{5.42}$$

and

$$\mathrm{C(dia)} = \{\nu_i C_{X_i} H_{Y_i}\}_{\rm dia}\,, \quad \Delta G = 0 \ (\text{constant } T,\ p), \tag{5.43}$$

where, $\{\nu_i C_{X_i} H_{Y_i}\}_{\rm gra*}$ represents hydrocarbons in the gas phase in dissipative states ($\Delta G = 0$) with the activated graphite phase. If the possible stabilization effect of superequilibrium atomic hydrogen on diamond can be neglected, the diamond gas–solid phase line in nonequilibrium (stationary) phase diagrams will be the same as that in equilibrium phase diagrams. Therefore, the key point is about the calculation of the gas–solid phase line with respect to activated graphite. According to the previously deduced thermodynamic data for activated graphite, there will no difficulty in calculating the gas–solid phase line with respect to activated graphite. Because the graphite phase has already been activated, activated graphite in a rather larger range of temperature has become a metastable phase. After the calculation it can be concluded that the concentrations of hydrocarbons with respect to the diamond gas–solid phase line do not equal the concentrations of hydrocarbons with respect to the activated graphite gas–solid phase line, i.e., $\{\nu_i C_{X_i} H_{Y_i}\}_{\text{dia}} \neq \{\nu_i C_{X_i} H_{Y_i}\}_{\text{gra*}}$. However, it should be noted that the expression $\Delta G = 0$ at constant temperature and pressure in (5.42) no longer means the system is in equilibrium, but only nonequilibrium nondissipation of the Gibbs free energy of the system. In the calculation procedure the Gibbs free energy of the reaction system will not be able to decrease any further, although a routine program based on the minimization principle of the Gibbs free energy can still be used. In fact, the practical meaning of such a calculation has been changed into the nonequilibrium non-(Gibbs free energy)-dissipation principle at constant temperature and pressure, i.e., the nonequilibrium nondissipation principle. The superequilibrium atomic hydrogen data included in the data for activated graphite have already been fixed at a special nonequilibrium concentration, so the whole system is limited in a stationary state. That is, the corresponding phase diagram is a nonequilibrium (stationary) phase diagram, or a stationary phase diagram, for short.

In addition, as discussed above, usually the stable and metastable solid phases do not appear simultaneously in an equilibrium gas–solid system in an interesting temperature range in low-pressure diamond vapor growth. Therefore, the drawing of both stable and metastable gas–solid phase lines on the same equilibrium phase diagram is more convenient only for discussion. However, that is not the same for a system of stationary states for the diamond process. The stable solid phase and the metastable solid phase may appear individually or simultaneously in a gas–solid system of stationary states. For instance, in the activated low-pressure CVD diamond growth system diamond growth and graphite etching may take place simultaneously, so the drawing of both stable and metastable gas–solid phase lines on the same phase diagram is not only allowed but also usually necessary. Moreover, the concentrations of gas species on the surfaces of stable and metastable solid phases and in the different parts of the gas phase may be different. All these characteristics are specific for the system in stationary states and for the nonequilibrium (stationary) phase diagram. Therefore, these characteristics of the calculated phase diagrams also confirm that these phase diagrams are not equilibrium phase diagrams but nonequilibrium (stationary) phase diagrams.

5.8 Some Results and Discussions

For comparison between the calculated results obtained after the fall of 1994 by the improved model and the previous calculated results during 1990–1994, all of these results for gas–solid phase lines together with the equilibrium phase lines of a carbon–hydrogen system are drawn in Fig. 5.9. It can be seen that the gas–solid phase line of activated graphite calculated by the improved method (bold solid lines in Fig. 5.9) naturally meets the ordinary equilibrium gas–solid phase line of graphite (thin solid line in Fig. 5.9) at the hot filament temperature. That is an important characteristic, which agrees very well with theoretical expectation. This excellent quantitative agreement had never been achieved before the fall of 1994. The previous calculated result of the "old activated graphite" gas–solid phase line (the short dashed line in Fig. 5.9) could not meet the equilibrium graphite line at the hot filament temperature.

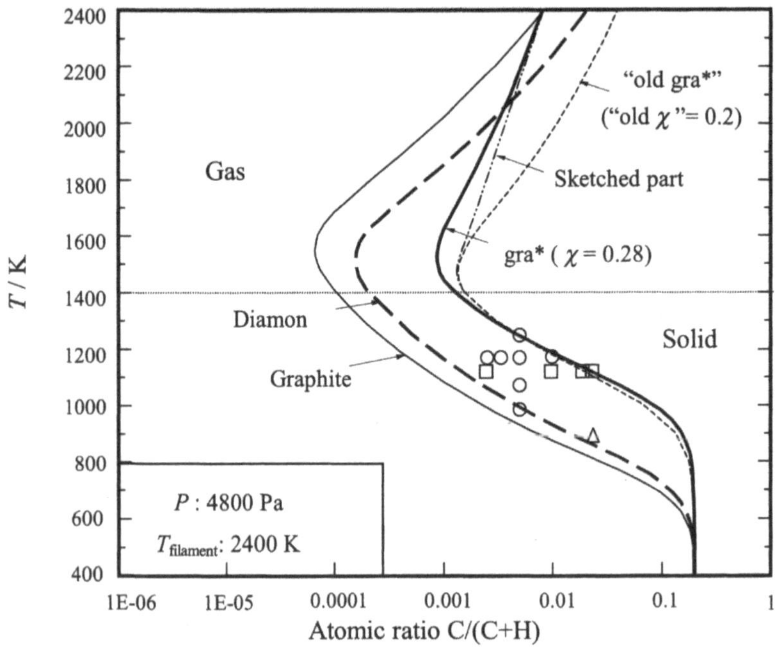

Fig. 5.9. Phase diagrams of the carbon–hydrogen system calculated before and after improvement. The related gas–solid phase lines and equilibrium phase lines are shown in one diagram for comparison between them. The symbols open circles, open squares and open triangles are experimental data reported by Matsumoto [5.10], Sato [5.30] and Southworth et al. [5.31], respectively

From the figure it can also be seen that the coupling parameter χ is different from the previous coupling parameter χ (denoted by "old χ") both in meaning and value. From the comparison between the improved theoretical calculated results and Matsumoto's experimental data it can be found that the value of χ should be 0.28, while the previous value of the "old χ" was 0.2.

Figure 5.9 shows an equilibrium phase diagram of a carbon–hydrogen (C-H) system consisting of the equilibrium graphite gas–solid phase line (thin solid line in Fig. 5.9) and the diamond gas–solid phase line (bold dashed line in Fig. 5.9). For clarity the equilibrium phase diagram of the C-H system has been separated from Fig. 5.9 and redrawn individually in Fig. 5.10. From the equilibrium phase diagram of Fig. 5.10 it can be found that graphite is a stable phase, and the graphite gas–solid phase line is drawn by a solid line, while diamond is a metastable phase, and the diamond gas–solid phase line is drawn by a dashed line. There are three phase regions: the gas region, the graphite deposition region (filled by a net-like pattern) and the carbon mixture deposition region. The graphite deposition region is located between the two gas–solid phase lines, that is, the graphite deposition region is inside the graphite solid region and outside the diamond solid region, so only graphite can deposit in this region. All of these agree very well with the conclusions from equilibrium thermodynamics that graphite is a stable phase and there is no possibility of transferring graphite to diamond under equilibrium low-pressure conditions.

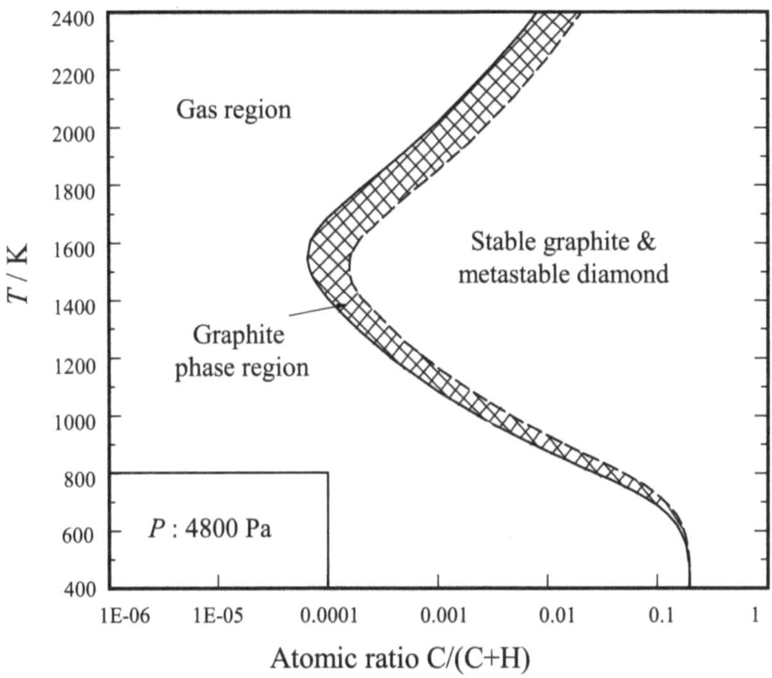

Fig. 5.10. Equilibrium phase diagram of carbon-hydrogen system

There is another phase diagram included in Fig. 5.9, which is a nonequilibrium (stationary) phase diagram for a C-H system with the existence of a superequilibrium concentration of atomic hydrogen and it consists of the activated graphite gas–solid phase line (bold solid line in Fig. 5.9) and diamond gas–solid phase line (bold dashed line in Fig. 5.9). For clarity the nonequilibrium (stationary) phase diagram for activated CVD diamond growth from the C-H system has also been separated from Fig. 5.9 and redrawn individually in Fig. 5.11 with different line styles. The diagram is divided into four regions by these two gas-solid phase lines and a horizontal line, that is (i) a gas phase region (no deposition region), (ii) a diamond phase region (diamond growth region), (iii) diamond stable and graphite metastable regions, and (iv) a graphite phase region at high temperature. For low-pressure diamond growth research regions (iii) and (iv) are usually called the nondiamond carbon deposition region.

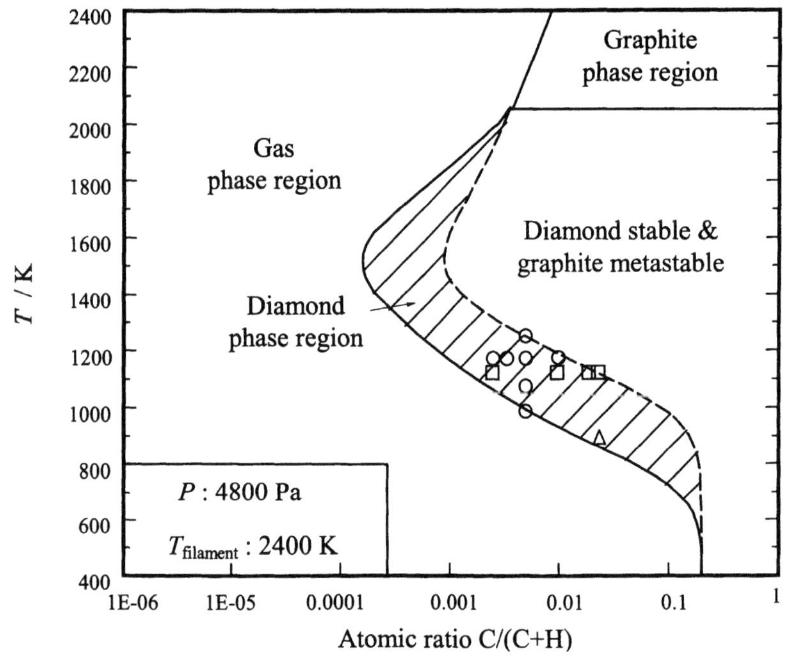

Fig. 5.11. Nonequilibrium (stationary) phase diagrams for C-H system with the existence of superequilibrium concentration atomic hydrogen. The symbols open circles, open squares and open triangles are experimental data reported by Matsumoto [5.10], Sato [5.30], and Southworth et al. [5.31], respectively

In Fig. 5.11 the temperature intersection point (about 2050 K) of the two gas–solid phase lines is a triple (three-phase) point, if the pressure is fixed. The gas phase, solid diamond phase and solid graphite phase may exist at the same time with or without a phase transformation between any two phases (supposing the graphitization of diamond at such a high temperature was negligible). At this point, all compositions of these three phases and temperature have already been fixed. For the two-phase line (stable diamond gas–solid phase line at the lower temperature or activated graphite gas–solid phase line at the high temperature) the gas phase and one solid phase (diamond or graphite) exist with or without a phase transformation between the phases, so one parameter (temperature or composition) can be changed. In the single-phase region (such as gas phase region) two parameters (such as temperature and gas composition) can be changed. All of those agree very well with the Gibbs phase rule for any binary system, that is, nonequilibrium (stationary) phase diagrams are still subordinated to the Gibbs phase rule.

It should be pointed out that in this book many phase diagrams are mainly focused on discussion of diamond growth. Region (iii) (diamond stable and graphite metastable region) and region (iv) (graphite phase region at high temperature) are usually called the nondiamond carbon deposition region. The line style (solid line or dashed line) usually is not dependent on the stable or metastable phase, but only dependent on the graphite line or diamond line. Sometime, all gas–solid lines are simply drawn as solid lines and no horizontal line between region (iii) and region (v) is drawn.

At high temperature (higher than the triple point) graphite is still a stable phase. This means that if the substrate temperature is too high close to the hot filament temperature, due to the insufficient concentration of superequilibrium atomic hydrogen activated by the hot filament, the relative stabilities between diamond and graphite are not inverted. This provides an explanation for the observation that if the hot filament temperature is too low and close to the substrate temperature, the relative stabilities between diamond and graphite are not inverted. Just as reported in the literature, if the hot filament is made of platinum, diamond cannot grow in the hot filament process because of the low melting point of platinum (the melting point of pure platinum is 1773 °C, and that of platinum alloy is about 1650 °C). In Fig. 5.11 below the temperature of the intersection point there is a diamond stable region (diamond growth region) between two gas–solid phase lines. Diamond has become a stable phase, while activated graphite has become a metastable phase. That is why diamond can grow from the vapor phase usually at a substrate temperature of 1000–1300 K, and why steady diamond growth together with simultaneous graphite etching can take place in the hot filament process. Experimental data for diamond growth from the vapor phase reported by Matsumoto [5.10], Sato [5.30] and Southworth et al. [5.31] in the literature are just dropped into the diamond growth (stable) phase region of Fig 5.11. All of these phenomena and data quantitatively agree with the calculated nonequilibrium (stationary) phase diagrams for activated CVD diamond growth.

To sum up, in modern thermodynamics steady low-pressure diamond growth from the vapor phase is a normal thermodynamic phenomenon. In other words, due to the generation of the external energy provided by the hot filament, plasma or other energy forms, atomic hydrogen remains in a superequilibrium concentration, so low-pressure diamond growth is possible. In different processes, the superequilibrium concentration of atomic hydrogen may be different, so the coupling parameter χ may also be different.

The diamond growth region will be larger with increase of χ value. That will be favorable for low-pressure diamond growth. The relationship between the diamond growth region and χ value is illustrated in Fig. 5.12. Comparison of the calculated curves of different χ values with Matsumoto's hot filament experimental data shows that a χ value equal to about 0.28 can be obtained. In Figs. 5.9 and 5.11 it has been seen that when $\chi = 0.28$, the experimental data for diamond growth from the vapor phase reported by Sato [5.30] and Southworth et al. [5.31] also agree well with the theoretical results. Experimental data are just dropped inside the diamond growth region of the theoretical calculated phase diagram. In the microwave process the temperature of neutral particles in the plasma is usually about 3000 K. That will be slightly favorable to increase the diamond growth region. The thermodynamic data for activated graphite can be deduced again based on the activated temperature of 3000 K, but the χ value can still be 0.28. For the normal pressure plasma torch process the temperature of neutral particles in the plasma is much higher, so the concentration of superequilibrium atomic hydrogen close to the surface of the substrate may be higher. That is more favorable for increasing the χ value, and low-pressure diamond growth from the vapor phase. The plasma torch process runs at normal pressure. From nonequilibrium (stationary) phase diagrams at different pressures it can be found that the corresponding saturated concentration of carbon increases with the pressure, that is, the higher pressure is more favorable to increase the growth rate from the vapor phase. The diamond growth rate at different places, even inside the same diamond growth region, is different. On the left-hand side close to the diamond phase line the supersaturation degree of carbon is lower. That is favorable for the growth of diamond single crystals, but unfavorable for the diamond nucleation. On the contrary, on the right-hand side close to the activated graphite phase line the supersaturation degree of carbon is higher. That is favorable for the diamond nucleation, but unfavorable for the growth of diamond single crystals. All of these theoretical predictions agree very well with the experimental observations. T-p-X nonequilibrium (stationary) phase diagrams of C-H system will be further discussed in the next chapter.

Finally, it should be kept in mind that the bulk materials or crystals of diamond are unstable above 1700 K (or above 1900 K, depending on the different conditions reported in the literature), and that thermodynamic data for diamond reported in the literature can be found usually below 1200 K. The data of diamond above 1200 K are obtained by extrapolation. Therefore, the corresponding high temperature part of the theoretical diamond curve is only for the theoretical prediction of change tendencies. In brief, according to the reaction coupling theorem the chemical pump model has been improved to be a nonequilibrium thermodynamic reaction coupling theoretical model (or a reaction coupling model, for short) for activated low-pressure diamond growth from the vapor phase. Meanwhile a new concept of nonequilibrium

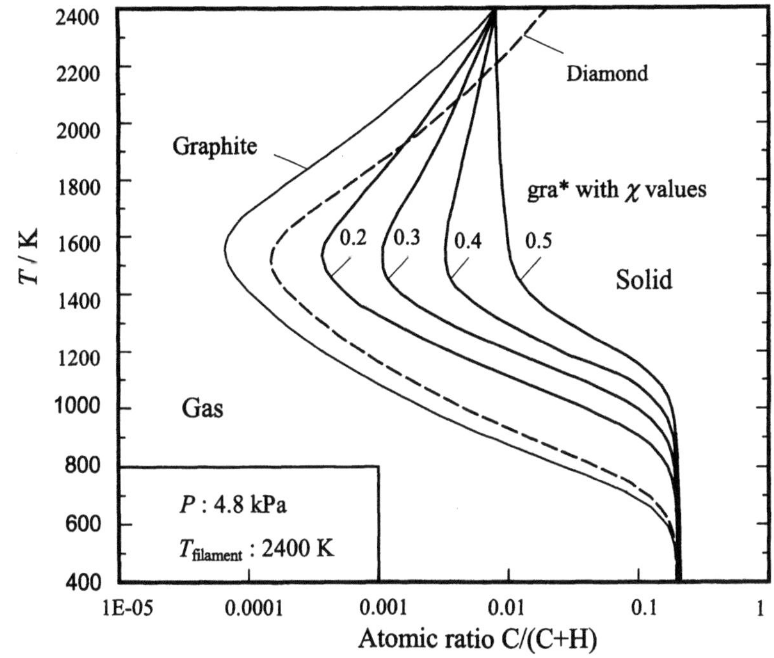

Fig. 5.12. Relationship between the diamond growth region and the coupling parameter χ in nonequilibrium (stationary) phase diagram

(stationary) phase diagrams has been proposed. Nonequilibrium (stationary) phase diagrams for the C-H system during the existence of superequilibrium atomic hydrogen have been calculated on the basis of rather strict deduction. Nonequilibrium (stationary) phase diagrams are still subordinated to the Gibbs phase rule. The theoretical model and nonequilibrium (stationary) phase diagrams are very useful not only for optimizing the low-pressure diamond process and but also for opening up a new field on materials and processes research on the basis of modern thermodynamics.

The discussions above focused on gas–solid phase diagrams. Gas–solid phase diagrams belong to one kind of phase diagram, which are mainly concerned with the saturation degree of each solid phase in the gas phase, such as the solid phase is undersaturated, saturated, or supersaturated in the gas phase. Therefore, in these phase diagrams there is often a single gas–solid phase line to divide the whole diagram into a gas phase region and a solid phase region for each solid phase. Such phase diagrams are usually not concerned with the detailed gas composition.

5.9 Morphology of Diamond Crystals and Orientation of Diamond Films

Another kind of phase diagram is mainly concerned with the gas phase composition. These are called gas phase composition phase diagrams. Both kinds of gas–solid phase diagrams and gas phase composition phase diagrams are complementary to each other. Figure 5.13 is the equilibrium gas phase composition phase diagram of a C-H system. The solid lines represent the saturation phase line of each gas species with respect to the graphite solid phase, while the dashed lines represent the saturation phase line of each gas species with respect to the diamond solid phase. In such phase diagrams there is always a saturation phase line for each gas species with respect to each solid phase. Therefore, there is a fixed value of saturated partial pressure for each gas species at different temperature and pressure for each solid phase. In this way the whole saturated gas phase composition is clearly known. In Fig. 5.13 all saturation phase lines of hydrocarbon gas species with respect to diamond (dashed lines) are higher than the corresponding saturation phase lines with respect to graphite, that is, diamond is a metastable phase, while graphite is a stable phase. That agrees with the conclusion of classical equilibrium thermodynamics from the gas–solid equilibrium phase diagrams of the C-H system at low pressure.

Figure 5.14 is a gas composition nonequilibrium (stationary) phase diagram for activated low-pressure diamond growth from the vapor phase of the C-H system at the activated temperature (such as the hot filament temperature) equal to 2400 K, and the total pressure equal to 10 kPa. In Fig. 5.14 at the substrate temperature of about 1200 K all saturation phase lines of hydrocarbon gas species with respect to diamond (dotted lines) are lower than the corresponding saturation phase lines with respect to graphite, that is, diamond has become a stable phase under the activated low-pressure diamond growth condition, while graphite has been activated into a metastable phase. Therefore, both diamond growth from a methane–hydrogen mixture, and from graphite via gaseous hydrocarbons under the activated low-pressure diamond growth conditions are thermodynamically reasonable. From gas composition nonequilibrium (stationary) phase diagrams more detailed theoretical information about the saturated concentration (or the saturated partial pressure) of each gas species can be obtained. That will be very favorable to know the main gas species, the growth mechanism, crystal orientation and so on.

It is also well known in the CVD diamond forum that the growth mechanisms of diamond crystals from different gas precursors are different. CH_3 and C_2H_2 are the dominant growth precursors during activated CVD diamond growth. It is known that the growth of diamond (111) and (100) facets is controlled by C_2H_2 and CH_3, respectively. The ratio will affect the facets of diamond crystals [5.32]. As a general rule, in the case of the growth of separate crystals, the facets with the maximal growth rate disappear and the

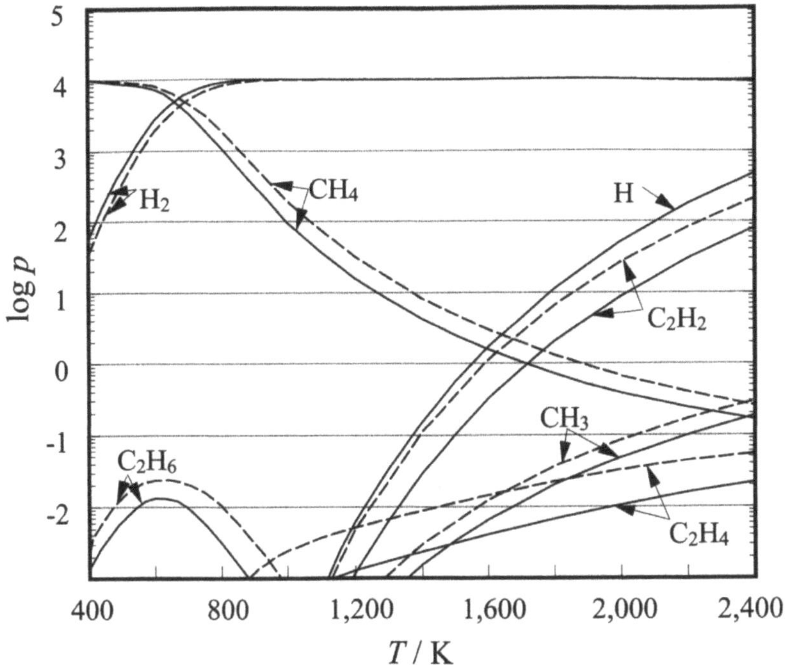

Fig. 5.13. Gas composition equilibrium phase diagrams for C-H system at $p = 10$ kPa (solid lines are phase lines of each gas species with respect to graphite, while dashed lines are phase lines of each gas species with respect to diamond)

crystal habit represents the crystallographic facets with the lowest growth rate (Wulff's rule). When the ratio C_2H_2 to CH_3 (i.e., $[C_2H_2]/[CH_3]$) exceeds a certain value, diamond (100) facets, whose growth is controlled by CH_3, well appear. Figure 5.15 shows the relationship between the concentration of C_2H_2 (or CH_3) and the substrate temperature on the basis of the nonequilibrium thermodynamic coupling model. The concentrations of C_2H_2 and CH_3 increase, when the substrate temperature rises. The concentration of C_2H_2 rises more quickly.

The relationship between the concentration ratio of $[C_2H_2]/[CH_3]$ and the substrate temperature is also calculated, as shown in Fig. 5.16. Three curves represent 1, 0.4 and 0.2 methane in the feeding gas, respectively. In the normal range of the deposition temperature (900–1400 K), the ratio of $[C_2H_2]/[CH_3]$ rises with increase of substrate temperature. Therefore, the growth ratio of diamond (100) to (111) facets (symbolized as $\alpha = r_{100}/r_{111}$) will fall with increase of the substrate temperature. Based on Wulff's theory, diamond (111) facets will appear when $\alpha \geq \sqrt{3}$, and when $\alpha \leq 1/\sqrt{3}$, diamond with (100) facets will be formed. It is concluded that the morphology of CVD separate diamond crystals shifts from initial (111) facets to (100) facets with the substrate temperature increasing, which is due to decrease of α with

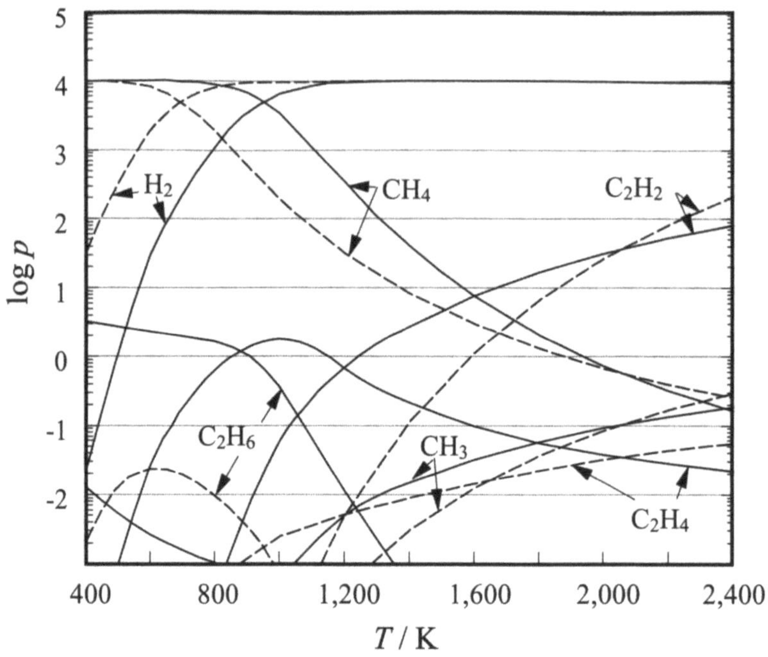

Fig. 5.14. Gas composition nonequilibrium (stationary) phase diagrams for activated low pressure diamond growth from a C-H system ($T_{\text{activated}} = 2400$ K, $p =$ 10 kPa) (solid lines are phase lines of each gas species with respect to activated graphite, while dashed lines are phase lines of each gas species with respect to diamond)

increase of $[C_2H_2]/[CH_3]$ value. Mastumoto's experiments [5.10] showed that diamond crystal facets changed from (111) to (100) with the increase of the substrate temperature from 610 to 970 °C. That is, the theoretical calculations are consistent well with experimental observations.

In Fig. 5.16 it is also shown that the ratio of $[C_2H_2]/[CH_3]$ changes with the methane concentration at the same substrate temperature. Figure 5.17 shows the dependence of $[C_2H_2]/[CH_3]$ on the methane concentration. The ratio of $[C_2H_2]/[CH_3]$ almost shows a linear increase with increase of the methane concentration. Because the ratio of $[C_2H_2]/[CH_3]$ has an opposite trend with the change of α, so when α decreases to a value smaller than $1/\sqrt{3}$, due to the ratio of $[C_2H_2]/[CH_3]$ increase with increase of the CH_4 concentration, the morphology of diamond crystals will be changed from (111) to (100). Kobashi et al. [5.33] used the different methane concentrations under 4 kPa pressure and 1100 K substrate temperature to growth diamond crystal. They found that when the methane concentration changed from 0.2% to 1.2%, the beginning morphology was (111) facet, subsequently, (100) facet appeared when the methane concentration was more than 0.4%. Their experimental

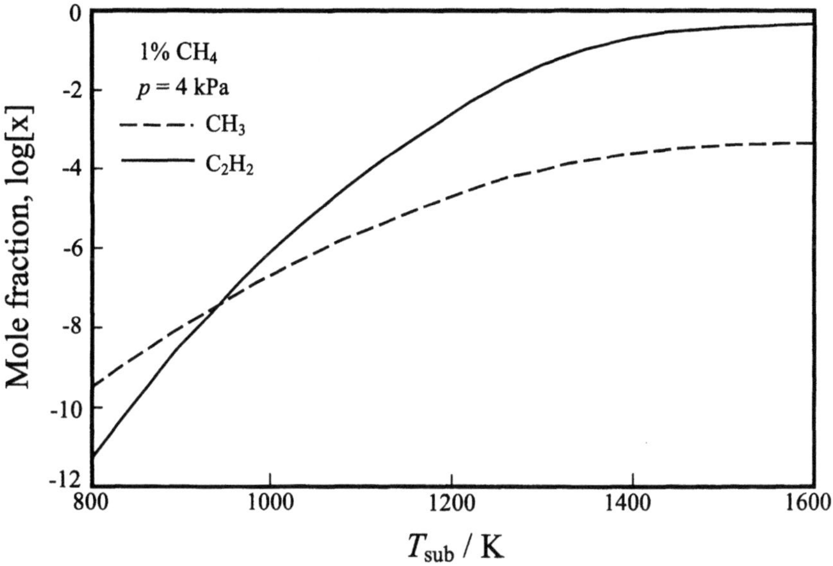

Fig. 5.15. The concentrations of CH_3 and C_2H_2 as the functions of substrate temperature during activated low-pressure diamond growth from the vapor phase

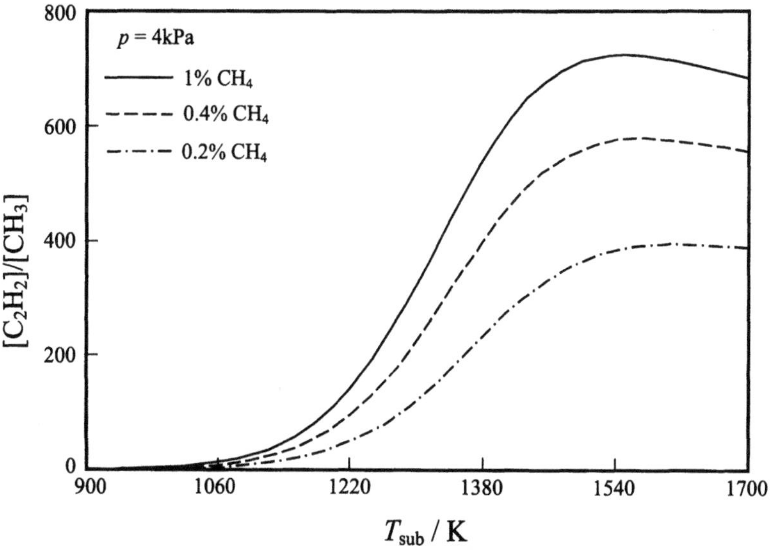

Fig. 5.16. The ratio of C_2H_2/CH_3 as a function of the substrate temperature during activated low-pressure diamond growth from the vapor phase

results indicate that our calculated results in Fig. 5.17 are reasonable. In addition, Sato et al. [5.30] also reported that when the methane concentration changed from low to high, the morphology changed from (111) facets to (100) facets at 0.5% methane concentration.

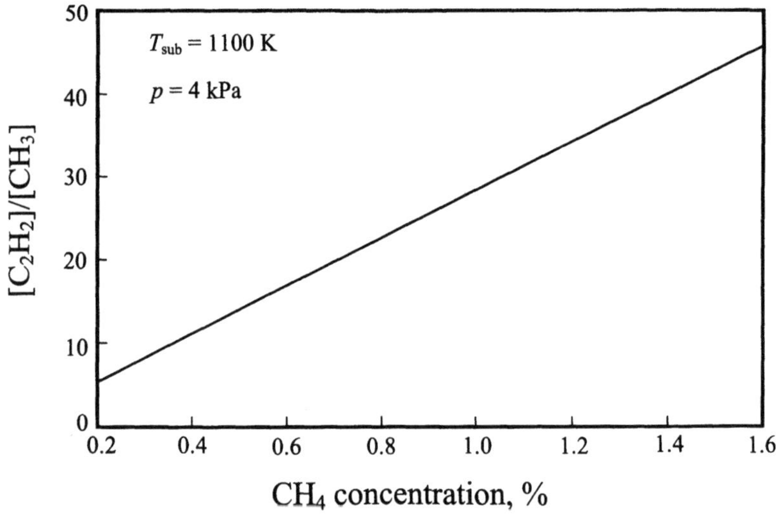

Fig. 5.17. The value of C_2H_2/CH_3 as a function of the methane concentration during activated low-pressure diamond growth from the vapor phase

However, in a polycrystalline diamond film, the growth of a community of microcrystals proceeds as a competition between separate crystals for feeding from the gas phase. In such a competition, prevalence and opportunity to be presented at the outer surface characterize the crystallites which have the maximal growth rate. Spitsyn [5.34] had summarized experimental observations of orientation or texture of diamond films, as shown in Fig. 5.18.

At the low crystallization temperature or at the early stages of the growth of diamond films nuclei consist of tiny polycrystalline diamond crystals, usually called "cauliflower-like" particles, as shown in the left of Fig. 5.18 and also in photographs of Figs. 4.5c and d. With increase of the substrate temperature the situation may change drastically, as shown in Fig. 5.18 and also in Fig. 4.4b a ⟨100⟩ orientated diamond film and in Fig. 4.4c a ⟨111⟩ orientated diamond film. ⟨100⟩ orientated diamond films should be grown at a little lower temperature (about 950 °C with $\alpha = r_{100}/r_{111}$) bigger than 1, while ⟨111⟩ orientated diamond films should be grown at a little higher temperature (about 1150 °C with $\alpha = r_{100}/r_{111}$) smaller than 1. These phenomena agree well with our theoretical calculation on the ratio of $[C_2H_2]/[CH_3]$, which has an opposite trend with the change of α. In such a way, a lot of experimental

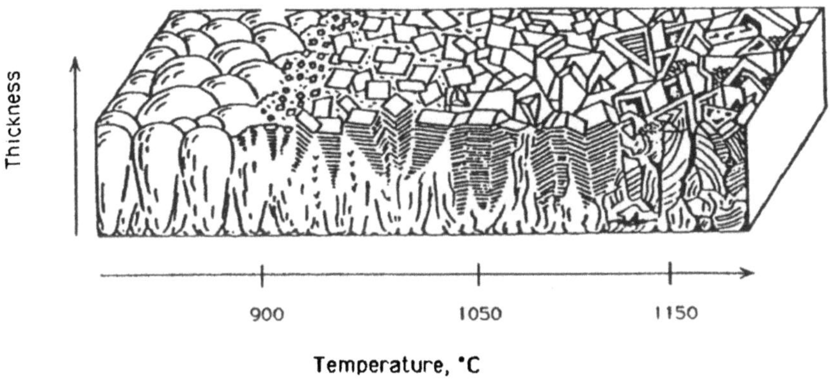

Fig. 5.18. Evolution of growth texture with diamond layer thickness and growth temperature. Reprinted from B.V. Spitsyn: 'Chap. 10 Growth of Diamond Films from the Vapor Phase'. In: *Handbook of Crystal Growth, vol.1*, ed. by D.T.J. Hurle (Elsevier Science, Amsterdam, 1994) pp. 403–456 with permission from Elsevier Science

observations on the morphology of diamond crystals and the orientation of diamond films in the literature can theoretically be summarized.

6. Nonequilibrium Phase Diagrams of C-H, C-O and Other Binary Systems

6.1 T-X Nonequilibrium Phase Diagrams for C-H Binary Systems

As mentioned in the last chapter, in the fall of 1994 the thermodynamic coupling model had made an important breakthrough [6.1–6.4]. Superequilibrium atomic hydrogen was noted by H*, instead of H. Moreover, the thermodynamic data for superequilibrium atomic hydrogen and activated graphite (activated by atomic hydrogen) were deduced, so the first quantitatively calculated nonequilibrium phase diagram for a C-H system was obtained by us, see Figs. 5.9 and 5.11 in the last chapter. H was the previous notation for superequilibrium atomic hydrogen (before the fall of 1994). In fact, there was confusion between equilibrium atomic hydrogen and superequilibrium atomic hydrogen before the improvement. After the improvement, a series of nonequilibrium phase diagrams of C-H systems at different pressures [such as at total pressures of 2.7 kPa (20 torr), 4.0 kPa (30 torr), 5.3 kPa (40 torr), 10.6 kPa (80 torr), 13.3 kPa (100 torr), 27 kPa (200 torr), 40 kPa (300 torr) and 53 kPa (400 torr), etc.] had been calculated. Here, in Figs. 6.1–6.4 nonequilibrium phase diagrams of C-H systems at four different pressures, i.e., 2.7 kPa, 4.0 kPa, 5.3 kPa and 10.6 kPa, are illustrated. From these phase diagrams, it can be seen that the diamond growth region shifts toward the high concentration of carbon (i.e., the right side of the phase diagrams) with increase of pressure. An empirical value of χ equal to 0.28 from the hot filament process was adopted during the calculation of these phase diagrams. Note that χ_{H}, χ_{O} and $\chi_{\mathrm{H,O}}$ will be used to represent coupling parameters in different systems, i.e., carbon–hydrogen (C-H), carbon–oxygen (C-O) and carbon–hydrogen–oxygen (C-H-O), respectively, in this chapter and in later chapters. gra*(H), gra*(O) and gra*(H, O) will also be used to represent activated graphite of in different systems, i.e., C-H, C-O, and C-H-O, respectively. Obviously, gra* $\equiv$ gra*(H), and $\chi \equiv \chi_{\mathrm{H}}$, here. There is always a diamond growth region in these nonequilibrium phase diagrams for the C-H system. The diamond growth region shifts with increase of pressure toward the right side in the diagram, i.e., toward the higher concentration of carbon. Inside the diamond growth region only diamond grows, but not graphite. That is why graphite can be etched away during simultaneous growth of diamond. These conclusions agree very well with a lot of experimental observations. During the 1994 Fall Meeting of the Chinese Materials Research Society (CMRS), in my oral presentation at the hard materials session the author made a comparison between experimental data concerned with activated CVD diamond growth data in the Abstracts of this meeting (see Table 6.1) and the theoretical phase diagrams calculated by us just before the meeting. The agreement were so excellent that nearly all the experimental data available had just dropped in the theoretical diamond growth phase region at the corresponding approximate pressure, see Figs. 6.1–6.3.

Table 6.1. Data on activated low-pressure diamond growth in the Abstract of 1994 Fall Meeting of CMRS

Reporters	CH_4 in H_2	p_{total}	$T_{substate}$	Ref.
C.-X. Gao et al.	0.5–0.8%	3.5 kPa	750–950 °C	[6.5]
M.-Y. Mao et al.	0.5–0.7%	20 torr (MW)	850–950 °C	[6.6]
Q.-Q. Yang et al.	1%	50 torr	950 °C	[6.7]
Y. Wang et al.	1.7%	4.8 kPa	900–950 °C	[6.8]
Y.-Z. Bai et al.	1%	3.7 kPa	850–900 °C	[6.9]

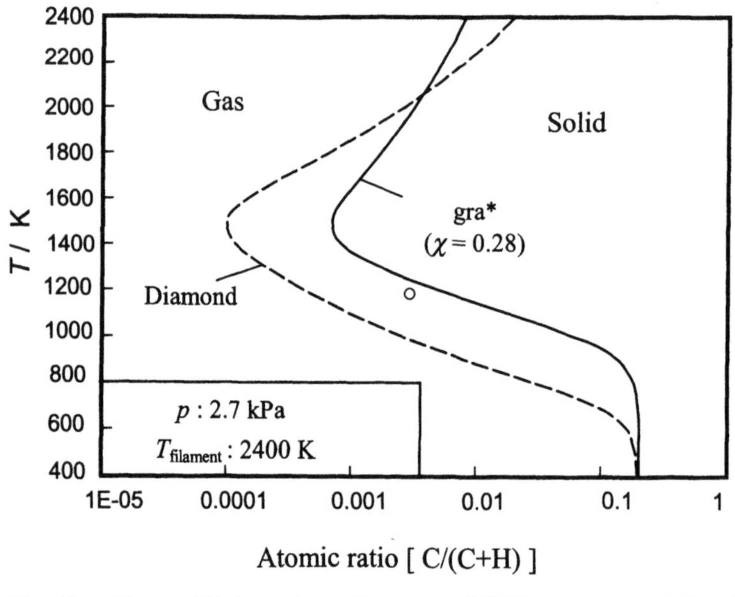

Fig. 6.1. Nonequilibrium phase diagrams of C-H system at 2.7 kPa (20 torr). "o" represents the diamond growth experimental data reported by M.-Y. Mao [6.6]

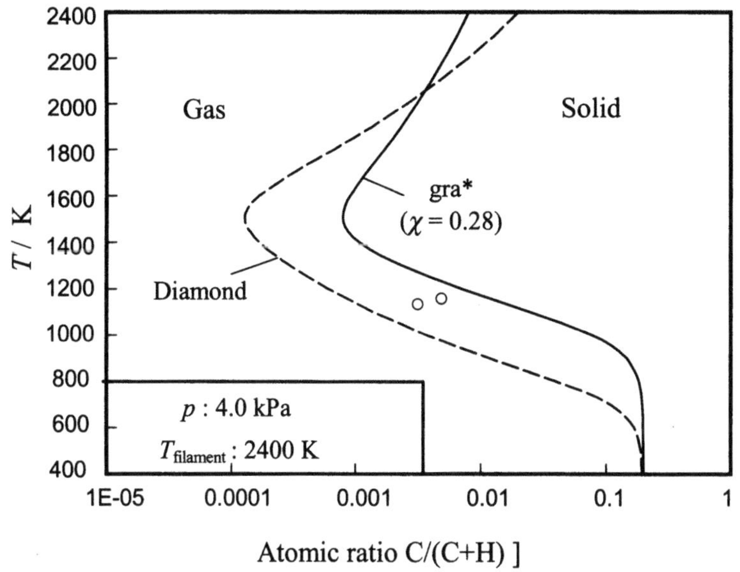

Fig. 6.2. Nonequilibrium phase diagrams of C-H system at 2.7 kPa (20 torr). The left "o" represents the diamond growth experimental data reported by C.-X. Gao [6.5], while the right "o" is that reported by Y.-Z. Bai [6.9]

In my presentation some empirical rules had also been theoretically explained. For instance, Qiao-Qin Yang et al. of Hunan University pointed out in the Abstracts of this meeting that if all other conditions (such as flow, temperature, pressure and so on) were kept unchanged, the growth rate of diamond was increased with the order of different carbon sources in the following order: 1% methane, alcohol, acetone, or isooctane and H_2 mixture [6.7]. Among them the growth rate was the lowest for the carbon source of methane, while the rate was the fastest for isooctane instead of methane. On the viewpoint of the phase diagram, the use of isooctane instead of methane increases the practical concentration of carbon significantly. (The carbon concentration increases about eight times, if the gas flow and other conditions remain unchanged.) This means the system shifts toward the right side along the X composition axis (or atomic ratio of [C/(C+H)]) in these phase diagrams. The system locates into the deeper part of the theoretical diamond growth region, i.e., with higher supersaturation degree, so the increase of the growth rate is a predicted result. Moreover, the substitution of isooctane for methane may even cause the deposition of nondiamond carbon. Another example had been found in the Abstract of this meeting. Zhen-Gang Zhang et al. of Lanzhou University reported their experiments about diamond growth from 1.58%, 1.66% and 1.86% CH_4-H_2 gas mixtures as the carbon sources [6.10]. They pointed out that the quality of the grown diamond crystals was better from

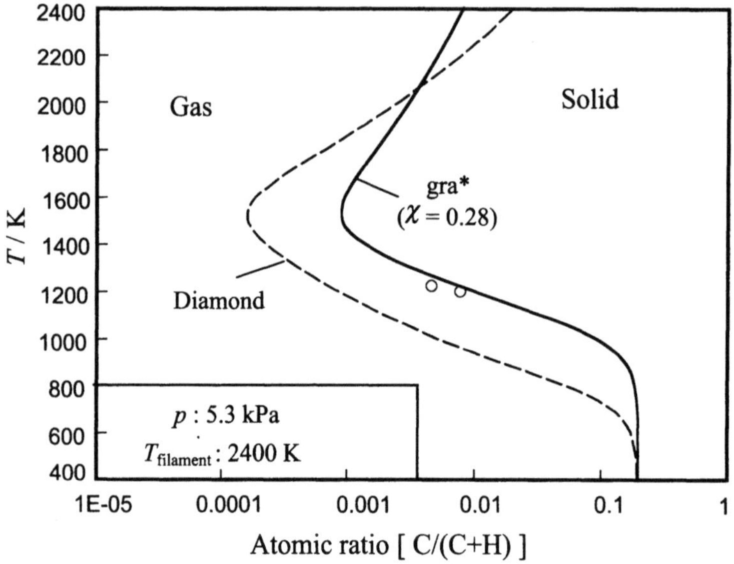

Fig. 6.3. Nonequilibrium phase diagrams of C-H system at 5.3 kPa (40 torr). The left "o" represents the diamond growth experimental data reported by Q.-Q. Yang [6.7], while the right "o" is that reported by Y. Wang [6.8]

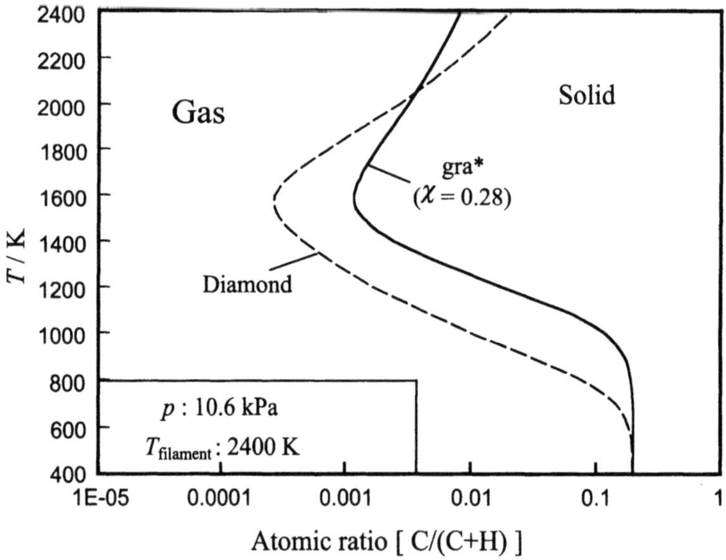

Fig. 6.4. Nonequilibrium phase diagrams of C-H system at 10.6 kPa (80 torr)

the lower carbon concentration. That agrees very well with the rules from our theoretical phase diagrams.

Later in the diamond session of the 1996 Fall Meeting of the CMRS Zhang-Da Lin of the Chinese Institute of Physics presented a paper concerned with nucleation density of diamond films [6.11]. There is a phenomenon that if all conditions remained unchanged except the lowering of the total pressure diamond nucleation would be easier and the density of the nucleation would be higher. This phenomenon is very easy to explain by our theoretical phase diagrams. For instance, in three T-X phase diagrams at 100 kPa, 10 kPa and 1 kPa, it is easy to find that the diamond growth phase region shifts toward the low concentration of carbon side with lowering of the total pressure, see Figs. 6.5–6.7 or Fig. 6.8. In Figs. 6.5–6.8 each diagram had been divided by two phase lines into four phase regions denoted by A, B, C and D, respectively. They are: Region A: gas phase region (no deposition region); Region B: graphite stable region; Region C: diamond stable region (diamond growth region); and Region D: nondiamond carbon deposition region. In Figs. 6.5–6.7 the diamond growth phase regions are represented by regions with oblique lines. For comparing the relationship between the diamond growth region and pressure, Figs. 6.5–6.7 can be redrawn into one diagram, as shown in Fig. 6.8. If the experimental conditions are at the carbon containing (expressed by mole ratio) in the gas source $[C/(C+H)] = 0.01$, and at the ordinary 1200 K substrate temperature (as shown by the intersection point of the two straight dash-and-dot lines in Fig. 6.8), then at a total pressure of 100 kPa the system is completely inside the gas phase region, in which diamond growth and the formation of diamond nuclei will be impossible. However, if the total pressure is lowered to 10 kPa but other conditions remain unchanged, then the experimental point has come deeply into the diamond growth region. That is, the supersaturation is very high. Under such conditions, not only the rate of the diamond growth but also the density of nucleation is very high. If the total pressure is further lowered to 1 kPa, then the experimental point has already come into the nondiamond carbon deposition region, and the nondiamond carbon will be deposited. Therefore, as the pressure is gradually lowered from 100 kPa to 1 kPa, the experimental point gradually comes from the gas region to the diamond growth region, and then into the nondiamond carbon deposition region. During the stage of moving deeper and deeper into the diamond growth region, the empirical rule for the relationship between the increase of nucleation density and the lowering of the pressure can be theoretically explained. In this way a lot of (maybe several hundred) experimental data of activated CVD diamond reported in the literature are dropped into our theoretical diamond growth region in our nonequilibrium phase diagrams under different conditions [6.12–6.24].

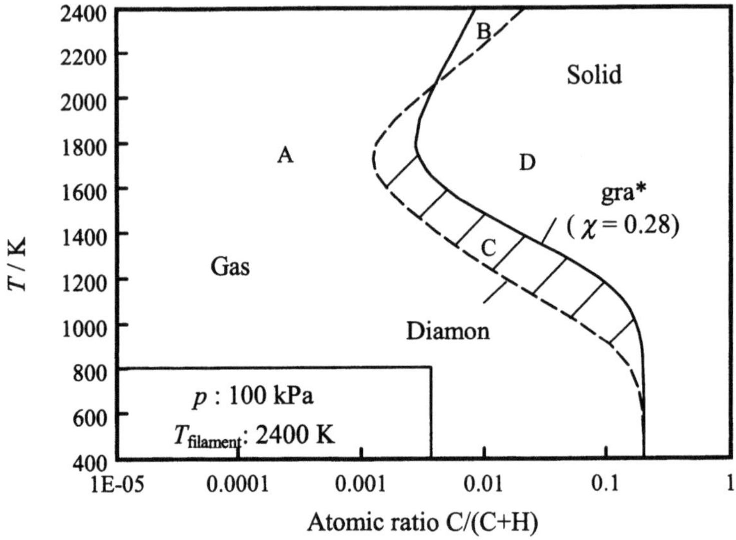

Fig. 6.5. *T-X* nonequilibrium phase diagram of C-H system at 100 kPa

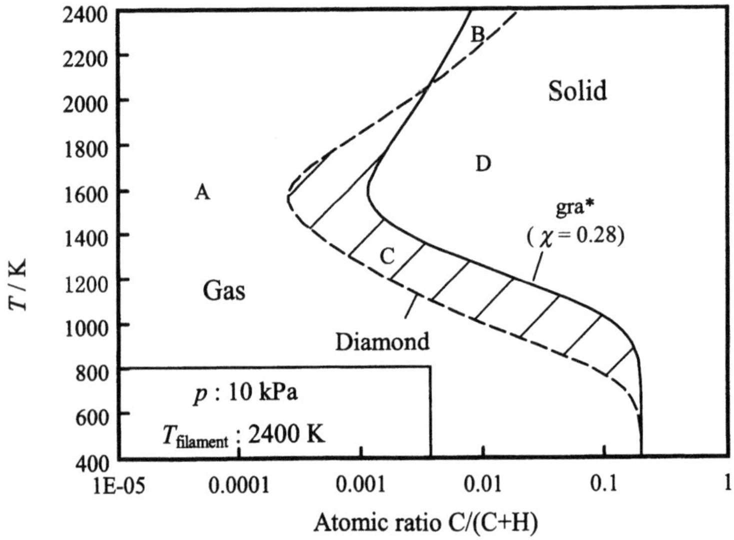

Fig. 6.6. *T-X* nonequilibrium phase diagram of C-H system at 10 kPa

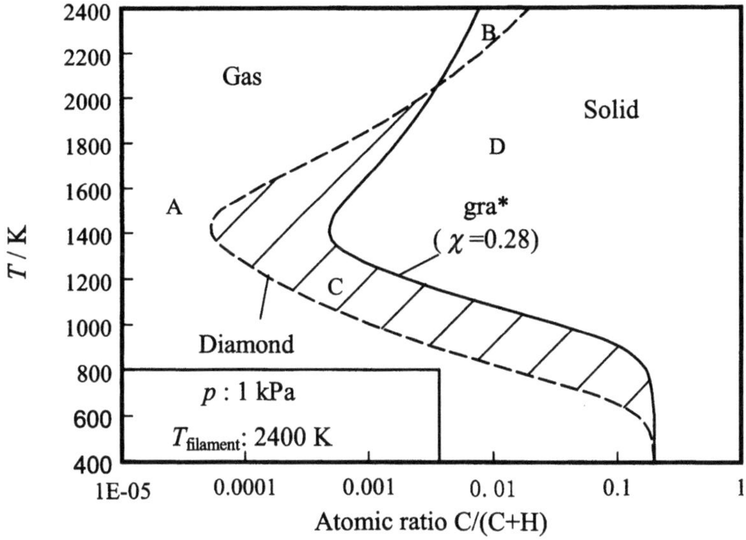

Fig. 6.7. *T-X* nonequilibrium phase diagram of C-H system at 1 kPa

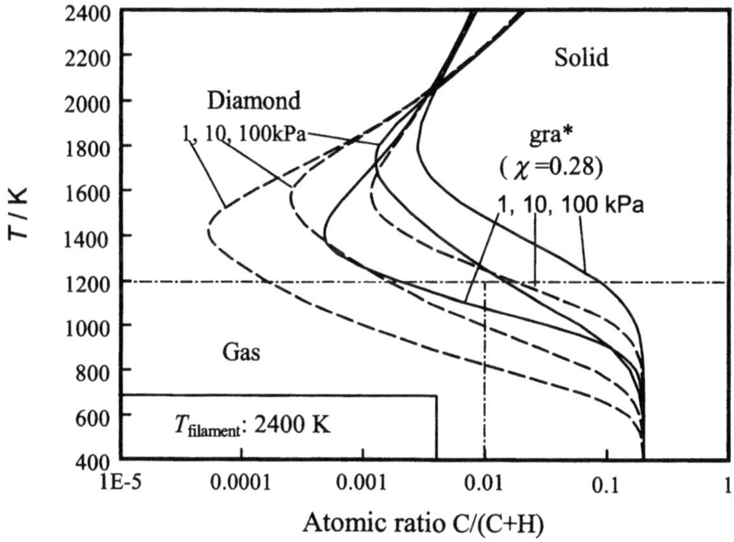

Fig. 6.8. *T-X* nonequilibrium phase diagrams of C-H system at three different pressures

6.2 *T-p-X* Nonequilibrium Phase Diagrams for C-H and C-O Binary Systems

It is well known that a complete phase diagram of the C-H binary system should be expressed in a three-dimensional space of temperature (T)–pressure (p)–composition (X). Figure 6.9 is the obtained T-p-X phase diagrams of the C-H system. Three isobaric cross-sections of T-X phase diagrams at three different pressures (0.01 kPa, 1 kPa and 100 kPa) are shown in the figure. There is a diamond stable phase region (also called the diamond growth region) in each cross-section of the T-X phase diagram in Fig. 6.9. The diamond growth phase region is represented by the region with oblique lines. If diamond growth is the purpose of our experiment, then the experiment should be run inside the diamond growth phase region. That provides a powerful theoretical tool for designing and optimizing the activated low-pressure CVD diamond process. During the calculation the empirical value of $\chi = 0.28$ from the hot filament process is used for the reaction rate ratio or the coupling parameter. The thermodynamic data for activated graphite for the C-H system has been discussed in the last chapter and listed in Tables 5.1 and 5.2. Those will not be repeated here.

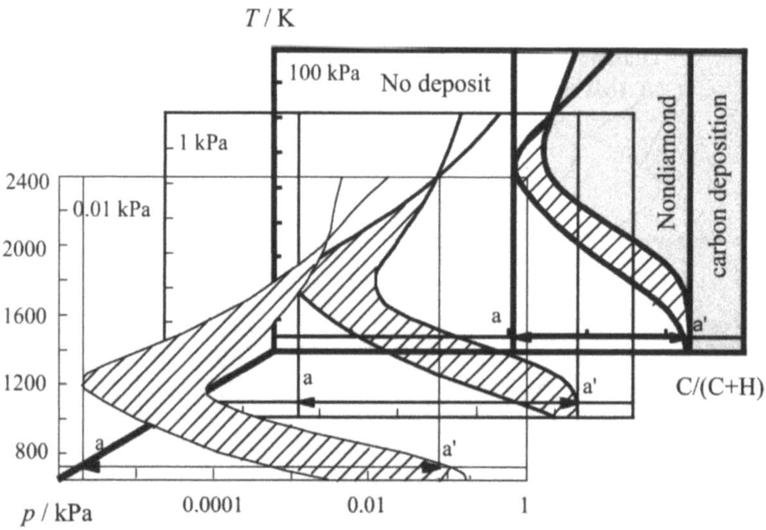

Fig. 6.9. T-p-X nonequilibrium phase diagrams of C-H system

The research method for nonequilibrium phase diagrams of the C-H system can be extended for the C-O system, for which a set of coupled reaction equations can be written as

$$C(gra) = C(dia), \ \Delta G_1 > 0 \ (T, \ p \leq 10^5 \text{ Pa}) \tag{6.1}$$

$$O* = 0.5O_2, \ \Delta G_2 \ll 0 \ (T_{\text{activated}} \gg T, \ p) \tag{6.2}$$

where, O* represents atomic oxygen with superequilibrium concentration. (6.3) = (6.1) + χ_O (6.2):

$$\mathrm{C(gra)} + \chi_O \ \mathrm{O*} = 0.5\ \chi_O\ \mathrm{O_2} + \mathrm{C(dia)},\ (T,\ p \leq 10^5\ \mathrm{Pa}) \tag{6.3}$$

or

$$\mathrm{C\,[gra*\,(O)]} = \mathrm{C(dia)},\ (T,\ p \leq 10^5\ \mathrm{Pa}) \tag{6.4}$$

where,

$$\mathrm{C\,[gra*\,(O)]} = [\mathrm{C(gra)} + \chi_O(\mathrm{O*} - 0.5\mathrm{O_2})] \tag{6.5}$$

$$\Delta G_4 = \Delta G_3 = \Delta G_1 + \chi_O \Delta G_2. \tag{6.6}$$

If the reaction rate ratio $\chi_O = r_2/r_1 >| \Delta G_1/\Delta G_2 |$ is not too small (i.e., atomic oxygen is of sufficient superequilibrium concentration), then we must have $\Delta G_3 < 0$, and the reaction (6.3) or (6.4) should be toward the right side. gra*(O) represents graphite activated by atomic oxygen. The corresponding thermodynamic data of superequilibrium atomic oxygen and activated graphite (activated by atomic oxygen) for the C-O system can be deduced by a similar method used for the C-H system. The obtained thermodynamic data for activated graphite (activated by superequilibrium atomic oxygen) are listed in Table 6.2. Then T-X phase diagrams of the C-O system at different pressures can be calculated, and finally the calculated T-p-X nonequilibrium phase diagram of the C-O system is illustrated in Fig. 6.10.

Table 6.2. Thermodynamic data for activated graphite (activated by superequilibrium atomic oxygen) [gra*(O)] with different χ_O values

$H_{\mathrm{gra*}}$(O) (kJ mol^{-1}) = $\chi_O \times 249.195$	
$S_{\mathrm{gra*}}$(O) (J mol^{-1} K^{-1}) = $(5.6861 + \chi_O \times 97.9819)$	
C_p (J mol^{-1} K^{-1}) $= A + B10^{-3}T + C10^5T^{-2} + D10^{-6}T^2$	
298–1000 K	1000–1550 K
$A = 0.1597 + \chi_O \times 8.4496$	$A = 25.5701 + \chi_O \times 8.4496$
$B = 38.8890 - \chi_O \times 7.3095$	$B = -0.2232 - \chi_O \times 7.3095$
$C = -1.5154 + \chi_O \times 0.7381$	$C = -39.1476 + \chi_O \times 0.7381$
$D = -17.4121 + \chi_O \times 2.2678$	$D = 0.0978 + \chi_O \times 2.2678$
1550–2000 K	2000–2400 K
$A = 25.5701 + \chi_O \times 5.5648$	$A = 25.5701 + \chi_O \times 5.5439$
$B = -0.2232 - \chi_O \times 2.1096$	$B = -0.2232 - \chi_O \times 2.0594$
$C = -39.1476 - \chi_O \times 5.4884$	$C = -39.1476 - \chi_O \times 6.4633$
$D = 0.0978 + \chi_O \times 0.1755$	$D = 0.0978 + \chi_O \times 0.1755$

Hot filament: 2400 K, Ref. Temp.: 298 K, Pressure: 100 kPa.
Definitions of B, C and D here are a little different from those in text

Although there are also three main regions (gas phase region, diamond growth region and nondiamond deposition region) in each T-X phase diagram of the C-O system, they are quite different from the corresponding ones both in shape and in position. The abscissa, i.e., the coordinate of the composition X, is selected by logarithms for the C-H system, but in the most general linear form for the C-O system. For the C-O system the temperature interval of the main part of the diamond growth region is obviously lower than the corresponding one for the C-H system. That agrees with a general empirical rule that diamond growth is usually at a somewhat lower temperature in an oxygen-containing system. The empirical rule can be illustrated by nonequilibrium phase diagrams for the C-O system very well. It should be pointed out that because experimental data of the pure C-O system were very few, the coupling parameter χ_O used for calculation was simply from the approximation $\chi_O \approx \chi_H = 0.28$. Besides, another gas allotrope of oxygen, i.e., ozone, has been neglected to simplify the treatment.

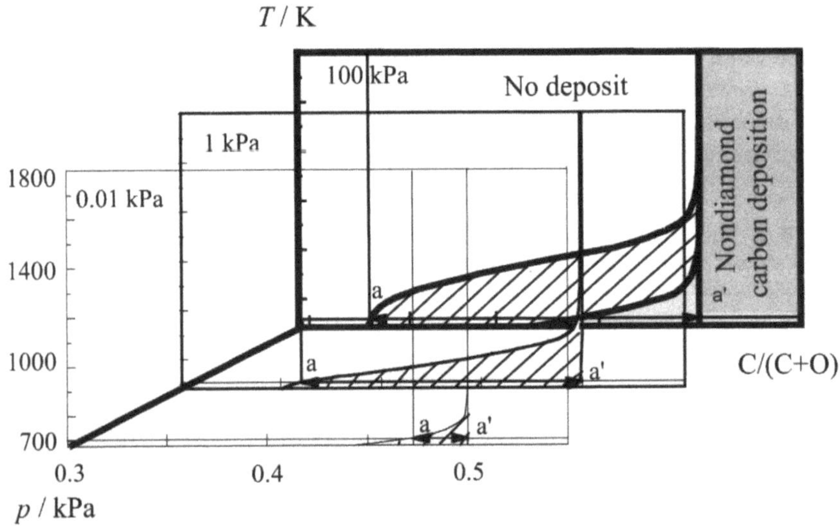

Fig. 6.10. T-p-X nonequilibrium phase diagrams of C-O system

As discussed for Fig. 5.11 in Chap. 5, the more suitable form of T-p-X nonequilibrium phase diagrams of a C-H system should be as shown in Fig. 6.11. Beside a gas phase region, there are three different phase regions including the solid phase, illustrated by different lines or net pattern styles. The total number of phase regions is four: (i) gas phase region, (ii) diamond phase region, (iii) diamond stable and graphite metastable phase region, and (iv) graphite phase region at high temperature. The gas phase region is a single-phase region. All of regions (ii)–(iv) are two-phase regions (a solid phase and a gas phase). Besides, at the end of the composition coordinate there is a single-phase line of 100% carbon. A metastable phase will not appear, if the stable phase exists. However, for low-pressure diamond growth Fig. 6.9 is clearer and convenient, because the most interesting phase region is only a diamond phase region, in which diamond can grow together with simultaneous graphite etching. Therefore, most of the other phase diagrams in this book concerned with the activated diamond vapor growth are still drawn in a simpler way, putting emphasis on the diamond growth region, as done in Fig. 6.9.

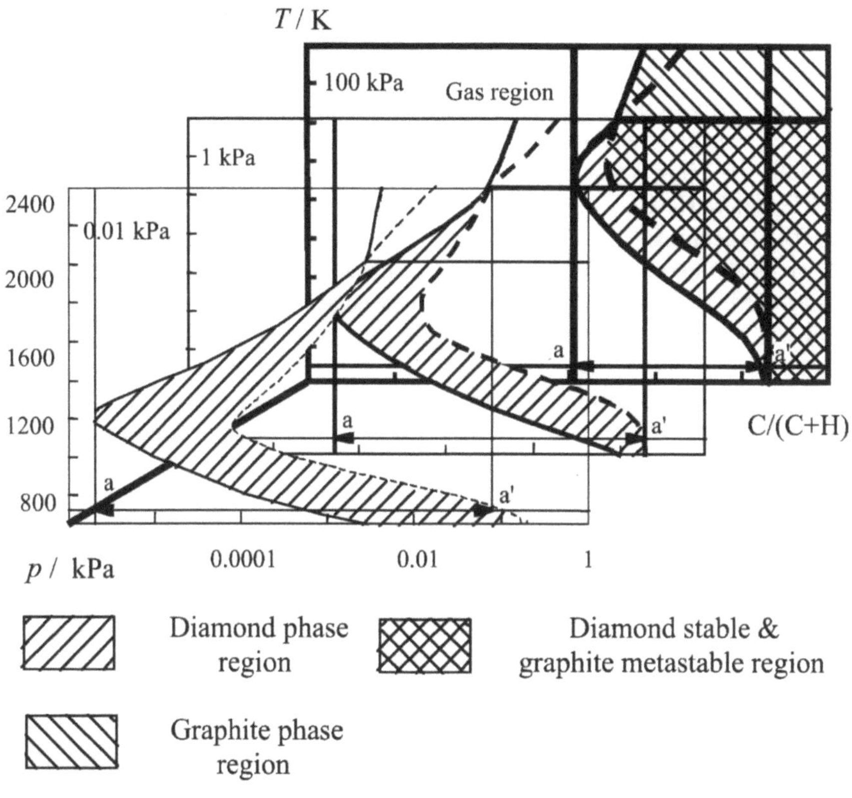

Fig. 6.11. More strict illustration of T-p-X nonequilibrium phase diagrams of C-H system

6.3 Nonequilibrium Phase Diagrams for Other Binary Systems

There are different expressions for nonequilibrium phase diagrams just as these are for equilibrium phase diagrams. For instance, nonequilibrium phase diagrams of C-H binary systems can also be expressed as nonequilibrium phase diagrams of the CH_4-H_2 binary system. The differences between them are only on the meaning of the binary components and the scale of the coordinate axes. However, the form of phase diagrams of the CH_4-H_2 system may be more convenient for practical applications [6.12, 6.25]. In phase diagrams of the CH_4-H_2 binary system the composition of the most generally used 1% CH_4-99% H_2 gas system equals $X' = [CH_4/(CH_4 + H_2)] = 0.01$, but in phase diagrams of the C-H binary system the composition should be calculated by the transformation equation $X = [C/(C{+}H)] = [1/(1{+}\ 4 + 2 \times 99)] \approx 0.005$. Such a transformation is still very simple, so in this book phase diagrams of the CH_4-H_2 binary system will not further be discussed. The phase diagrams of CO-H_2, C_2H_6-CO_2, C_2H_4-CO_2, C_2H_2-CO_2, etc., binary systems also have practical meaning [?,6.13–6.16,6.18,6.19], these phase diagrams are also parts of the phase diagrams of the C-H-O ternary system. However, these binary phase diagrams can be transformed into each other. The following discussion will focus on nonequilibrium phase diagrams of a general pseudobinary system with a fixed molar ratio [$\lambda = O/(H{+}O)$]. In fact, these systems are part of the C-H-O ternary system. A nonequilibrium phase diagram of the C-H-O ternary system can be constructed on the basis of a series of T-p-X phase diagrams of a pseudobinary system with different $\lambda = 0.0$, 0.1, 0.2, 0.3, ..., 0.9, and 1.0, respectively. The nonequilibrium phase diagram of the C-H-O ternary system will be discussed in the next chapter. To limit the length of the book T-p-X phase diagrams of a pseudobinary system with C-[$O/(O{+}H) = \lambda$] are illustrated only at three different pressures of 100 kPa, 1 kPa and 10 Pa, respectively in Figs. 6.12–6.14. Moreover, a series of T-X phase diagrams of the pseudobinary system for $\lambda = 0.0$, 0.1, 0.2, 0.3, ..., 0.9 and 1.0, respectively, at the same pressure are drawn into one of Figs. 6.12–6.14. It should be noted that if all the phase lines for $\lambda = 0.0$ are taken from Figs. 6.12, 6.13 and 6.14, respectively, then a three-dimensional T-p-X phase diagram of a C-H binary system similar to Fig. 6.9 can be constructed. The difference between the new constructed phase diagram and Fig. 6.9 is on the abscissa. In the new constructed phase diagram the abscissa of the composition is linear, while in Fig. 6.9 the abscissa of the composition is logarithm it. In this way, a series of three-dimensional T-p-X phase diagrams of the C-(O+H) pseudobinary system with different λ values can be constructed. A three-dimensional T-p-X phase diagram of the C-(O+H) pseudobinary system with $\lambda = 1$ is just the same as the three-dimensional T-p-X phase diagram of the C-O pure binary system, i.e., Fig. 6.10. Details about how to construct these T-p-X nonequilibrium phase diagrams of C-H-O ternary systems will be discussed in the next chapter.

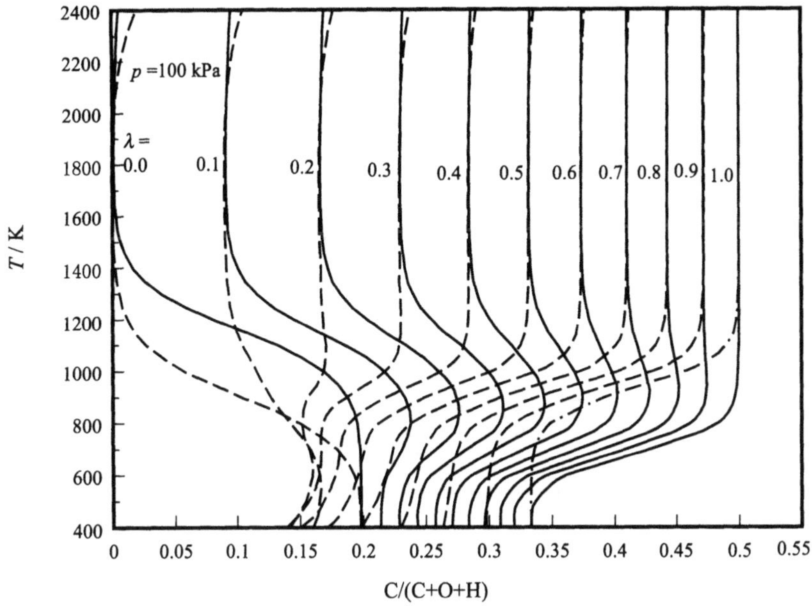

Fig. 6.12. *T-p-X* phase diagrams of the [C-(O+H)] pseudobinary system at 100 kPa

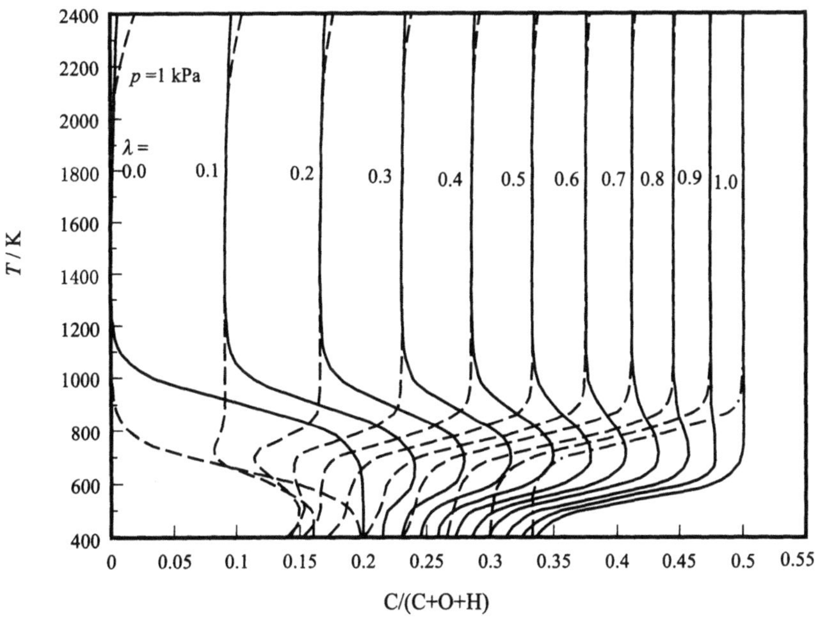

Fig. 6.13. *T-p-X* phase diagrams of the [C-(O+H)] pseudobinary system at 1 kPa

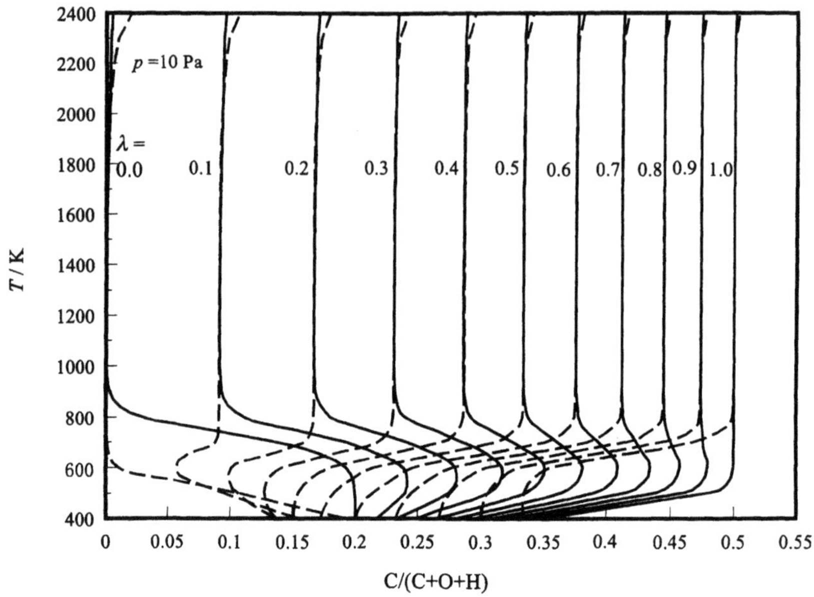

Fig. 6.14. *T-p-X* phase diagrams of the [C-(O+H)] pseudobinary system at 0.010 kPa

7. Nonequilibrium Phase Diagrams of C-H-O and Other Ternary Systems

Since 1986 a great upsurge in research on activated low-pressure diamond growth from the vapor phase has unfolded throughout the world. Many scientists coming from different academic fields have joined in this research and research papers have increased very quickly. At the early stage, the gas sources of carbon were mainly a CH_4-H_2 mixture, and other hydrocarbons in a hydrogen mixture. The general expression could be written as C_xH_y-H_2. A little later alcohol-H_2, methanol-H_2, acetone-H_2, ether-H_2, and other oxygen-containing organic compounds in a hydrogen mixture were also used as reaction sources. The CH_4-H_2 mixture together with direct addition of a little oxygen was also used for diamond growth. It was observed that the addition of oxygen in the reaction sources is often favorable to get diamond thin films with higher quality and to lower the temperature for diamond growth. In the oxygen–acetylene torch combustion process for diamond thin film growth oxygen plays a more important role. Although as pointed out in the last chapter a C-H-O system with a fixed ratio between oxygen and hydrogen or consisting of two reactants can be treated as a binary or pseudobinary system, from the viewpoint of thermodynamic analysis, these systems are C-H-O ternary systems. It is well known that a ternary system has two independent component variables, which may be denoted by X_1 and X_2. The change of composition of the system may be illustrated by a plane triangular phase diagram. Each sideline of the plane triangular phase diagram and each straight line inside the triangular phase diagram can be treated as a binary or pseudobinary system. For instance, two sidelines of C-H and C-O of the triangular phase diagram of the C-H-O system are just C-H and C-O binary systems for diamond growth from the vapor phase, respectively; and the tie line between the apex of H and CO can also be treated as a H-CO binary or pseudobinary system. However, since many experiments on activated low-pressure diamond growth from the vapor phase reported in the literature run in the C-H-O ternary system, so the phase diagrams for the C-H-O ternary system should be studied and discussed in detail.

7.1 Bachmann's Empirical Phase Diagram for the C-H-O System

In 1991 P.K. Bachmann et al. summarized the experimental data for activated CVD diamond processes for C-H-O systems reported in the literature [7.1, 7.2]. This summary included all kinds of different processes, such as the hot filament process, combustion process, microwave process, DC or AC plasma process, plasma torch process, and so on, without distinction. In this summary the experimental data (i.e., the C, H and O compositions of the gas sources) were divided into three classes: no growth (or no deposition), diamond growth, and nondiamond carbon deposition. The experimental points where nothing at all was deposited were labeled by the number n of data points, the experimental points of diamond growth were marked by open diamond symbols, and the experimental points where only nondiamond carbon was deposited were marked by full circles. The system composition (i.e., carbon, hydrogen, oxygen concentrations) can be represented as usual in a planar triangular phase diagram. It was found that the three classes of experimental points were concentrated into three corresponding regions inside the triangular ternary phase diagrams. These regions were called the no growth region, the diamond growth region (Bachmann called it the "diamond domain"), and the nondiamond carbon growth region. In this way an atomic compositional C-H-O diamond deposition phase diagram with "diamond domain" was obtained, see Fig. 7.1. In fact, this phase diagram was an empirical phase diagram for activated low-pressure diamond from the vapor phase of the C-H-O system and comprised more than 80 deposition experiments from over 25 references. Obviously, the system composition position might be independent of the initial raw materials inside the phase diagram. For instance, the position of CH_3OH is located at the same position as the 33.3% CO and 66.6% H_2 mixture, and also at the same position as the 66.6% CH_4 and 33.3% O_2 mixture. Moreover, the composition position inside the phase diagram for an experimental point is also independent of the process used, that is, all the different processes, such as the hot filament process, microwave process, combustion process, and plasma torch process, are treated the same. This characteristic that the phase diagram depends only on the total composition of the system but not on the initial raw materials and the processes used clearly showed that activated low pressure diamond growth must be subordinated to thermodynamic rules. The phase diagram itself is a concept and a term of thermodynamics. There was a viewpoint for a long time that activated low-pressure diamond growth from the vapor phase was controlled only by kinetics but not by thermodynamics. The reason might be based on a conclusion of equilibrium thermodynamics that "diamond is a metastable phase at low pressure". Now the existence of the diamond growth phase region clearly shows that such a viewpoint is incorrect for the activated low-pressure diamond process. The existence of the diamond growth phase region also forced some scientists who had the incorrect kinetic control viewpoint to consider the thermodynamics aspects. That was the advantage of Bachmann's empirical phase diagram.

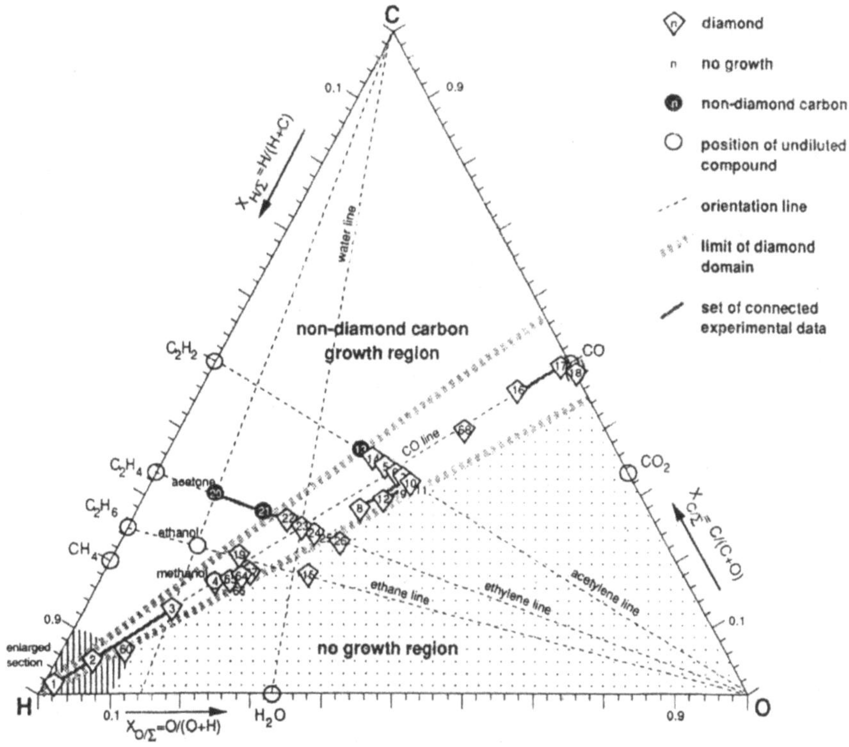

Fig. 7.1. Bachmann's empirical phase diagram for diamond deposition from C-H-O systems. Reprinted from P.K. Bachmann, D. Leers and H. Lydtin [7.1] with permission from Elsevier Science

However, the most important disadvantage was that the influences of temperature and pressure are not included in Bachmann's empirical phase diagram, and thus the second important disadvantage follows. This is the inexplicability of the nature of a projective phase diagram. It is well-known that in phase diagram theory that a planar ternary phase diagram must be an incomplete phase diagram, because the coordinates of temperature and pressure are missed. If all experimental data are obtained at a certain temperature and pressure, that will be a cross sectional phase diagram. There may be some clear and fixed phase line(s) in it. If the experimental data are obtained at different temperatures and pressures, as done by Bachmann, that must be a projective phase diagram in a certain ranges of temperatures and pressures. In a projective phase diagram in a certain ranges of temperatures and pressures phase regions may be overlapped each other. However,

Bachmann had made effort to exclude experimental points of no deposit and nondiamond carbon deposition out of the diamond growth phase region, i.e., Bachmann's "diamond domain". Because the influences of temperature and pressure were not considered, the shape of Bachmann's diamond growth region must be incorrect. After a critical test in 1994 Marinelli et al. said that "Quantitatively, however, the diamond domain deduced from our data considerably differs from the one indicated by Bachmann and co-workers ..." [7.3]. Bachmann et al. revised their empirical phase diagram in 1995, as shown in Fig. 7.2 [7.4, 7.5]. However, from basic knowledge of phase diagram any composition phase diagram must connected with temperature and pressure, so Bachmann's revised empirical phase diagram is still incorrect. The theoretical aspect has been talked about a little above. Further discussion in both empirical and theoretical aspects on C-H-O and other ternary phase diagrams for activated low-pressure diamond synthesis will be going on in the following sections of this chapter.

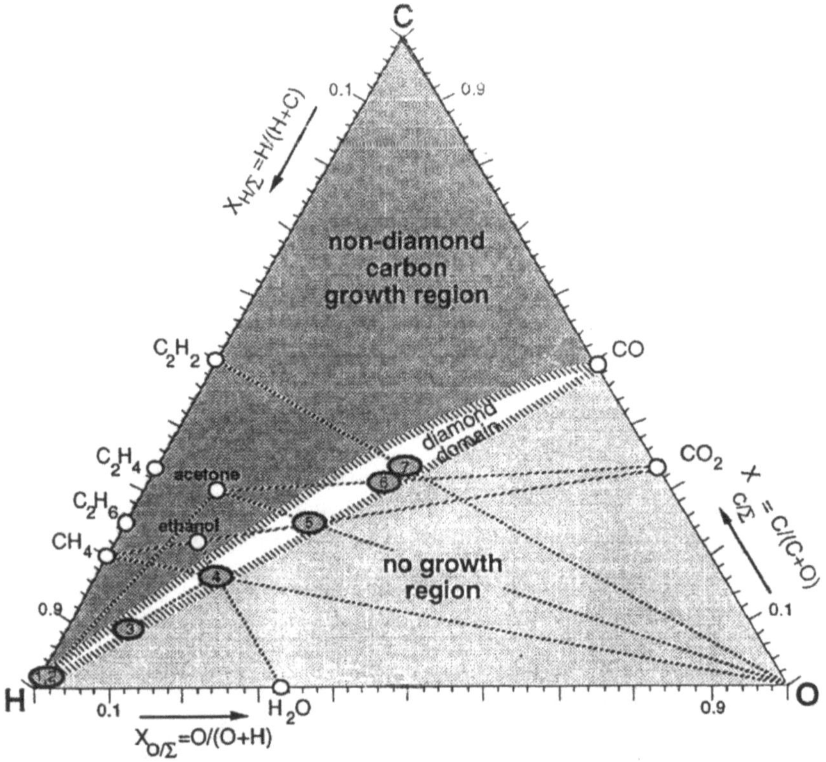

Fig. 7.2. Revised Bachmann's empirical phase diagram for diamond deposition from C-H-O systems. Reprinted from P.K. Bachmann, H.J. Hagemann, H. Lade, Leers, D.U. Wiechert, H. Wilson, D. Fournier and K. Plamann [7.5] with permission from Elsevier Science

7.2 Projective Nonequilibrium Phase Diagrams for C-H-O Systems

As mentioned above, the composition of a ternary system in a phase diagram should be expressed by two independent variables, i.e., the composition should be expressed in a planar phase diagram. The substrate temperature T and the total pressure p need two additional coordinates. Therefore, a four-dimensional T-p-X_1-X_2 phase diagram of a ternary system cannot be completely expressed in ordinary three-dimensional space. To express such a ternary system in a three-dimensional diagram it is usual to fix one of these variables, such as pressure or temperature, and then a phase diagram of T-X_1-X_2 or p-X_1-X_2 can be drawn in three-dimensional space. Even so, such three-dimensional phase diagrams are not very convenient for quantitative applications. It is usual to introduce projective phase diagrams or cross-sectional phase diagrams. The common advantage of these two kinds of planar phase diagrams is easy to compare between theory and experiments, and to offer a theoretical guide for practical work. However, these two different planar phase diagrams have their own advantages and disadvantages. For example, the projective planar diagram can include a lot of information for an overview. Projective planar diagrams may still be divided into three subclasses: (i) projective diagrams along the temperature coordinate within a certain temperature range at a fixed pressure; (ii) projective diagrams along the pressure coordinate within a certain pressure range at a fixed temperature; and (iii) projective diagrams both along the temperature coordinate within a certain temperature range and along the pressure coordinate within a certain pressure range. Among theses diagrams the third subclass of projective planar phase diagrams includes the largest amount of information. Figure 7.3 is such a projective nonequilibrium phase diagram of a C-H-O system calculated by us on the basis of the nonequilibrium nondissipation principle within the temperature range 700–1900 K and within the pressure range 0.01–100 kPa [7.6–7.9]. The ranges are selected so that most of the experimental data for activated low-pressure diamond growth from the vapor phase reported in the literature can be included. The selected ranges are also close to the ranges selected by Bachmann (¿ 400 °C, 0.1–760 torr), so that a comparison between our theoretically calculated phase diagram and Bachmann's empirical phase diagram can be made. Therefore, Bachmann's empirical phase lines (dotted lines) are also drawn in our theoretically calculated phase diagrams, see Fig. 7.3.

There are also three phase regions in the theoretically calculated projective phase diagram of the C-H-O ternary system: (i) no deposition region (also called the gas region); (ii) possible diamond growth region (note that in projective phase diagrams the name "possible diamond growth region" should be used: if the temperature and/or pressure are not suitable, then diamond growth may not occur); (iii) nondiamond carbon deposition region. These phase regions are divided by the carbon concentration limitation phase lines.

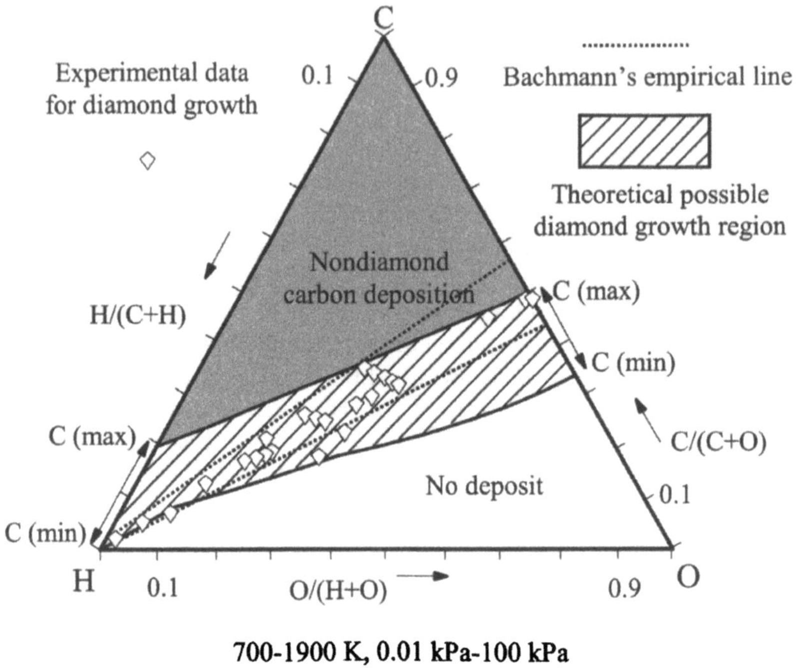

Fig. 7.3. Theoretically calculated projective nonequilibrium phase diagram of C-H-O ternary systems under 700–1900 K, 0.01–100 kPa

The low concentration limitation line is located between the no deposition region and the possible diamond growth region, while the high concentration limitation line is located between the possible diamond growth region and the nondiamond carbon deposition region. The theoretically calculated low carbon concentration limitation line is much lower than that of Bachmann's empirical line, so the theoretical possible diamond growth region includes more experimental data than that in Bachmann's empirical phase diagram. Therefore, all the relatively reliable experimental data are included in the possible diamond growth region, but the empirical phase diagram cannot be so. In the high carbon concentration limit the theoretical line obviously crosses the empirical line, so that there is a small triangular region at each of the right and the left sides. The right triangle is located below Bachmann's empirical phase line (inside Bachmann's diamond growth region) but above the theoretical phase line (outside our theoretical possible diamond growth region), that is, in the right-hand small triangular region, i.e., a molar ratio greater than C/O = 1, diamond growth is impossible based on our theoretical phase diagram, but diamond growth is possible based on Bachmann's empirical phase diagram. However, up to now no diamond growth experiment has yet been reported in the literature. On the contrary, the left triangle is located above Bachmann's empirical phase line (outside Bachmann's diamond growth region) but below the theoretical phase line (inside our theoretical possible diamond growth region), that is, in the left small triangular region diamond growth is possible based on the theoretical phase diagram, but diamond growth is impossible based on Bachmann's empirical phase diagram. In recent years, experiments have confirmed that in the left small triangular region diamond growth is possible, and diamond nucleation is also possible on a nondiamond substrate [7.10]. In this way the correctness of our theoretical phase diagram has been further confirmed. The author met Bachmann himself during an international conference, and discussed this with him. Bachmann said he had already found this problem, and had shifted the high limitation phase line of the left part approach to the position C/O = 1. His related paper was published in the proceedings of the conference (i.e., Fig. 7.2 of this book). It should be noted that the tendency of the revision may be correct, but even after revision such a phase line did not agree very well with the theoretical line and was still empirical without the necessary theoretical basis.

There is an obvious disadvantage for all projective phase diagrams: that is, during the projection different phase regions in space may partially overlapp each other on the projective plane. For instance, the boundary of the projective diamond growth phase region is selected as a maximum borderline. Therefore, all diamond growth experimental points should be dropped into the projective diamond growth composition phase region, but maybe some parts of the no deposition phase region (the gas region) and nondiamond carbon deposition region have also been projected into the projective diamond growth composition phase region during projection; that is, the temperature and/or the pressure may be unsuitable for diamond growth. Due to this uncertainty the projective diamond growth phase region should be called the "possible diamond growth phase region". Therefore, a few experimental points with no deposition and a few experimental points of nondiamond carbon deposition drop inside the possible diamond growth region. That is normal and reasonable. In cross-sectional phase diagrams all the phase lines and phase boundaries are clear and fixed. This is different from projective phase diagrams. However, there are also other disadvantages for cross-sectional phase diagrams. For instance, for each practical experimental condition a special cross-sectional phase diagram for the corresponding temperature and pressure should be drawn. That is not favorable to give an overview of comparison between theoretical predictions and experiments. In this book the projective phase diagrams for C-H-O systems will be discussed first, and then the discussion on cross-sectional phase diagrams will be given. The calculation method is the same for both the projective phase diagram and the cross-sectional phase diagram, but the calculation range and treatment of data are different. The following discussion on Fig. 7.3 will be taken as an example of projective phase diagrams for C-H-O systems.

There are two sidelines of C-H and C-O in the C-H-O triangular composition phase diagram. The different phase ranges on these two sidelines (C-H and C-O) are obtained by two projections from T-p-X phase diagrams of C-H and C-O systems, respectively. For instance, if the T-p-X phase diagrams of the C-H system of Fig. 6.9 (or the C-O system of Fig. 6.10) are projected along the T coordinate from 1900 K onto the plane of 700 K, then after the projection the diamond growth region (represented by regions with oblique lines) of each T-X phase diagram at different pressures will be a line segment on the composition X coordinate. These line segments are composition ranges for possible diamond growth (i.e., the possible diamond growth composition ranges), which are denoted as **a**-**a**′ line segments at all the selected pressures (0.01 kPa, 1 kPa and 100 kPa) in Fig. 6.9 (or Fig. 6.10). In the next step, both the minimum value of **a** in the selected pressure range (0.01–100 kPa), i.e., $\mathbf{a}_{min}$, and the maximum value of **a**′ in the selected pressure range (0.01–100 kPa), i.e., $\mathbf{a}_{max}$, are drawn on the corresponding sideline of C-H (or C-O) of the triangular phase diagram, as shown in Fig. 7.3, and denoted as C (min) and C (max), respectively. For the calculation of the inner part of the C-H-O triangular phase diagram of the ternary system, a linear combination approximation is adopted for getting the thermodynamic data for graphite simultaneously activated by superequilibrium atomic hydrogen and superequilibrium atomic oxygen. For example, the thermodynamic data for graphite simultaneously activated by both superequilibrium atomic hydrogen and superequilibrium atomic oxygen with the atomic ratio H/O = 1 for the C-H-O system can be taken as the average of both the corresponding thermodynamic data for graphite activated by superequilibrium atomic hydrogen [gra*(H)] (see Table 5.1 and note that gra* ≡ gra*(H), $\chi \equiv \chi_H$) and the corresponding thermodynamic data for graphite activated by superequilibrium atomic oxygen [gra*(O)] (see Table 6.2). Thermodynamic data for graphite simultaneously activated by superequilibrium atomic hydrogen and superequilibrium atomic oxygen with other H/O atomic ratios of the C-H-O system can be obtained by a linear insertion method. After getting the thermodynamic data for graphite simultaneously activated by superequilibrium atomic hydrogen and superequilibrium atomic oxygen with any fixed H/O atomic ratio, the T-X or T-p-X nonequilibrium phase diagrams of the C-(H/O) pseudobinary system (refer to Figs. 6.12–6.14 in Chap. 6) with any fixed H/O atomic ratio can be calculated. After two projections along the T coordinate and along the p coordinate, respectively, the line segment $\mathbf{a}_{min}$-$\mathbf{a}_{max}$ can be obtained. All these $\mathbf{a}_{min}$-$\mathbf{a}_{max}$ line segments with a fixed H/O ratio are next drawn into the planar triangular C-H-O composition phase diagram. The locus of all these $\mathbf{a}_{min}$ points is just the low carbon concentration limitation line C (min)-C (min) for possible diamond growth, while the locus of all these $\mathbf{a}_{max}$ points is just the high carbon concentration limitation line C (max)-C (max) for possible diamond growth. In this way Fig. 7.3 of the projective phase diagram of C-H-O ternary systems is obtained. According to the construction of the phase diagram mentioned above, it is clear that the planar phase diagram of Fig. 7.3 is a theoretical ternary projective composition phase diagram, while the corresponding Bachmann phase diagram is an empirical ternary projective composition phase diagram. It is also clear that the projective composition nonequilibrium phase diagram of a ternary system, as shown in Fig. 7.3, is obtained on the basis of the nonequilibrium nondissipation principle, and is suitable for activated low-pressure diamond growth from the vapor phase.

7.3 Influences of T and p Ranges on C-H-O Ternary Phase Diagrams

As mentioned above, a complete T-p-X phase diagram of a C-H-O ternary system must be in four-dimensional space. In the last section, the discussion focused only on a projective planar composition phase diagram for the 0.01–100 kPa pressure range and 700–1900 K temperature range. Obviously, projective planar composition phase diagrams for different pressure ranges and different temperature ranges must be different. Figure 7.4 shows projective phase diagrams for a fixed 0.01–100 kPa pressure range but for three different 700–1900 K, 900–1900 K and 1100–1900 K temperature ranges [7.11–7.19]. Bachmann's empirical phase lines are still included for comparison. From Fig. 7.4, it is clearly seen that the possible diamond growth phase region obviously shrinks with the increase of the lower temperature. In particular, close to the C-O sideline of the triangle the shrinkage is more remarkable. The low carbon concentration phase line of the possible diamond growth region (i.e., the low carbon concentration limitation for diamond growth) rises quickly close to the tie line of H-CO with an increase of the low-temperature limitation of the substrate temperature range. The high carbon concentration line of the possible diamond growth region (i.e., the high carbon concentration limitation for diamond growth) lowers with increase of the low-temperature limitation of the temperature range. However, the right end (on the C-O sideline of the triangle) of lines remains nearly unchanged at the fixed point of the CO position. In this way, the possible diamond growth phase region of the projective phase diagram for the 1300–1900 K temperature range is nearly a very narrow tapered shaped triangle with a vertex on the C-O sideline of the ternary phase diagram. The lower carbon concentration border is very close to the H-CO tie line. The narrow tapered shaped triangle for the 1300–1900 K temperature range is not drawn in Fig. 7.4, but can be found from the change tendency. In comparison with Bachmann's empirical phase region such a theoretically possible diamond growth region is much narrower in shape than the empirical phase region. Moreover, the vertex of the narrow triangle is just in the opposite direction. That clearly reflects the most important disadvantage of Bachmann's empirical phase diagram, which does not represent the influences of temperature and pressure.

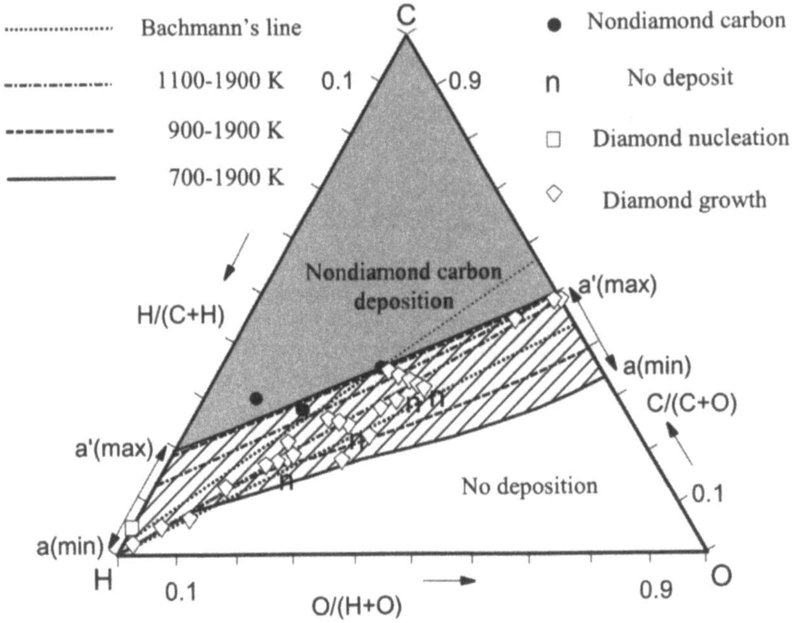

Fig. 7.4. Theoretical projective nonequilibrium phase diagrams for C-H-O ternary systems under 0.01–100 kPa

In 1994, Marinelli et al. made a detailed comparison between their critical experiments and Bachmann's empirical phase region [7.3]. Marinelli's results confirmed the existence of a single compositional "diamond domain" in a C-H-O triangular diagram. Bachmann correctly divided the triangular phase diagram into three phase regions, i.e., (i) no growth region, (ii) diamond growth region, also called the "diamond domain" and (iii) nondiamond carbon growth region. However, Marinelli et al. also concluded that the shape and the borders of their experimental diamond growth region are quantitatively rather different from those assumed in the work of Bachmann and coworkers. According to Marinelli's study of the plasma optical emission spectra the border from the no growth region into the diamond growth region is independent of the identity of the initial chemical species. That shows the presence of the same mechanisms of diamond growth from the vapor phase for all the gas mixtures investigated. The growth gas mixtures investigated were CH_4-CO_2, C_2H_6-CO_2, C_2H_4-CO_2 and C_2H_2-CO_2. The concentration of hydrocarbons increased gradually, so that the total composition of the system is consequent to moving along the corresponding tie lines of CH_4-CO_2, C_2H_6-CO_2, C_2H_4-CO_2 and C_2H_2-CO_2, respectively. They found that (i) the compositional border between the no growth region and the diamond growth region is extremely sharp; (ii) just beyond this border good quality diamond films are obtained; SEM photos show the perfection of crystals; (iii) moving into the diamond growth region, the diamond quality worsens, and nonfaceted materials start appearing between the diamond crystallites; (iv) the border between the diamond growth region and the nondiamond carbon deposition region is rather fuzzy. Around this border the amount of nondiamond material increases continuously, and the diamond facets are gradually lost. Experiments were run at a substrate temperature of 900 °C (1173K) and a pressure of 50 torr (6.6 kPa). Marinelli's experiments did confirm the existence of the diamond domain (the diamond growth region) in the C-H-O triangular diagram proposed by Bachmann. However, quantitatively, the diamond growth region deduced from Marinelli's experimental data considerably differs from Bachmann's one in that: (i) the very sharp border between the no growth region (gas phase region) and the diamond growth region occurs for higher carbon concentrations, and coincides (within experimental uncertainty) with the H-CO tie line itself; (ii) the width of the diamond growth region shrinks while moving away from the H vertex. In the case of a C_2H_2-CO_2 mixture, Marinelli found only a single composition (C_2H_2 67.3%, CO_2 32.7%) which results in diamond deposition. Even deviations in gas composition of the order of 1% will make the working point fall either in the no growth region or in the nondiamond carbon deposition region. The results obtained for all of the different gas mixtures are summarized in Table 7.1.

Table 7.1. Relationship between Marinelli's deposition results and hydrocarbon concentrations in CO_2 [7.3] with permission from the American Institue of Physics

	CH_4	C_2H_6	C_2H_4	C_2H_2
no deposit	47.7	32.5	32.1	32.0
diamond	49.1	34.2	35.1	32.7
nondiamond carbon	56.3	39.4	37.4	33.4

According to our coupling model, the projective nonequilibrium phase diagram with narrow temperature and pressure ranges (1100–1500 K and 1–10 kPa) can be theoretically calculated, as shown in Fig. 7.5. For comparison with experiments Marinelli's data and some experimental data reported by Ramesham [7.20] and Cassidy et al. [7.21] under nearly the same temperature and pressure conditions are also drawn in Fig. 7.5a. It can be seen that the agreement between theory and experiment is excellent. Even if the local part including Marinelli's experimental points is amplified, the agreement between our theory and Marinelli's experiments is still very satisfactory, see Fig. 7.5b.

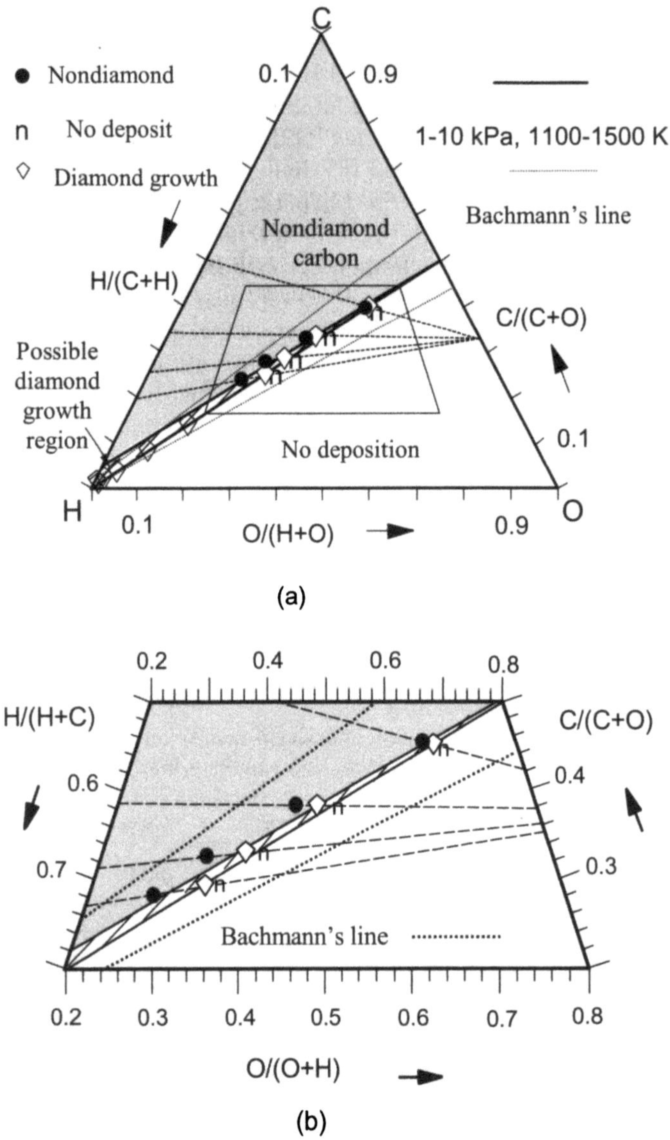

Fig. 7.5. Excellent agreement between our theoretical projective nonequilibrium phase diagram and experimental data reported by Marinelli [7.3], Ramesham [7.20] and Cassidy et al. [7.21] (**a**) Projective nonequilibrium phase diagram calculated in the 1100–1500 K and 1–10 kPa ranges. (**b**) Amplified picture of the ladder-shaped part in (a)

7.4 Cross-Sectional Nonequilibrium Phase Diagrams for C-H-O Ternary Systems

Because of the complexity of *T-p-X* phase diagrams of ternary systems, phase diagrams of a four-dimensional character can also be studied by way of isothermal, isobaric cross-sectional phase diagrams and/or cross-sectional phase diagrams with a fixed H/O composition ratio. For instance, Fig. 7.6 shows isothermal and isobaric cross-sectional phase diagrams at a fixed total pressure of 10 kPa. The isothermal cross-sectional phase diagrams of different temperature are drawn in Fig. 7.6 for 400 K, Fig. 7.7 for 600 K, Fig. 7.8 for 800 K and Fig. 7.9 for 1000 K, respectively [7.11–7.14, 7.22, 7.23]. For convenience, the different isothermal and isobaric phase diagrams may also be drawn together. The different temperatures may be represented by different kinds of lines. The region between the high and the low carbon concentration phase lines at the same temperature is the diamond growth region. From these phase diagrams it can be found that even at the same pressure the diamond growth region at different temperatures may still change a lot. It should be noted that in these isothermal and isobaric cross sectional phase diagrams all variables of temperature, pressure and composition are completely fixed. There is no possibility of finding an overlap between different phase regions in each of these isothermal and isobaric cross-sectional phase diagrams. The overlap between different phase regions usually occurs in projective phase diagrams. Therefore, the names of the phase regions here are: (i) no deposition region (i.e., gas phase region), (ii) diamond growth region and (iii) nondiamond carbon deposition region. Among them phase region (ii) is called the diamond growth region in cross-sectional phase diagrams, while the similar phase region (ii) is called the possible diamond growth region in projective phase diagrams. This means that if the experimental conditions (i.e., composition, temperature and pressure of the system) are inside the diamond growth region of the isothermal and isobaric cross-sectional phase diagrams, the thermodynamic possibility is only for diamond growth, but not for no growth or nondiamond carbon deposition.

Figure 7.10 is a theoretically calculated isothermal and isobaric cross-sectional composition nonequilibrium phase diagram at exactly the same temperature (1170 K) and pressure (6.67 kPa) as reported by Marinelli et al. in their critical experiments [7.3]. In comparison with the projective composition nonequilibrium phase diagram in the temperature range of 1100–1500 K and in the pressure range of 1–10 kPa (i.e., Fig. 7.5) it can be found that both of them are very similar, because of the very narrow ranges used in Fig. 7.5. In fact, if the projective ranges of temperature (1100–1500 K) and pressure (1–10 kPa) are shrunk into a certain temperature point and a certain pressure point, then the projective phase diagram will be exactly the same as the isothermal and isobaric cross-sectional phase diagram. The phase lines have stricter meaning in cross-sectional phase diagrams for the division of phase regions.

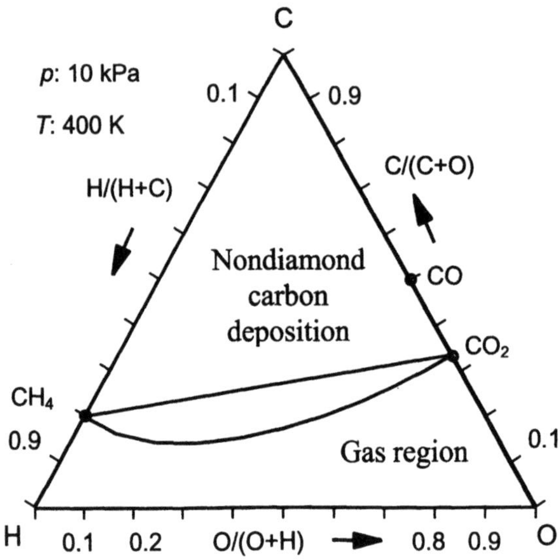

Fig. 7.6. Isothermal and isobaric cross-sectional nonequilibrium phase diagrams at 400 K and 10 kPa

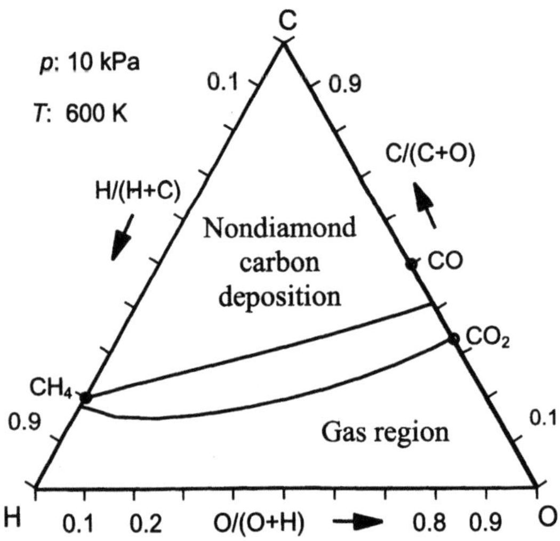

Fig. 7.7. Isothermal and isobaric cross-sectional nonequilibrium phase diagrams at 600 K and 10 kPa

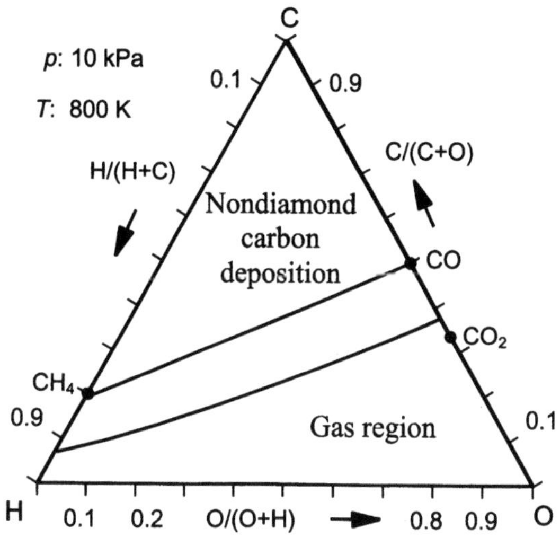

Fig. 7.8. Isothermal and isobaric cross-sectional nonequilibrium phase diagrams at 800 K and 10 kPa

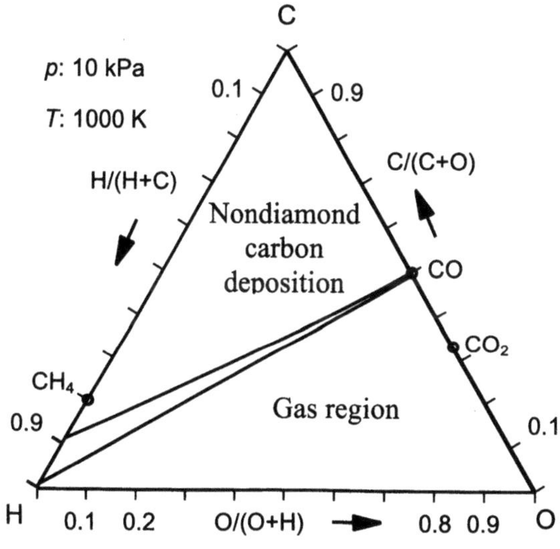

Fig. 7.9. Isothermal and isobaric cross-sectional nonequilibrium phase diagrams at 1000 K and 10 kPa

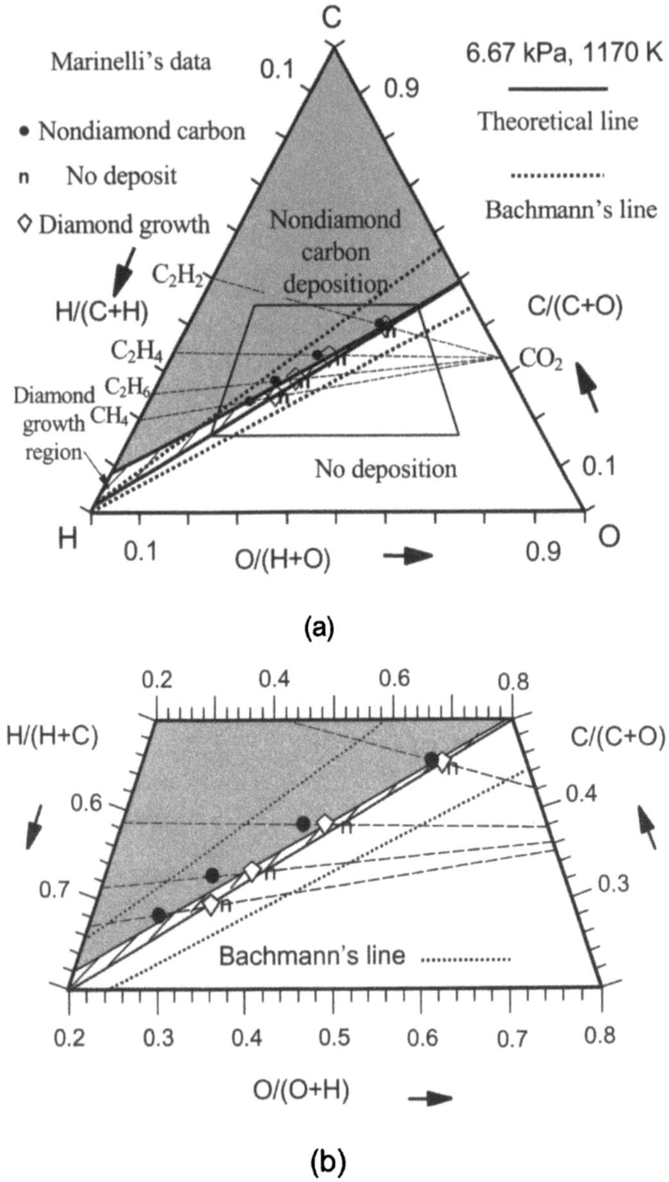

Fig. 7.10. Excellent agreement between theoretical cross-sectional nonequilibrium phase diagram and Marinelli's critical experiments [7.3]. (**a**) Cross-sectional nonequilibrium phase diagram calculated at 1173 K and 6.6 kPa. (**b**) Amplified phase diagram of the ladder-shaped part in (a)

Figure 7.11 is Marinelli's empirical phase diagram for activated diamond growth from the vapor phase based on their critical experiments at 900 °C (1173K) and 50 torr (6.6 kPa) [7.3]. Comparison between Figs. 7.10 and 7.11 clearly shows that the agreement between our theory and experiments is really excellent. It can also be found that based on the same critical experimental data if the vertex of the diamond growth region in Marinelli's empirical phase diagram were a little extended to the right C-O sideline, the empirical phase diagram, Fig. 7.11, would be exactly thc same as the theoretical phase diagram, Fig. 7.10. This confirms either that Marinelli's experiments are really critical experiments with very nice accuracy and or that the author's theoretical nonequilibrium phase diagram agrees excellently with experiments. The most important characteristic of both the theoretical planar ternary phase diagram Fig. 7.10 and Marinelli's phase diagram Fig. 7.11 is their relation to the temperature and pressure. In general, any planar ternary compositional phase diagram should only correspond with a certain temperature and pressure (for cross-sectional phase diagrams) or with a certain temperature and pressure range (for projective phase diagrams).

7.5 Nonequilibrium Phase Diagrams for C-H-X Ternary Systems

Similar to the calculation method used for the C-H-O system, nonequilibrium phase diagrams of C-H-F, C-H-Cl, C-H-N ternary systems have also been calculated for activated low-pressure diamond growth from the vapor phase [7.24–7.38]. In this book nonequilibrium phase diagrams of C-H-F, and C-H-Cl systems will only be illustrated, as shown in Figs. 7.12 and 7.13, respectively. The experimental data reported in the literature are also illustrated. The agreement between theoretically calculated results and experimental data is also very nice.

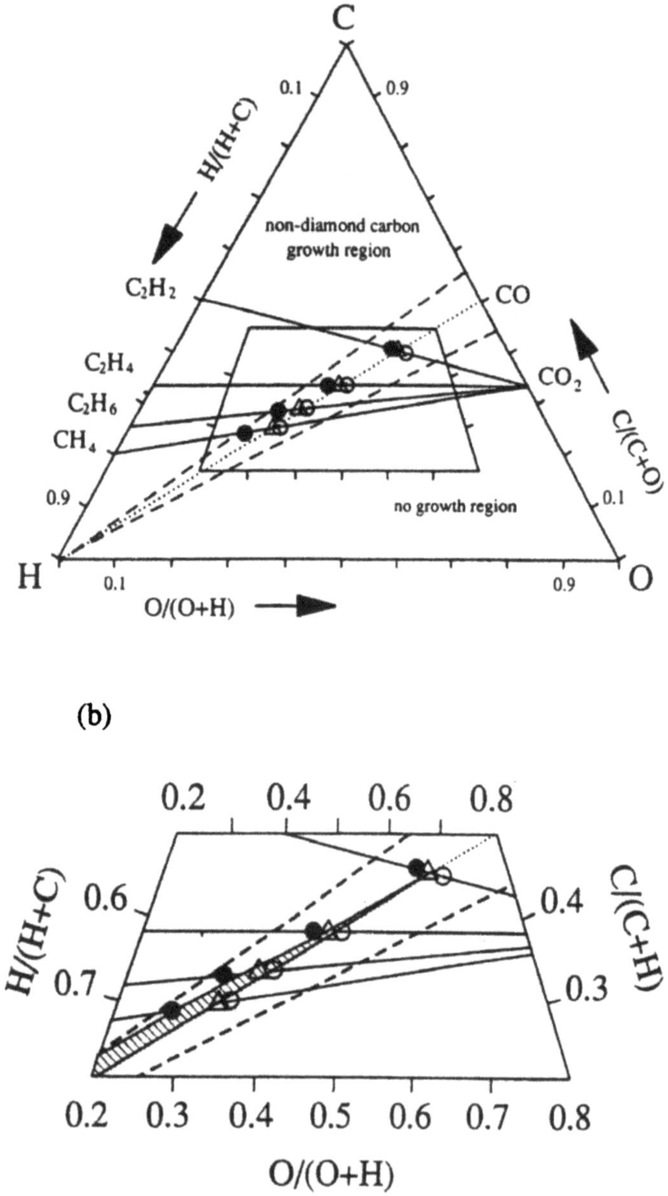

Fig. 7.11. Marinelli's original empirical phase diagram at 1173 K and 6.6 kPa. (o) no growth, (Δ) diamond, and ($\bullet$) graphite deposition. Reprinted from M. Marinelli, E. Milani, M. Montuori, A. Paoletti, A. Tebano, G. Balestrino and P. Paroli [7.3] with permission from American Institute of Physics

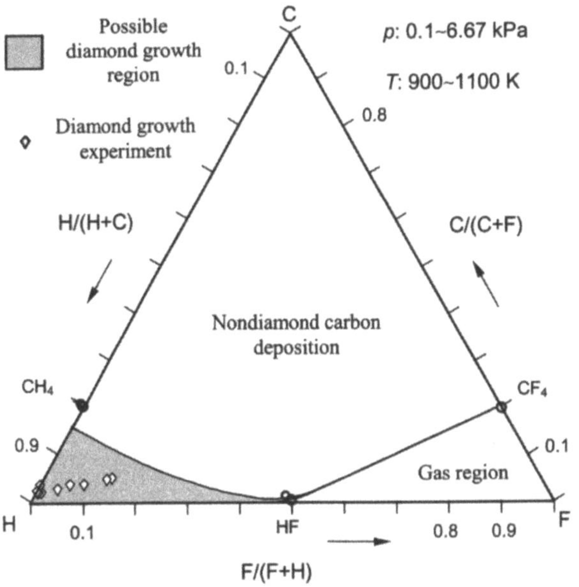

Fig. 7.12. Comparison between calculated projective composition nonequilibrium phase diagrams and experiments for diamond growth from C-H-F systems

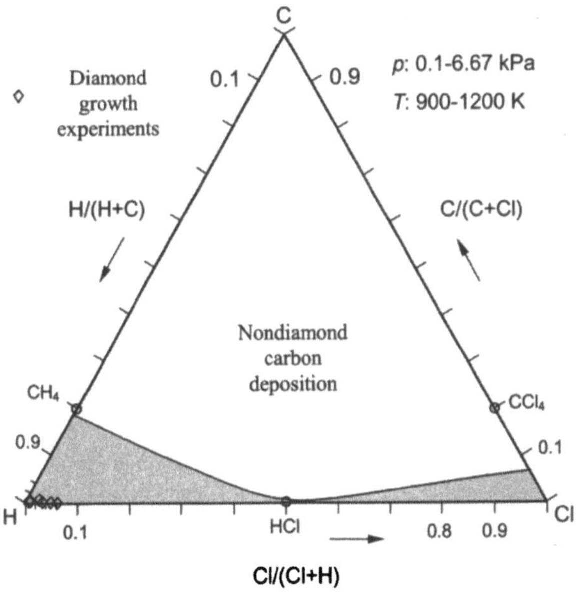

Fig. 7.13. Comparison between calculated projective composition nonequilibrium phase diagrams and experiments for diamond growth from C-H-Cl systems

8. Further Discussions on Some Debates

Up to now everything mentioned in this book might seem reasonable. Suppose that:

1. If Sect. 2.5. "Systematization of Modern Thermodynamics" were deleted, perhaps the essential idea of the book would be limited only to the use of the coupling of two different reactions: one of them is the transfer from graphite to diamond, see (5.5), which at low pressure goes against the usual direction implied by equilibrium thermodynamics, and the association of superequilibrium atomic hydrogen to produce molecular hydrogen, see (5.6). The second reaction acts as a chemical pump (or say, through a thermodynamic reaction coupling) driving the former reaction in the experimentally observed (but initially counterintuitive) direction. It may seem that the activated low-pressure diamond process may be only a nice example of the "well known" situation of reaction coupling or coupled reaction, see the set of equations (2.19), i.e.,

$$A_1 r_1 < 0, \ A_2 r_2 > 0 \text{ and } (A_1 r_1 + A_2 r_2) \geq 0$$

with A_i and r_i the affinities and the rates of reaction i, respectively (reaction 2 being spontaneous and driving reaction 1 against its spontaneous tendency).

2. If Sect. 2.5. "Systematization of Modern Thermodynamics" were deleted and all discussion of the activated CVD diamond process were simply and directly put on the basis of the molar free energy change formulation of a coupled reaction, as done by D. Burk in 1931 for biochemical systems (see the next section of this chapter) [8.1], there would be little concerned with modern thermodynamics. The molar Gibbs free energy formulation or criterion of coupled reaction, see the set of equations (2.20), i.e.,

$$(\Delta G_1)_{T,p} > 0, \ (\Delta G_2)_{T,p} < 0 \text{ and } \left[(\Delta G_1)_{T,p} + \chi\,(\Delta G_2)_{T,p}\right] \leq 0$$

is so simple and can be deduced directly from the second law of thermodynamics. Many sections in the book could be deleted, and all conclusions could easily be accepted. Here, χ is the rate ratio of the two reactions, i.e., $\chi = \frac{r_2}{r_1}$. $(\Delta G_1)_{T,p}$, $(\Delta G_2)_{T,p}$, and $\left[(\Delta G_1)_{T,p} + \chi\,(\Delta G_2)_{T,p}\right]$ are the molar Gibbs free energy changes of reaction 1, reaction 2 and the whole reaction (or system) at constant temperature and pressure, respectively.

However, neither of these suggestions are the real situation. First, the new development of thermodynamics or the essential idea in this book is mainly on the "Systematization of Modern Thermodynamics", which provides the theoretical basis for the new concepts, quantitative calculations and a series of results of nonequilibrium (stationary) phase diagrams in previous chapters. Second, if the Sect. 2.5 were denied, there would be no room for nonequilibrium (stationary) phase diagrams in thermodynamics. Nonequilibrium (stationary) phase diagrams do not belong to classical equilibrium thermodynamics, or to classical or traditional nonequilibrium thermodynamics, either. Third, the "Systematization of Modern Thermodynamics" is a natural theoretical conclusion of the reaction coupling model for activated CVD diamond processes, if the concept of reaction coupling or coupled reaction has really been accepted.

The concept of coupled reaction is really important, but the introduction and discussion of the concept made by D. Burk directly on the basis of free energy changes of biochemical reactions as early as in 1931 did not play a sufficiently important role in the development of thermodynamics in the last 70 years or so [8.1]. Another affinity formulation of reaction coupling had also been proposed by De Donder in 1936 [8.2], but the concept of coupled reaction or reaction coupling had still been bogged down in a 70-year controversy. That together with other explanations or debates for the activated CVD diamond processes will be discussed in this chapter.

8.1 Coupled Reaction in Biochemistry

Dean Burk made a detailed discussion on coupled reaction in his paper "*The Reversibility of Coupled Reactions in Biological Systems and The Second Law of Thermodynamics*" in 1931 [8.1]. He treated the nonphotochemical reduction of carbon dioxide by the autotrophic hydrogen bacterium, *Bacillus pycnoticus*, as an example of a coupled reaction. The whole coupled reaction of glucose biosynthesis was under consideration. Note that his reaction equation numbers have been changed to fit this book style. He wrote:

"The coupled reaction in the standard states

$$CO_2(g) + 2H_2(g) = \frac{1}{6}C_6H_{12}O_6(s) + H_2O(l) \quad (8.1)$$

$$\Delta F^0 = 1184 \quad (8.2)$$

may be divided into its two components and the actual states of the reacting substances indicated.

$$2\ H_2(0.35\ \text{atm}) + O_2(0.06\ \text{atm}) = 2\ H_2O(l) \quad (8.3)$$

$$\Delta F = \Delta F^0 - RT\ln(0.35)^2 - RT\ln(0.06) - RT\Delta N \qquad (8.4)$$
$$= (2 \times -56560) + 1250 + 1630 + 1780 = -108460$$

where $\Delta F^0 = (2 \times -56560)$; $RT\Delta N$ is a factor correcting for the circumstance that the reaction takes place at constant volume (i.e., without external work), N being the mol volume change, in this instance −3; and the stated pressure of the gases are the averages between the initial and final pressures in each of the eleven experiments. (Note: refer to Ruhland [8.3])

$$CO_2(0.06\ \text{atm}) + H_2O(l) = \frac{1}{6}C_6H_{12}O_6(10^{-6}M) + O_2(0.06\ \text{atm}) \qquad (8.5)$$

$$\Delta F = \Delta F^0 - RT\ln(0.06) + RT\ln(0.06) + RT\ln(10^{-6})^{1/6} \qquad (8.6)$$
$$= 114300 + 1630 - 1630 - 1365 = 112933$$

where $\Delta F^0 = 114300$, and the activity of glucose is considered to be that of about 1/1000 the average cell carbon in 50 cc. of medium (and the maximum solubility of glucose 5M), i.e., (1/1000) (10 cc./2) (1/22400) (1000/50 cc.) (1/5 M) = ca. 10^{-6} M per ($C_6H_{12}O_6$).

However, from the experimental data, the empirical formula of the reduced CO_2 as in (8.5) is not ($C_6H_{12}O_6$) but ($C_6H_{12x}O_6$) where x is slightly less than unity. (1196.2 − (2 × 455.21) − 134.02)/134.02, or 0.931, i.e., (cc. H_2 actually consumed, minus cc. H_2 appearing as H_2O as a result of combining with the O_2 lost from the gas phase of the system, minus cc. H_2 appearing as H_2O as a result of combining with one half the O_2 in CO_2 as in equation (8.1))/ (cc. H_2 theoretically combined with C as in glucose). This gives for the true free energy of reduction of CO_2, 0.931 × 112935 = 105 140. This assumption involving the method of deriving a conversion factor for obtaining the free energy of reduction of $C_6H_{0.931\times12}O_6$ from $C_6H_{12}O_6$ is the most uncertain element in the second law efficiency calculated. What is indicated literally is that (8.5) should read

$$CO_2(0.06\ \text{atm}) + 0.931\ H_2O(l) = \frac{1}{6}C_6H_{0.931\times12}O_6(10^{-6}M) + \frac{1.931}{2}O_2(0.06\ \text{atm}) \qquad (8.7)$$

$$\Delta F = 105140 \qquad (8.8)$$

i.e., that (100 × (1.000 − 0.931)), or 6.9%, less than H_2O must be decomposed to yield the necessary H_2 to reduce the CO_2. "

Here, the second law efficiency was defined by Burk as the work done in any specific reaction divided by the free energy dissipated by that reaction only.

He continues:

"In accordance with (8.4) the oxidation of 2 mole H_2 can furnish 108 460 cal. of free energy. In accordance with (8.7) the reduction of one mole CO_2 requires a slightly smaller amount. Hence we should expect the following relation between the numbers of mols (or cc.) involved,

$$\frac{(\text{No. of molsH}_2\text{ needed})}{2\times(\text{No. ofmols CO}_2\text{ reduced})}=\frac{105140}{108460}=0.970. \tag{8.9}$$

On the other hand, as has been shown, in accordance with Table I Column 7 (Note: i.e.,Table 8.1 Column 7 here), the experiments gave

$$\frac{(\text{No. of mols H}_2\text{ used})}{2\times(\text{No. of mols CO}_2\text{ reduced})}=0.966\pm0.012 \tag{8.10}$$

Hence there was apparently very slightly less hydrogen oxidized than was necessary to produce the available energy needed for the reduced of CO_2, the efficiency of the process being

$$100\times\frac{0.970\pm0.000}{0.966\pm0.012}=100.4\pm1.2\%$$

if one assign a probable error (in distinction to any other type of error) of ± 0.000 to the thermodynamic data."

Table 8.1. Ratio of gases consumed by *Bacillus Pycnoticus* (i.e., Table I of Burk's paper and original from Ruhland [8.3]). From D. Burk [8.1] with permission from the American Chemical Society

Column	1	2	3	4	5	6	7
Exp. No.	H_2	O_2	CO_2^a	H_2/CO_2	H_2/O_2	O_2:CO_2	$(H_2-2O_2)/2CO_2$
2	137.82	52.75	16.90	8.15	2.63	3.12	0.956
3	111.48	44.72	10.40	10.72	2.50	4.30	1.060
4	89.48	39.06	6.20	14.42	2.29	6.30	0.916
15	87.92	41.15	3.00	29.31	2.14	13.7	0.937
19	91.21	40.71	5.30	17.20	2.24	7.68	0.924
22	103.20	39.39	13.10	7.88	2.62	3.01	0.932
28	29.51	10.58	4.31	6.84	2.79	2.45	0.969
29	85.20	30.64	10.89	7.84	2.78	2.82	1.098
30	113.13	40.71	17.01	6.65	2.78	2.39	0.932
31	94.93	34.12	13.87	6.85	2.78	2.46	0.962
32	225.34	81.38	33.04	6.83	2.78	2.46	0.947
Total	1169.22	455.21	134.02				
Average	106.29	41.38	12.18	8.71	2.56	3.40	0.966±0.012
Relative	8.71	3.40	1				

[a] This refers to the C in the cells; i.e., reduced CO_2

Burk concluded,

"the second law is not transgressed, since the free energy efficiency is not significantly greatly than 100 %."

Based on Ruhland's experimental data Burk also pointed out that

"Under the most favorable conditions only six and eight-tenths volumes of hydrogen are consumed for every volume of carbon dioxide finally changed to organic carbon. Calculations show that the ratio corresponds to a maximum machine free energy efficiency of (105 140 × 100 %)/(54 230 × 6.8), or 28.4 %, where 105 140 and − 54 230 are the respective molar free energies of reduction of carbon dioxide and oxidation of hydrogen under the conditions of Ruhland's experiments."

Here, the machine free energy efficiency was defined by Burk as the work done in any specific reaction occurring in a biological system divided by the free energy consumed by the system as a whole.

Burk had correctly introduced the concept of reaction coupling or coupled reaction and he had correctly concluded that "the second law is not transgressed" in biological systems. However, it is difficult to accept that the 100% (i.e., 100.4 ± 1.2 %) reversibility of the coupled reaction, could really occur in biological systems, based on "the second law efficiency" defined by Burk.

Now it can be found that based on Ruhland's average experimental data in Table 8.1, i.e., $H_2/CO_2 = 8.71 \approx 8.8$ and $O_2/CO_2 = 3.4$, the practical overall coupled reaction should almost consist of two components of a non-spontaneous reaction with a free energy change increase, i.e.,

$$CO_2(0.06\text{ atm}) + H_2O(l) = \frac{1}{6}C_6H_{12}O_6(10^{-6}M) + O_2(0.06\text{ atm}), \quad (8.11)$$

$$\Delta F_1 = 112933 \quad (8.12)$$

and another spontaneous reaction with a free energy change decrease, i.e.,

$$2\ H_2(0.35\text{ atm}) + O_2(0.06\text{ atm}) = 2\ H_2O(l), \quad (8.13)$$

$$\Delta F_2 = -108460 \quad (8.14)$$

with the reaction rate ratio $\chi = \frac{r_2}{r_1} = \frac{8.8}{2} = 4.4$. The practical overall coupled reaction is nearly

$$CO_2(0.06\text{ atm}) + 8.8\ H_2(0.35\text{ atm}) + 3.4\ O_2(0.06\text{ atm}) \quad (8.15)$$
$$= \frac{1}{6}C_6H_{12}O_6(s) + 7.8\ H_2O(l),$$

$$\Delta F = (\Delta F_1 + \chi \Delta F_2) = 112\ 933 + 4.4 \times (-108\ 460) = -364\ 291. \quad (8.16)$$

All these values (in cal. in Burk's paper and in this section) agree very well with the criterion or definition of reaction coupling, as shown in the set of equations (2.21), i.e.,

$$(\Delta F_1)_{T,V} > 0, \ (\Delta F_2)_{T,V} < 0, \ and \ \left[(\Delta F_1)_{T,V} + \chi (\Delta F_2)_{T,V}\right] \leq 0$$

that is, this biological reaction agrees very well with the thermodynamic coupling theorem, as mentioned in Chap.2.

8.2 The Seventy-Year Controversy About Reaction Coupling in Chemistry

As shown in (2.19) another formulation of reaction coupling, i.e., $A_1 r_1 < 0$, $A_2 r_2 > 0$ and $(A_1 r_1 + A_2 r_2) \geq 0$, was proposed in the 1936 by De Donder. However, the visiting scientist of Th. De Donder's group, P. Van Rysselberghe, had ambiguously introduced an approximate or incorrect criterion nearly at the same time (1936–1937) [8.4–8.6]. Prigogine in his book *Chemical Thermodynamics* (1954) cited Van Rysselberghe's work and wrote that [8.7]:

> "Thermodynamics coupling is an essential feature of living systems. For example the synthesis of urea is coupled with the combustion of glucose and other substances in the liver. This synthesis may be represented schematically by
>
> $$2\ NH_3 + CO_2 \rightarrow (NH_2)_2CO + H_2O \text{(coupled reaction)}$$
>
> $$\frac{1}{8}C_6H_{12}O_6 + O_2 \rightarrow CO_2 + H_2O \ \text{(coupling reaction)}$$
>
> The affinities of these reactions under standard conditions are found to be
>
> $$A_1 = -11,000 \text{ cal.}$$
>
> $$A_2 = +115,000 \text{ cal."}$$

In fact, the phrase "under standard conditions" means that A_1 and A_2 here should be represented by A_1^0 and A_2^0, respectively, that is, the criterion of reaction coupling was ambiguously changed into its approximate or incorrect form, i.e.,

$$A_1^0 r_1 < 0, A_2^0 r_2 > 0 \text{ and } \left(A_1^0 r_1 + A_2^0 r_2\right) \geq 0 \tag{8.17}$$

Up to now, this seems to have been adopted in all textbooks of bioenergetics and biochemistry which we have found, that is, the approximate or incorrect criterion of reaction coupling was practically written as follows.

$$\Delta G_1^0 > 0, \ \Delta G_2^0 < 0 \text{ and } \Delta G^0 = \Delta G_1^0 + \Delta G_2^0 \leq 0 \tag{8.18}$$

or

$$\Delta G_1'^{\;0} > 0,\; \Delta G_2'^{\;0} < 0 \text{ and } \Delta G'^{\;0} = \Delta G_1'^{\;0} + \Delta G_2'^{\;0} \leq 0 \tag{8.19}$$

where, $\Delta G'^{\;0}$ represents the change of the Gibbs free energy of a reaction taking place in a solution, in which all the reactants and products are in the standard state [The standard state of a substance in solution is defined as a solution of substance at a concentration of 1 M, i.e., 1 mol kg^{-1} or 1 mol dm^{-3} (if the H^+ ion participates in a reaction, then the pH value equals 7) and at 25 °C]. $\chi = 1$ is also approximately or unsuitably adopted. In this way ΔG approximately equal to ΔG^0 or $\Delta G'^{\;0}$ with the same sign has been ambiguous assumed. It can be generally found that in textbooks of physical chemistry, biochemistry and other related books, the description of these kinds of biochemical reactions are similar:

"$ATP + H_2O \rightarrow ADP + P_i$ $\quad \Delta G'^{\;0} = -30.5$ kJ

glucose + Pi → glucose – 6 – phosphate + H_2O $\quad \Delta G'^{\;0} = +13.8$ kJ

- -

ATP + glucose → ADP + glucose – 6 – phosphate $\quad \Delta G'^{\;0} = -16.7$ kJ"

In some books, such as in Southerland's *Biochemistry* [8.8], there was a notation that "Usually, ΔG for biologic reactions has the same sign as ΔG^0, and the two often do not differ greatly in value" although such an approximation might be still incorrect.

Due to progress in modern thermodynamics, in the 1980s the terms of "nonequilibrium thermodynamics coupling", "reaction coupling" and "coupling reaction" became fashionable. The approximate but not strict or not correct criteria (8.17)–(8.19) were also introduced into the inanimate chemistry field. For instance, the paragraph entitled COUPLED REACTIONS is cited as follows from *Physical Chemistry* by Castellan [8.9,8.10].

"⋆11.18 COUPLED REACTIONS

...

The reaction scheme is

coupled reactions:
$TiO_2(s) + 2Cl_2(g) \rightarrow TiCl_4(l) + CO_2(g)$,
$\Delta G^0_{298} = 152.3$ kJ mol^{-1},
$C(s) + O_2(g) \rightarrow CO_2(g)$,
$\Delta G^0_{298} = -394.4$ kJ mol^{-1},

and the overall reaction is

$$C(s) + TiO_2(s) + 2Cl_2(g) \rightarrow TiCl_4(l) + CO_2(g),$$

$$\Delta G^0_{298} = -242.1 \text{ kJ mol}^{-1}.$$

Since the overall reaction has a highly negative ΔG^0, it is spontaneous. As a general rule metal oxides cannot be converted to chlorides by simple replacement; in the presence of carbon, the chlorination proceeds easily.

Coupled reactions have great importance in biological systems ...".

Note that ΔG^0 had been unsuitably used as a criterion of a spontaneous reaction. In fact, the overall reaction practically occurs at about 1200 °C and 100 kPa. Under conditions of the existence of white-hot carbon the concentration or partial pressure of O_2 is very low. Therefore, the ΔG s of the two so-called "coupled reactions" are practically of small negative value, i.e., both of these two reactions are spontaneous. There is nothing concerned with thermodynamic coupling at all. The introduction of carbon at high temperature is only based on the well-known La Chatelier principle, the mass action law, the equilibrium shift principle, also called simultaneous equilibrium. Carbon is introduced only to remove the product O_2 of the reaction between TiO_2 and Cl_2.

The confusion between "equilibrium shift" and "reaction coupling" had already been felt to a certain extent even at that time. Therefore, two parallel definitions of "reaction coupling" could be found in some textbooks. For instance, Leidler and Meiser in their book *Physical Chemistry* wrote [8.11]:

"**4.7 Coupling of Reactions**

For a chemical reaction

1.

$$\mathrm{A + B \longleftrightarrow X + Y}$$

removal of either of the products X or Y will lead to a shift of equilibrium to the right. Frequently a product is removed by the occurrence of another reaction; for example, X might isomerize into Z:

2.

$$\mathrm{X \longleftrightarrow Z}$$

We can write an eqilibrium constant for reaction 1,

$$K_1 = \left(\frac{[\mathrm{X}]\,[\mathrm{Y}]}{[\mathrm{A}]\,[\mathrm{B}]}\right)_{\mathrm{eq}}$$

and a corresponding standard Gibbs energy change:

$$\Delta G_1^0 = -RT \ln K_1 .$$

Similarly for the second reaction

$$K_2 = \left(\frac{[\mathrm{Z}]}{[\mathrm{X}]}\right)_{\mathrm{eq}}$$

and

$$\Delta G_2^0 = -RT \ln K_2 .$$

We can add reaction 1 and 2 to obtain
3.

$$A + B = Y + Z$$

The eqilibrium constant K_3 for the combined reaction 3 is

$$K_3 = \left(\frac{[Y]\,[Z]}{[A]\,[B]}\right)_{eq}.$$

The same result is obtained by multiplying K_1 and K_2 :

$$K_1 K_2 = \left(\frac{[X]\,[Y]}{[A]\,[B]}\right)_{eq} \times \left(\frac{[Z]}{[X]}\right)_{eq} = \left(\frac{[Y]\,[Z]}{[A]\,[B]}\right)_{eq} = K_3.$$

We can take natural logarithms of both sides of this equation, and then multiply by $-RT$:

$$-RT \ln K_3 = -RT \ln K_1 - RT \ln K_2$$

so that

$$\Delta G_3^0 = \Delta G_1^0 + \Delta G_2^0.$$

The result is easily shown to be quite general; if we add reactions, the ΔG^0 of the resulting reaction is the sum of the ΔG^0 values of the component reactions ... .

As a result of this coupling of chemical reactions, it is quite possible for a reaction to occur to a considerable extent even though it has a positive value of ΔG^0. Thus for the scheme just given, suppose that reaction 1 has a positive value:

$$\Delta G_1^0 > 0.$$

Reaction 1 by itself will therefore not occur to a considerable extent (K_1 ¿ 1). Suppose, however, that reaction 2 has a negative value. It is then possible for ΔG_3^0 to be negative, so that A + B may react to a considerable extent to give X+Y. In terms of the Le Chatelier principle, this simply means that X is constantly removed by the isomerization reaction 2, so that the equilibrium of reaction 1 is shifted to the right."

Leidler and Meiser continue:

"There is another way in which chemical reactions can be coupled. In the cases we have considered, the product of one reaction is removed in a second reaction. Alternatively, two reactions might be coupled by a catalyst. Suppose, for example, that we have two independent reactions, having no common reactants or products:
1.

$$A + B \longleftrightarrow X + Y$$

2.

$$C \longleftrightarrow Z$$

There might exist some catalyst that brings about the reaction

3.

$$A + B + C \longleftrightarrow X + Y + Z$$

and this catalyst would couple the reactions 1 and 2. This type of coupling of reactions is quite common in biological systems."

Similar incorrect descriptions concerned with the term "vreaction coupling" can also be found in many chemistry and thermodynamics books. Some papers and books on irreversible thermodynamics incorrectly emphasized that at least one of the reactants and products must be the same in coupled reactions, and was called a "coupling substance". Such an incorrect term was based only on unsuitable speculation and was not deduced from the second law of thermodynamics. Van Rysselberghe in his paper (1937) wrote that "It seems fairly certain that coupling of a reaction with negative affinity by a reaction with a positive affinity is possible only if at least one reactant or product is common to both reactions." [8.6, 8.12]

In 1983 M. Boudart in his paper *Thermodynamic and Kinetic Coupling of Chain and Catalytic Reactions* wrote [8.13]:

"Any guidance from thermodynamics in the art and science of chemical kinetics has a value which is particularly appreciated as it is limited but general. An intriguing example of such thermodynamic assistance to chemical kinetics comes from the De Donder's inequality

$$Ar \geq 0.$$

linking the affinity $A = -\Delta G$ of a reaction and its rate r. Indeed, if two reactions (subscripts 1 and 2) occur simultaneously, it is conceivable that the following inequalities

$$A_1 r_1 < 0, \quad A_2 r_2 > 0$$

might hold provided that

$$A_1 r_1 + A_2 r_2 > 0$$

It is said that the first reaction is coupled to the second one, 'thermodynamics coupling thus allows the coupled reactions to proceed in a direction opposite that dictated by its affinity' [8.14]. Unfortunately, the coupling mechanism, clearly a notion from chemical kinetics, needs to be known before the inequality $A_1 r_1 < 0$ can be accepted. It is quite possible that, when the coupling mechanism is understood, the need for the thermodynamic coupling as defined above disappears, together with the uncomfortable inequality $A_1 r_1 < 0$, which is purely formal in the sense that the coupled reaction does not take place at all but is only a stoichiometric statement devoid of kinetic (i.e., r_1) meaning."

Boudart continued discussing examples of the so-called "thermodynamic reaction coupling" found in the literature. The first example was a chain reaction. Note that his reaction numbers have been changed to fit this book style, and the symbol for the rate v has been changed into r.

"Consider first the two gas-phase reactions

$$H_2 = 2\,H \tag{8.20}$$

$$2\,H_2 + O_2 = 2\,H_2O \tag{8.21}$$

taking place at 1000 K between reactants and products at their standard state (1 atm ≡ 101.3 kPa). The standard affinity for the first one is $A_1^0 = -331$ kJ mol^{-1} while $A_2^0 = 385$ kJ mol^{-1}. For the overall reaction

$$3\,H_2 + O_2 = 2\,H_2O + 2\,H \tag{8.22}$$

The standard affinity is 385 − 331 = 54 kJ mol^{-1}, so that it would appear that reaction 1 [Note: i.e., (8.20) here] is the reaction coupled by reaction 2 [Note: i.e., (8.21) here], the coupling reaction, and that indeed, for reaction 1, $A_1 r_1 < 0$. But this is not so, because, in fact, reaction 1 does not take place *at all*. Rather, according to Semenov [8.15], the branching chain reaction propagates through three elementary processes as follows:

$$OH + H_2 \rightarrow\ H_2O + H \tag{8.23}$$

$$H + O_2 \rightarrow\ OH + O \tag{8.24}$$

$$O + H_2 \rightarrow\ OH + H \tag{8.25}$$

If we sum up these three elementary processes, taking the first one twice, so that its stoichiometric number σ is given a value equal to 2, we reproduce the overall reaction 3 [Note: i.e., (8.22) here]. The overall De Donder's inequality applies to it, but 'reaction 1' has nothing to do with what happens at the mechanistic level. Indeed, reaction 1 does not take place at all. There is no thermodynamic coupling in the sense mentioned above. The multiplication of hydrogen atoms is due not to reaction 1 but to the branching chain reaction $H + O_2 \rightarrow OH + O$. Thus, the concept of thermodynamic coupling disappears at the kinetic level of interpretation."

Boudart had also pointed out that the synthesis of HBr from H_2 and Br_2 is only a chain reaction in kinetics and there was nothing concerned with thermodynamic coupling.

It is obvious that Boudart's disagreements are reasonable. Boudart's paper was mainly on kinetics and did not concern with the nature of thermodynamic coupling, so even sometime in recent years the synthesis of HBr from H_2 and Br_2 had still been discussed under the title of coupled reactions [8.16]. In fact, there is nothing concerned with reaction coupling, but all mistakes may be only due to the incorrect criterion of reaction coupling [such as (8.17) or (8.18)] used.

From thermodynamic data book, it is easy to find the standard Gibbs free energy values of all concerned species of the synthesis of HBr from H_2 and Br_2. The reaction usually occurs at the temperature between 500 K and 600 K, so the values of standard Gibbs free energy, i.e., G^0 at 500 K and G^0 at 600 K, are listed in Table 8.2

Table 8.2. Standard Gibbs free energies of the species concerned

Species	G^0 at 298 K	G^0 at 500 K	G^0 at 600 K
H_2	-38.93 kJ mol^{-1}	-66.94 kJ mol^{-1}	-81.77 kJ mol^{-1}
Br_2	-42.25 kJ mol^{-1}	-93.85 kJ mol^{-1}	-120.63 kJ mol^{-1}
HBr	-95.65 kJ mol^{-1}	-137.39 kJ mol^{-1}	-159.04 kJ mol^{-1}
H	183.82 kJ mol^{-1}	159.51 kJ mol^{-1}	146.77 kJ mol^{-1}
Br	59.73 kJ mol^{-1}	23.25 kJ mol^{-1}	4.49 kJ mol^{-1}

Therefore, the main steps of the chain reaction together with the corresponding standard Gibbs free energy changes of each step are shown in (8.26)–(8.28).

$$\mathrm{Br_2 = 2\ Br} \tag{8.26}$$

$$\Delta G^0_{298\ \mathrm{K}} = 161.72\ \mathrm{kJ\ mol^{-1}}$$

$$\Delta G^0_{500\ \mathrm{K}} = 140.35\ \mathrm{kJ\ mol^{-1}}$$

$$\Delta G^0_{600\ \mathrm{K}} = 129.63\ \mathrm{kJ\ mol^{-1}}$$

$$\mathrm{Br + H_2 = HBr + H} \tag{8.27}$$

$$\Delta G^0_{298\ \mathrm{K}} = 67.37\ \mathrm{kJ\ mol^{-1}}$$

$$\Delta G^0_{500\ \mathrm{K}} = 65.81\ \mathrm{kJ\ mol^{-1}}$$

$$\Delta G^0_{600\ \mathrm{K}} = 65.01\ \mathrm{kJ\ mol^{-1}}$$

$$\mathrm{Br_2 + H = HBr + Br} \tag{8.28}$$

$$\Delta G^0_{298\ \mathrm{K}} = -177.49\ \mathrm{kJ\ mol^{-1}}$$

$$\Delta G^0_{500\ \mathrm{K}} = -179.80\ \mathrm{kJ\ mol^{-1}}$$

$$\Delta G^0_{600\ \mathrm{K}} = -180.69\ \mathrm{kJ\ mol^{-1}}$$

The overall reaction of the synthesis of HBr from H_2 and Br_2 together with its standard Gibbs free energy changes are shown in (8.29).

$$\mathrm{H_2 + Br_2 = 2\ HBr} \tag{8.29}$$

$$\Delta G^0_{298\ \mathrm{K}} = -110.12\ \mathrm{kJ\ mol^{-1}}$$

$$\Delta G^0_{500\ \mathrm{K}} = -113.99\ \mathrm{kJ\ mol^{-1}}$$

$$\Delta G^0_{600\ \mathrm{K}} = -115.68\ \mathrm{kJ\ mol^{-1}}$$

The overall reaction (8.29) does really consist of (8.27) and (8.28). However, the incorrect criterion of reaction coupling (8.18), i.e., $\Delta G_1^0 > 0$, $\Delta G_2^0 < 0$ and $\Delta G^0 = \Delta G_1^0 + \Delta G_2^0 \leq 0$ should never be used, because ΔG^0 is different from ΔG not only in value but also often in sign. For instance, it is well-known that the reaction (8.26) is a fast reaction and practically in equilibrium. so ΔG of the reaction equals zero and not a big positive value, as shown in (8.26) for ΔG^0. It is also known in kinetics study that the practical concentration of H is about one tenth of the equilibrium concentration of atomic hydrogen, so the Gibbs free energy changes of both reactions (8.27) and (8.28) have negative values. Boudart had made calculation and concluded that even for the "thermodynamic unfavorable" reaction (8.27) $\Delta G_{298\ \mathrm{K}} = -1.43$ kcal mol^{-1} while $\Delta G^0_{298\ \mathrm{K}} = +15.7$ kcal mol^{-1}. [Note: A little difference between Boudart's data and data of this book in (8.27) is due to a little difference in thermodynamic data books] Therefore, there is nothing concerned with reaction coupling in the synthesis of HBr from H_2 and Br_2.

In 1989 R.-S. Li had also lodged an objection against the abuse of the conception of reaction coupling or thermodynamic coupling [8.17]. After his derivation he pointed out that "the thermodynamic coupling is impossible for those reactions which obey kinetically the mass action law." In this way, not only in Leidler and Meiser's and Castellan's books but also in all other relevant ordinary chemistry textbooks at that time the examples of reaction coupling might be incorrect [8.18].

8.3 Quantitative Verification of Reaction Coupling in Inanimate Systems

It might be unsuitable to simply say the cause of the confusion mentioned in the previous section was only due to the confusion between ΔG and ΔG^0. The most important puzzle might be because no practical example in inanimate systems, which really did agree with the criterion of coupled reaction or reaction coupling, had been found.

An opportunity finally came due to the appearance of activated low pressure CVD diamond processes [8.19, 8.20], which can be represented by three modeling reactors, as shown in Figs. 8.1a–c. In all of theses modeling reactors steady diamond growth together with simultaneous graphite etching may take place. There are only two elements, carbon and hydrogen included in the reaction system. The standard thermodynamic data of all species can be found in data books or in the literature. These are favorable factors for finding a quantitative theoretical explanation.

In activated low-pressure processes both the practical growth rate of diamond and the etching rate of graphite are rather slow (about 10 μm h^{-1}), so all of the three modeling reactor systems in Fig. 8.1 can be treated as stationary states. Suppose that the growth rate on the diamond surface in the reactors and the etching rate on the graphite surface are uniform, the diamond part in the modeling reactors is drawn out by the same rate of the growth rate r_{d}, and the graphite part is pushed into the reactors by the same rate of the etching rate r_{g}. Then the modeling reactor systems in Fig. 8.1 become systems in ideal stationary states, as shown in Fig. 8.2. In brief, heat emission and other sources of energy dissipation are out of consideration here.

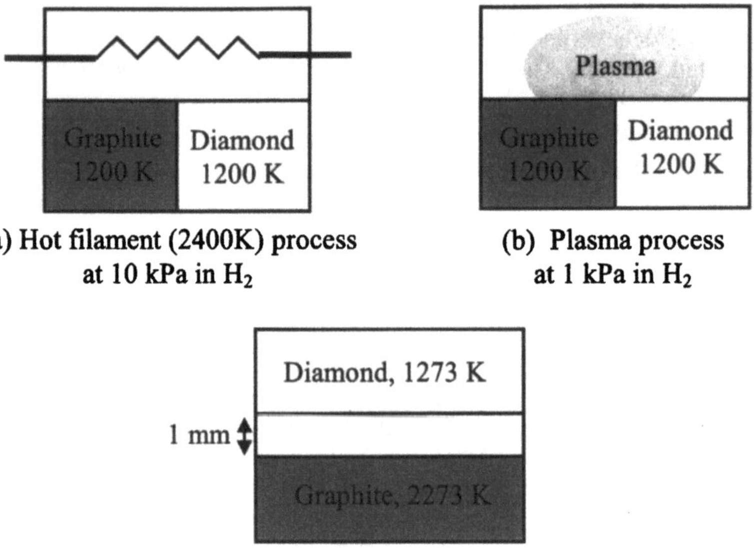

(a) Hot filament (2400K) process at 10 kPa in H_2

(b) Plasma process at 1 kPa in H_2

(c) High gradient chemical transport reaction (HGCTR) process at 1.6 kPa in H_2

Fig. 8.1. Three activated diamond modeling reactors

A series of theoretically calculated nonequilibrium (stationary) phase diagrams have been illustrated in Chaps. 5–7 of this book. The excellent agreement with a lot of experimental data reported in the literature is the most powerful evidence for the correctness of reaction coupling and the coupling model itself. All of these phase diagrams do not correspond to equilibrium systems, but only to nonequilibrium nondissipative states of systems, and usually to special stationary states, such as for activated low-pressure diamond growth. That is also the reason for using the name of nonequilibrium stationary phase diagram, or phase diagram of stationary states. A set of quantitative data will be listed as follows for further understanding of the negative entropy production reaction, reaction coupling and the model itself.

The typical experimental conditions for activated low-pressure diamond growth from the vapor phase are: substrate temperature T = 1200 K, to-

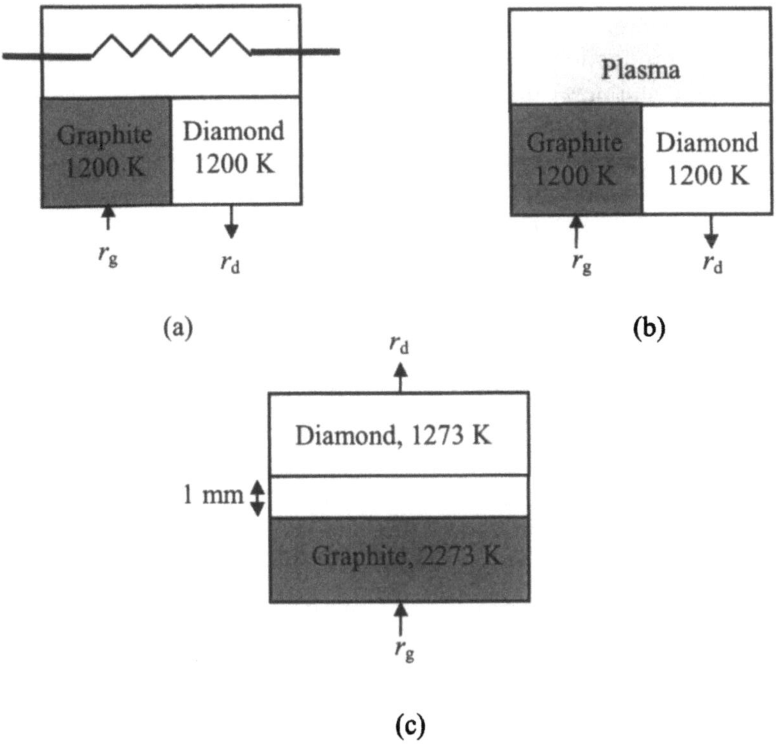

Fig. 8.2. Three ideal diamond growth systems of stationary states

tal pressure $p = 10$ kPa and hot filament temperature $T_{\text{activation}} = 2400$ K. From handbooks of thermodynamic data the standard thermodynamic data for graphite and diamond (Gibbs free energy G, enthalpy H, entropy S at 298 K and 100 kPa; and isobaric heat capacity C_p in a certain temperature range) can be found. According to the relationship between these thermodynamic functions and temperature mentioned in Chap. 1, the calculated results for diamond and graphite are listed in Tables 8.3 and 8.4, respectively. Thermodynamic data of a solid under low pressure can be treated as a constant independent of pressure. In this way from Table 8.3, Table 8.4 and (2.22), i.e., $\Delta_i S = -\Delta G/T$, the Gibbs free energy change and the entropy production of the transformation from graphite to diamond of reaction (5.5) (denoted by subscript 1), i.e.,

$$\text{C(gra)} = \text{C(dia)} \ (T = 1200 \text{ K}, \ p = 10 \text{ kPa})$$

are

$$\Delta G_1 = 6.96 \text{ kJ and } \Delta_i S_1 = -5.80 \text{ J K}^{-1} \tag{8.30}$$

respectively. Therefore, the transformation from graphite to diamond is a negative entropy production reaction or a positive Gibbs free energy change reaction, which should never occur spontaneously.

Table 8.3. Thermodynamic data of diamond

T K	G kJ mol^{-1}	H kJ mol^{-1}	S J $mol^{-1}K^{-1}$	C_p J $mol^{-1}K^{-1}$
298.15	1.1863	1.8950	2.3770	6.0971
300.00	1.1819	1.9064	2.4151	6.2072
400.00	0.82278	2.7652	4.8561	10.539
500.00	0.20421	3.9627	7.5169	13.255
600.00	−0.67862	5.3956	10.124	15.334
700.00	−1.8169	7.0196	12.624	17.112
800.00	−3.1997	8.8128	15.016	18.731
900.00	−4.8168	10.763	17.311	20.256
1000.00	−6.6590	12.862	19.521	21.723
1100.00	−8.7186	15.106	21.659	23.153
1200.00	−10.989	17.492	23.734	24.557

$p = 101.325$ kPa (1 atm)

T (K)	A	B	C	D
298–1200	9.1211	13.2214	−6.1923	0

$C_p = A + 10^{-3}BT + 10^5CT^{-2} + 10^{-6}DT^2$

Similarly, the standard thermodynamic data for atomic hydrogen and molecular hydrogen can also be found from handbooks of thermodynamic data. Using the partial pressure of superequilibrium atomic hydrogen [i.e., 0.49500 kPa (0.0048853 atm), equal to the equilibrium partial pressure of atomic hydrogen at the hot filament temperature 2400 K], the total pressure, and the relationship between thermodynamic functions and temperature or pressure mentioned in Chap. 1, the calculated results for superequilibrium atomic hydrogen and molecular hydrogen are listed in Tables 8.5 and 8.6, respectively. In this way from Table 8.5, Table 8.6 and (2.22), i.e., $\Delta_i S = -\Delta G/T$, the Gibbs free energy change and entropy production of the association of superequilibrium atomic hydrogen of reaction (5.6) (denoted by subscript 2), i.e.,

$$H* = 0.5H_2 \ (T = 1200 \text{ K}, \ p = 10 \text{ kPa and } T_{\text{activated}} = 2400 \text{ K})$$

are

$$\Delta G_2 = -112.72 \text{ kJ and } \Delta_i S_1 = 93.93 \text{ J K}^{-1} \tag{8.31}$$

Table 8.4. Thermodynamic data of graphite

T K	G kJ mol^{-1}	H kJ mol^{-1}	S J mol^{-1}K^{-1}	C_p J mol^{-1}K^{-1}
298.15	– 1.6953	0.0000	5.6860	8.5019
300.00	– 1.7058	0.015797	5.7388	8.5755
400.00	– 2.4265	1.0519	8.6959	11.982
500.00	– 3.4452	2.3880	11.666	14.645
600.00	– 4.7563	3.9642	14.534	16.804
700.00	– 6.3473	5.7348	17.260	18.541
800.00	– 8.2030	7.6595	19.828	19.890
900.00	– 10.307	9.7005	22.231	20.869
1000.00	– 12.644	11.821	24.465	21.485
1100.00	– 15.195	14.010	26.550	22.208
1200.00	– 17.949	16.257	28.506	22.725
1300.00	– 20.893	18.551	30.341	23.129
1400.00	– 24.014	20.880	32.067	23.452
1500.00	– 27.303	23.239	33.695	23.715
1600.00	– 30.750	25.622	35.232	23.934
1700.00	– 34.347	28.025	36.689	24.119
1800.00	– 38.085	30.445	38.072	24.277
1900.00	– 41.959	32.880	39.389	24.415
2000.00	– 45.961	35.327	40.644	24.536
2100.00	– 50.086	37.786	41.844	24.645
2200.00	– 54.328	40.256	42.993	24.744
2300.00	– 58.683	42.735	44.095	24.834
2400.00	– 63.145	45.223	45.153	24.918

$p = 101.325$ kPa (1 atm)

T (K)	A	B	C	D
298–1000	0.1597	38.889	– 1.5154	– 17.4121
1000–2400	25.5701	– 0.2232	– 39.1476	0.0978

$C_p = A + 10^{-3}BT + 10^5CT^{-2} + 10^{-6}DT^2$

respectively. Here, H* represents atomic hydrogen of superequilibrium concentration. H* is defined here as the equilibrium concentration at the activated temperature and of superequilibrium concentration at the substrate temperature. This reaction is a positive entropy production reaction or a negative Gibbs free energy change reaction, so it is spontaneous.

Table 8.5. Thermodynamic data of superequilibrium atomic hydrogen

T K	G kJ mol^{-1}	H kJ mol^{-1}	S J mol^{-1}K^{-1}	C_p J mol^{-1}K^{-1}
298.15	170.62	217.99	158.85	20.786
300.00	170.33	218.02	158.98	20.786
400.00	154.12	220.10	164.96	20.786
500.00	137.38	222.18	169.60	20.786
600.00	120.23	224.26	173.39	20.786
700.00	102.72	226.34	176.59	20.786
800.00	84.921	228.42	179.37	20.786
900.00	66.860	230.50	181.82	20.786
1000.00	48.566	232.57	184.01	20.786
1100.00	30.065	234.65	185.99	20.786
1200.00	11.374	236.73	187.80	20.786
1300.00	−7.4898	238.81	189.46	20.786
1400.00	−26.514	240.89	191.00	20.786
1500.00	−45.687	242.97	192.44	20.786
1600.00	−64.998	245.05	193.78	20.786
1700.00	−84.440	247.13	195.04	20.786
1800.00	−104.00	249.20	196.23	20.786
1900.00	−123.68	251.28	197.35	20.786
2000.00	−143.47	253.36	198.42	20.786
2100.00	−163.36	255.44	199.43	20.786
2200.00	−183.36	257.52	200.40	20.786
2300.00	−203.44	259.60	201.32	20.786
2400.00	−223.62	261.68	202.21	20.786

p = 10.1325 kPa (0.1 atm), $T_{\text{activation}}$ = 2400 K, partial pressure of superequilibrium atomic hydrogen p_{H*} = 0.49500 kPa (0.0048853 atm)

T (K)	A	B	C	D
298–2400	20.7861	0	0	0

$C_p = A + 10^{-3}BT + 10^5CT^{-2} + 10^{-6}DT^2$

Table 8.6. Thermodynamic data of molecular hydrogen

T K	G kJ mol^{-1}	H kJ mol^{-1}	S J mol^{-1}K^{-1}	C_p J mol^{-1}K^{-1}
298.15	−44.766	0.0000	150.14	28.872
300.00	−45.043	0.053420	150.32	28.879
400.00	−60.512	2.9545	158.67	29.109
500.00	−76.716	5.8721	165.18	29.237
600.00	−93.509	8.8017	170.52	29.357
700.00	−110.79	11.744	175.05	29.502
800.00	−128.50	14.703	179.00	29.685
900.00	−146.58	17.683	182.51	29.913
1000.00	−164.99	20.688	185.68	30.189
1100.00	−183.71	23.722	188.57	30.515
1200.00	−202.70	26.792	191.24	30.892
1300.00	−221.95	29.902	193.73	31.321
1400.00	−241.44	33.058	196.07	31.802
1500.00	−261.16	36.265	198.28	32.336
1600.00	−281.09	39.527	200.39	32.923
1700.00	−301.23	42.815	202.38	33.175
1800.00	−321.57	46.153	204.29	33.581
1900.00	−342.09	49.530	206.11	33.960
2000.00	−362.79	52.944	207.87	34.314
2100.00	−383.66	56.393	209.55	34.649
2200.00	−404.69	59.874	211.17	34.967
2300.00	−425.89	63.386	212.73	35.270
2400.00	−447.24	66.927	214.24	35.560

p = 10.1325 kPa (0.1 atm), $T_{\text{activation}}$ = 2400 K, partial pressure of molecular hydrogen p_{H_2} = 9.6375 kPa (0.095115 atm)

T (K)	A	B	C	D
298–1550	30.0453	−2.4799	−0.5975	2.6832
1550–2400	29.6687	3.2702	−41.3714	−0.2151

$C_p = A + 10^{-3}BT + 10^5CT^{-2} + 10^{-6}DT^2$

According to the empirical value (from comparison with experiments) of the reaction rate ratio $\chi = r_2/r_1 = 0.28$, the Gibbs free energy change and the entropy production of the overall coupled reaction (5.7) (denoted by subscript 3), i.e., (5.7) = (5.5) + χ (5.6),

$$\mathrm{C(gra)} + \chi\mathrm{H}* = 0.5\chi\mathrm{H_2} + \mathrm{C(dia)}\ (T = 1200\ \mathrm{K\ and}\ p = 10\ \mathrm{kPa})$$

or reaction (5.8)(denoted by subscript 4), i.e.,

$$\mathrm{C(gra*)} = \mathrm{C(dia)},\ \Delta G_4 = \Delta G_3\ (T = 1200\ \mathrm{K\ and}\ p = 10\ \mathrm{kPa})$$

with

$$\mathrm{C(gra*)} = [\mathrm{C(gra)} + \chi\,(\mathrm{H}* - 0.5\mathrm{H_2})]$$

are

$$\Delta G_3 = 6.96 - 0.28 \times 112.72 = -24.60\ \mathrm{kJ} < 0 \tag{8.32}$$

and

$$\Delta_i S_3 = -5.80 + 0.28 \times 93.93 = 20.50\ \mathrm{J\ K^{-1}} > 0 \tag{8.33}$$

respectively.

That is, at low pressure the reaction between graphite and superequilibrium atomic hydrogen to produce diamond and molecular hydrogen is completely in agreement with the second law of thermodynamics. Such a quantitative verification in an inanimate system cannot only dispel nearly 30 year of doubts that activated low-pressure diamond growth might be violating thermodynamic laws, but also remove a 70-year controversy surrounding reaction coupling and negative entropy production reactions.

It should be noted that these discussions are based on the viewpoint of thermodynamics. In fact, the thermodynamic viewpoint is the same as the viewpoint of energy at a certain condition. In particular, for coupling reactions at constant temperature and pressure in a close system the energy relationship is clearly on the basis of the Gibbs free energy. The key effect of reaction coupling is that a nonspontaneous reaction (also called a coupled reaction) is driven forward by another reaction process (also called a coupling reaction), which is more aggressively spontaneous. In other words, an endergonic reaction with a positive free energy ΔG_1 is forced to occur by virtue of its coupling to an exergonic reaction with a negative free energy ΔG_2 together with a negative sum of $(\Delta G_1 + \chi\Delta G_2)$. However, the limitation of thermodynamics is only able to provide information on the reaction possibility and the reaction direction, but not on the practical inevitability. The practical inevitability is still dependent on reasonable kinetic mechanisms and paths. Therefore, after the above discussion on thermodynamic aspects the kinetic mechanism of the reaction coupling should also be discussed. Due to the complexity of living systems nearly all sets of coupled reactions in living systems are represented simply by a scheme of two arrows. Similarly, the mechanism of the activated low-pressure diamond growth can also be represented in this way, as shown in Fig. 8.3a. The mechanism of the reaction coupling in inanimate systems is much simpler than that in living systems, that is to say, the mechanism can be illustrated in detail, as shown in Fig. 8.3b.

Under activated low-pressure conditions diamond but not graphite may grow up from a gas mixture of hydrocarbon, such as methane in hydrogen, and even diamond may grow up from graphite via gaseous hydrocarbon intermediates. All these diamond growth processes are typical organic reaction processes. Due to the high reactivity between superequilibrium atomic hydrogen and unsaturated graphite clusters the graphite clusters will be decomposed by the addition reaction of atomic hydrogen, and result in a corresponding higher concentration of methane, acetylene, ethylene, methyl, ethyl and other small molecule hydrocarbons (the detailed gas composition can be found at the gas–solid phase line of activated graphite in its gas phase composition nonequilibrium stationary phase diagram). The concentrations of these gaseous hydrocarbons is already higher than the saturated concentration for diamond growth (the detailed gas composition can be found at the gas–solid phase line of diamond in its gas phase composition nonequilibrium stationary phase diagram). Therefore, diamond clusters can form and grow up on the diamond surfaces, and that results in a release of hydrogen molecules. The opposite process is not easy, because on the diamond surface carbon atoms are saturated in an sp^3 configuration and are difficult to react with superequilibrium atomic hydrogen. There are many experimental demonstrations reported in the literature to show the influence of superequilibrium atomic hydrogen. The higher concentration of atomic hydrogen does not show any obvious influence on the etching rate of diamond, but shows an obvious influence on increasing the etching rate of graphite. Therefore, during the existence of superequilibrium atomic hydrogen it may be impossible to form graphite nuclei with an unsaturated sp^2 configuration. The overall coupling process is a vector in a single direction, so that during the existence of superequilibrium atomic hydrogen both the transformation from graphite to diamond via gas-phase intermediates or the pyrolysis of hydrocarbons (such as methane) to grow only diamond are kinetically reasonable. In comparison with the discussion in Sect. 5.1 of Chapter V an emphasis on the key role of superequilibrium atomic hydrogen has been made. In Fig. 8.3b superequilibrium atomic hydrogen is denoted by H^*. If atomic hydrogen is in equilibrium concentration, then ΔG_2 equals zero, so there is no nonequilibrium thermodynamic coupling, and no possibility of continuously growing diamond at low pressure. Figure 8.3b may be the most detailed discussion for mechanisms of the reaction coupling. The detailed mechanism may help to understand the reaction coupling and negative entropy production reactions.

From the discussion on the mechanism of reaction coupling, it can clearly be found that the so-called "coupling substance", which was supposed to be a common reactant or product in both the coupling reaction and the coupled reaction, is not a necessary condition for reaction coupling. In spite of no common substance in formulations of the coupled reaction and the coupling reaction in the activated low-pressure diamond process, superequilibrium atomic hydrogen does really react with graphite at the microscopic

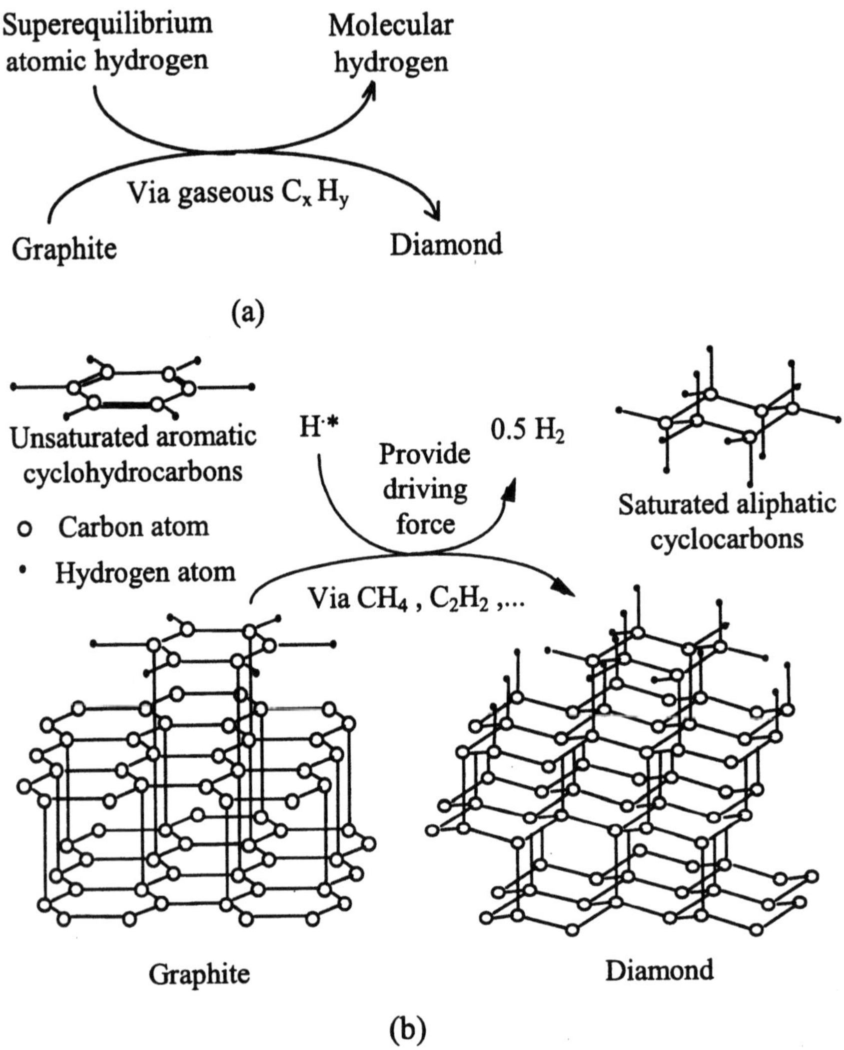

Fig. 8.3. Mechanism for diamond growth at activated low pressure (**a**) is a simple expression of (**b**)

level, that is, the coupling between the coupled reaction and the coupling reaction can in fact occur, even without the so-called "coupling substance" in the formulation of these reactions.

8.4 Reaction Barrier in Synthetic Diamond Processes

8.4.1 "Unified Barrier" Model

In 1994 Ya-Fei Zhang et al. proposed a "unified barrier" model, "which is thought to be a general mechanism for all these experiments" of synthetic diamond processes including super-high-pressure high-temperature (HPHT) processes, high-pressure catalyst-assisted (HP catalyst) processes, and activated low-pressure chemical vapor deposition (activated CVD) processes [8.21]. The barrier of this model is illustrated in Fig. 8.4. The equilibrium point between graphite and diamond was estimated by an approximate calculation of diamond and graphite growth probabilities. However, an obvious mistake like "1 + 1 = 1" (in fact, a mistake of $\Delta G_T^p + \Delta G_T^p = \Delta G_T^p$) had been made during his derivation. Therefore, the results of that model are incorrect and violate the basic laws of thermodynamics. The model might conclude that a catalyst could only accelerate the forward reaction rate but not the reverse reaction rate.

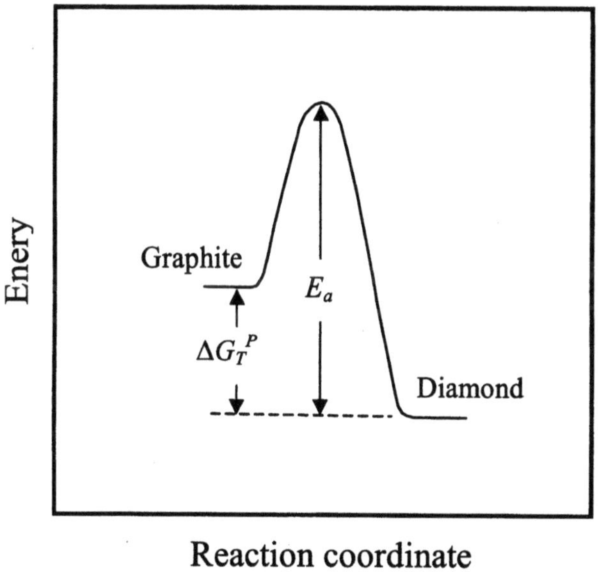

Fig. 8.4. The barrier illustrated in the original paper of the unified barrier model [8.21]

In textbooks of physical chemistry the incorrectness of such a conclusion had already been pointed out. For instance, Moore in his book *Physical Chemistry* (1976) wrote that:

"Wilhelm Ostwald was the first to emphasize that a catalyst influences the rate of a chemical reaction but has no effect on the position of equilibrium. His famous definition was, 'A catalyst is a substance that changes the velocity of a chemical reaction without itself appearing in the end products.' Ostwood's proof was based on the First Law of Thermodynamics. Consider a gas reaction that proceeds with a change in volume. The gas is confined in a cylinder fitted with a piston; the catalyst is in a small receptacle within the cylinder, and can be alternately exposed and covered. If the equilibrium position were altered by exposing the catalyst, the pressure would change, the piston would move up and down, and a perpetual-motion machine would be available. Since a catalyst changes rate but not equilibrium, *it must accelerate the forward and reverse reactions in the same proportion.*" (Note: The italic emphasis was made by Moore himself.) [8.22]

Moreover, the unified barrier model created more confusion between the high-pressure catalyst-assisted diamond process and the activated low-pressure CVD diamond process by introducing the special term "the metal and hydrogen catalyst-assisted diamond synthesis". The high-pressure catalyst-assisted diamond process completely differs from the activated low-pressure CVD diamond process, so that the derivation and the results of the unified barrier model were in a mess. We had to calculate them again on the basis of correct barriers for different processes, as shown in Figs. 8.5, 8.6 and 8.7. For comparison barriers practically used in the unified barrier model are drawn in Figs. 8.8 and 8.9. Due to these errors the barriers in Figs. 8.8 and 8.9 cannot be drawn by continuous curves. If the right side of the barrier line in Fig. 8.8 were lowered for ΔG_T^p, then the energy difference between graphite and diamond in Fig. 8.8 would not be ΔG_T^p, but $2\times\Delta G_T^p$. That does not agree with the original definition of ΔG_T^p in the unified barrier model. Discussion of Fig. 8.9 is more difficult. Even if the right side of the barrier line in Fig. 8.9 is lowered for ΔG_T^p, a continuous curve will not be obtained. For clearing up the misunderstanding, we have to make a complete derivation again, and we published our paper in the same *Journal of Materials Research* in 1997 [8.23, 8.24].

8.4.2 Thermodynamic Coupling Effect and Catalytic Effect

There are mainly three processes for diamond syntheses: (1) the super-high-pressure high-temperature process, (2) the high-pressure catalyst-assisted process, and (3) the activated low-pressure CVD process. The first and second processes take place in the diamond stable region of the equilibrium phase diagram of carbon, so these processes can be explained by equilibrium thermodynamics. However, the activated low-pressure CVD process runs under nonequilibrium conditions, so that this process should be explained by nonequilibrium thermodynamics or modern thermodynamics. In the high-pressure catalyst-assisted process, due to the existence of the catalyst for the same growth rate the pressure (p) and the temperature (T) may be lower

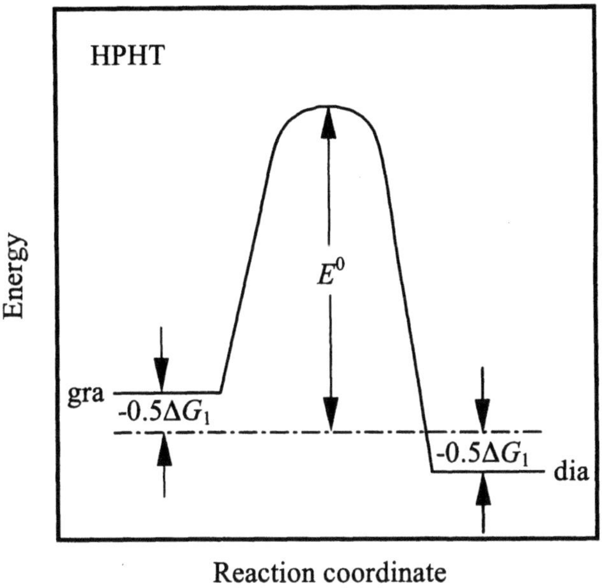

Fig. 8.5. Correct barriers for the super-high-pressure high-temperature synthetic diamond process

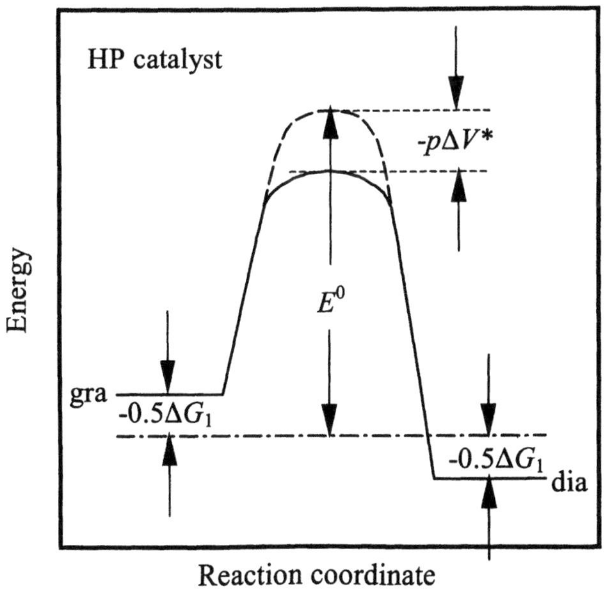

Fig. 8.6. Correct barriers for the high-pressure catalyst-assisted synthetic diamond process

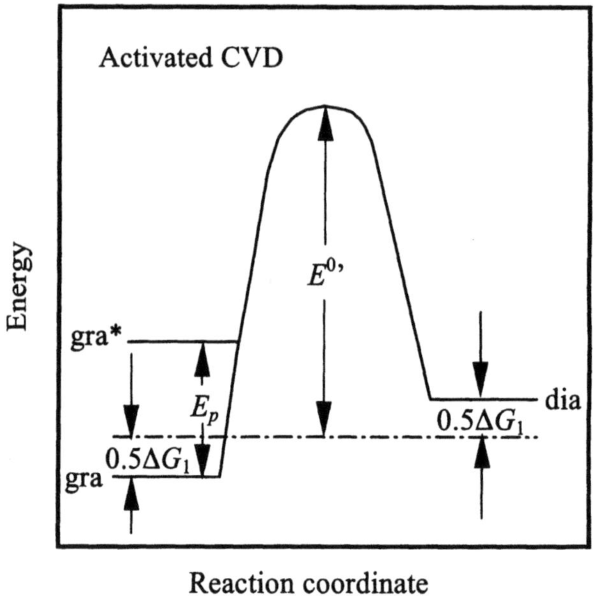

Fig. 8.7. Correct barriers for activated low-pressure CVD synthetic diamond process

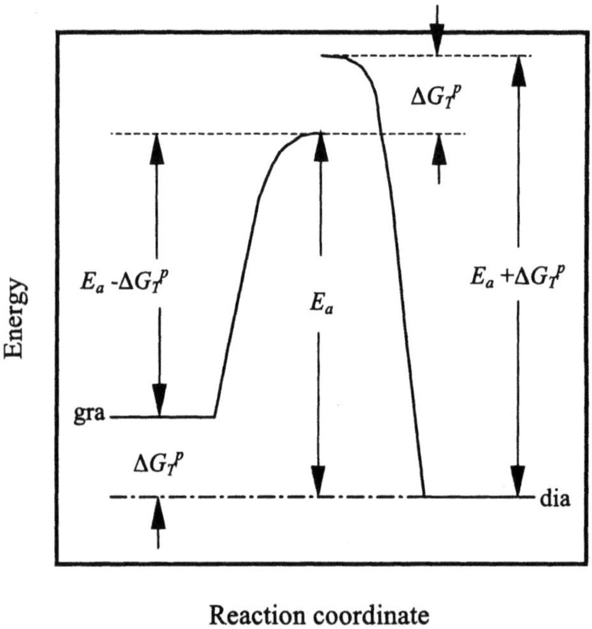

Fig. 8.8. Illustration of the incorrect barrier used in practical calculations by the unified barrier model for the super-high-pressure high-temperature process

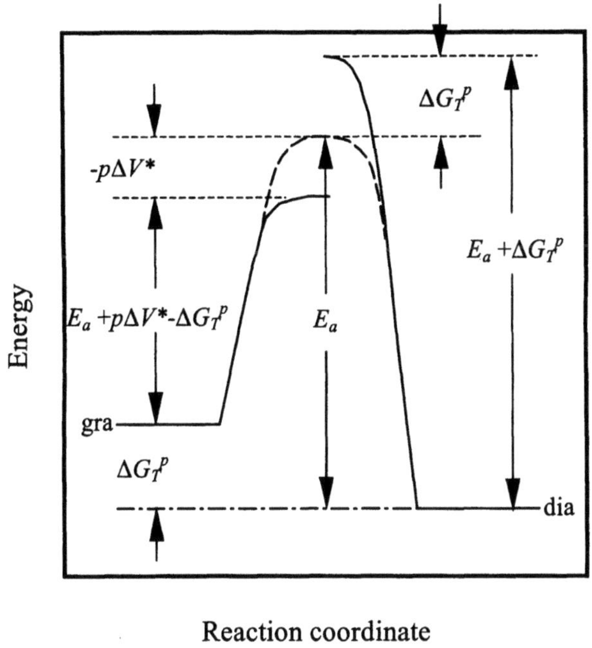

Fig. 8.9. Illustration of the incorrect barrier used in practical calculations by the unified barrier model for the high-pressure catalyst-assisted process and the activated low-pressure CVD process (these two processes are confused in the unified barrier model as "metal and hydrogen catalyst-assisted process")

than those in the super-high-pressure high-temperature process without a catalyst.

"Hence *a catalyst cannot alter the equilibrium point of a reaction.* This being so, a catalyst for a forward reaction must be a catalyst for the reverse reaction also." (note: the italic emphasis is original), see *Physical Chemistry* by Levine (1983) [8.25].

Therefore, equilibrium point between diamond and graphite in either the super-high-pressure high-temperature process or high-pressure catalyst-assisted process is always on the Berman–Simon line of the phase diagram of carbon, that is

$$p^e(\mathrm{Pa}) = 2.73 \times 10^6 T\ (\mathrm{K}) + 7.23 \times 10^8. \tag{8.34}$$

For the activated low-pressure CVD process, the reaction equation may be written as

$$[\mathrm{C(graphite)} + \chi(\mathrm{H}* - 0.5\mathrm{H_2})] = \mathrm{C(diamond)} \tag{8.35}$$

where, χ is the coupling coefficient and H* denotes superequilibrium atomic hydrogen. χ (H*−0.5 H_2) is the influence of superequilibrium atomic hydrogen on the transformation from graphite to diamond. This influence can be called the thermodynamic coupling effect or the chemical pump effect. The left side of the reaction equation can be treated as a whole, and is called activated graphite. If the coupling coefficient χ is not too small (i.e., the concentration of superequilibrium atomic hydrogen is not too low), the Gibbs free energy change ΔG must have a negative value, that is, the reaction should be toward the right side. Therefore, due to the thermodynamic coupling effect, diamond is more stable than activated graphite, that is to say, diamond becomes a stable phase under activated low-pressure conditions. According to the thermodynamic coupling model, the relationship of the nondissipation phase line (nondissipation pressure vs. temperature, $p^{e'} - T$) can be calculated for different χ values at the hot filament temperature of 2400 K in the activated CVD process. The relationship of the nondissipation phase line between the diamond phase and activated graphite is listed in Table 8.7. (heat capacity at constant pressure $C_p = A + BT \times 10^{-3} + CT^{-2} \times 10^5 + DT^2 \times 10^{-6}$). Note that the term "nondissipation" should be used in nonequilibrium phase diagrams instead of term "equilibrium" in equilibrium phase diagrams.

Table 8.7. $p^{e'}$ vs. T relationship between diamond and activated graphite in the hot filament (2400 K) activated CVD process with different χ values

For temp. range: 300–1550 K

χ	α	β	γ	δ	ε	ζ
0.01	0.8441	−4.992	2.984	3.503	−1.263	3.256
0.02	1.688	−17.21	3.239	7.006	−2.527	6.512
0.04	3.376	−41.66	3.748	14.01	−5.053	13.02
0.08	6.753	−90.54	4.766	28.02	−10.11	26.05

For temp. range: 1550–2400 K

χ	α	β	γ	δ	ε	ζ
0.01	58.43	−5.143	2.993	−4.619	0.1013	3.363
0.02	116.9	−17.52	3.257	−9.238	0.2026	6.725
0.04	233.7	−42.26	3.783	−18.48	0.4053	13.45
0.08	467.5	−91.75	4.837	−36.95	0.8105	26.90

$p^{e'}$ (Pa) $= \alpha 10^8 T^{-1} + \beta 10^8 + \gamma 10^6 T\,(\mathrm{K}) + \delta T^2 + \varepsilon 10^{-3} T^3 + \zeta 10^4 T \ln T$

Figure 8.10 illustrates the $p^{e'}$–T relationship between diamond and activated graphite for different χ value at the hot filament temperature of 2400 K in the activated CVD process. From Fig. 8.10 it can be seen that all these nondissipation lines are much lower than the Berman–Simon line in the equilibrium phase diagram. The corresponding nondissipation line is lower, χ is bigger. If the value of χ is bigger than 0.12 (in the ordinary hot filament method, the empirical value of χ is about 0.28), diamond may be grown under low pressures at temperature below 1600 K. It has been confirmed by experiments that diamond can really grow in the wide range of 800–1500 K and 10^3–10^5 Pa. At low temperature below 700 K the growth rate of diamond is too low, and at high temperature above 1600 K diamond is easy to graphitize. All of these phenomena agree with our theoretical prediction. The following discussion is based on the correct reaction barriers for different processes as shown in Figs. 8.5–8.7.

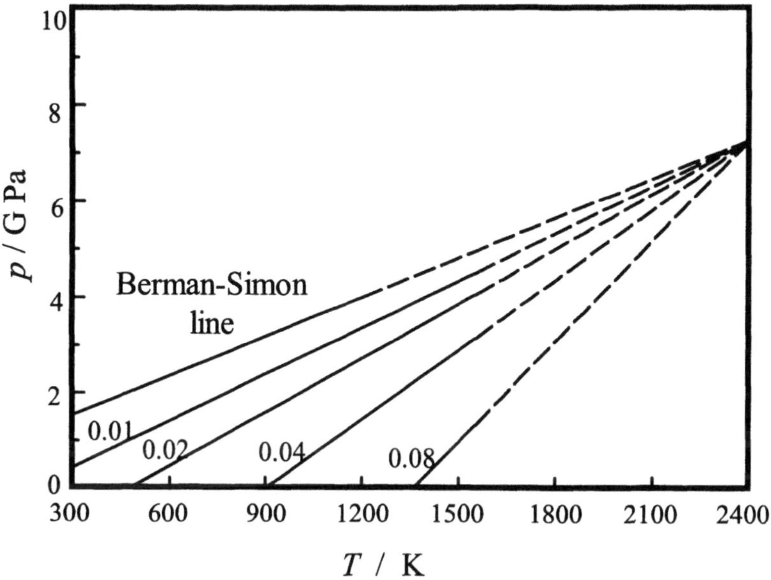

Fig. 8.10. Nondissipative phase lines between diamond phase and activated graphite in the hot filament (2400 K) activated CVD process with different χ values

The basic reaction discussed here is the transformation from graphite (denoted by gra) to diamond (denoted by dia), i.e., reaction (1), C (gra) = C (dia). At a certain temperature (T) and pressure (p) the Gibbs free energy change of the reaction is denoted by ΔG_1. In this chapter, the transformation from graphite to diamond is called the forward reaction, while the transformation from diamond to graphite is called the reverse reaction.

The illustration of the barrier for the HPHT process is the simplest one. According to different zero points and different names of symbols the barrier for the HPHT process can be drawn as shown in Fig. 8.4 or Fig. 8.5. Here, $E_a = E^{\text{o}} - 0.5\ \Delta G_1$, $G_T^p = -\Delta G_1$. Now, the discussion here will be based on these barriers illustrated in Fig. 8.5. From Fig. 8.5 it can be seen that the barriers for the forward and the reverse reactions in the HPHT process are E^o+0.5 ΔG_1 and E^o−0.5 ΔG_1, respectively. At super high pressures ΔG_1 has a negative value, so the reverse barrier is bigger than the forward barrier. The forward reaction occurs easily. That agrees with experimental observations, that is, the transformation from graphite to diamond is spontaneous under super-high-pressure high-temperature conditions.

In the HP catalyst process, due to the existence of the catalyst, the catalyst effect can lower the reaction barrier and correspondingly increase the transformation rate or decrease the transformation temperature. As mentioned above a catalyst cannot change the equilibrium point of a reaction. The catalyst must accelerate the forward and reverse reactions in the same proportion. Therefore, in the HP catalyst process the reaction barrier can be drawn as shown in Fig. 8.6, that is, the barrier for the forward and the reverse reactions in the HP catalyst process are $E^o + p\Delta V^* + 0.5\ \Delta G_1$ and $E^o + p\Delta V^* - 0.5\ \Delta G_1$, respectively, where, ΔV^* is called the activated volume, and has a negative value. In comparison with the HPHT process, both the forward and the reverse reaction barriers decrease in energy level by $-p\Delta V^*$. Correspondingly, the transformation rate increases, or the reaction temperature decreases. That agrees with experimental observations.

For the activated low-pressure CVD process the hot filament or microwave methods can be taken as an example. The process runs at typical nonequilibrium conditions, so it, of course, should be treated by nonequilibrium thermodynamics (or thermodynamics for irreversible processes). In the activated low-pressure process, during the transformation from graphite to diamond there is another irreversible reaction process, usually an association of superequilibrium atomic hydrogen (denoted by H*) to produce hydrogen molecules, that is reaction (2), H* = 0.5 H_2. At the same conditions of T and p together with a sufficiently high activation temperature, the Gibbs free energy change of reaction (2), ΔG_2, has a greater negative value. The reaction (1) occurs only when driven by the reaction (2). The overall reaction is reaction (3) = (1) + χ(2), C (gra) + χ H* = 0.5 χ H_2 + C (dia). The Gibbs free energy change of reaction (3) is $\Delta G_3 = \Delta G_1 + \chi \Delta G_2$. Here, χ is the rate ratio of reaction (2) to reaction (1), i.e., $\chi = \frac{r_2}{r_1}$. If $-\chi\ \Delta G_1$ is represented by E_p, then the forward and reverse reaction barriers are $E^o + 0.5\ \Delta G_1 - E_p$, and $E^o - 0.5\ \Delta G_1$, respectively. The reaction barrier for the activated CVD process can be illustrated by Fig. 8.7. Obviously, at ordinary low pressures the transformation from graphite to diamond cannot occur spontaneously. Driven by the coupling effect, reaction (1) can take place toward the right-hand side. Prigogine said: "Thermodynamic coupling allows one of the reactions to progress in a direction contrary to that prescribed by its own affinity."

To sum up, all of these forward and reverse reaction barriers for different processes discussed in this chapter can be listed in Table 8.8. Obviously, the forward and reverse reaction barriers in different processes are different, and therefore the mechanisms in different processes are different as well.

Table 8.8. Reaction barriers in the thermodynamic coupling model

Processes	Forward barrier	Reverse barrier
HTHP	$E^{\circ} + 0.5\ \Delta G_1$	$E^{\circ} - 0.5\ \Delta G_1$
HP catalyst	$E^{\circ} + p\Delta V^* + 0.5\ \Delta G_1$	$E^{\circ} + p\Delta V^* - 0.5\ \Delta G_1$
Low-pressure CVD	$E^{\circ} + 0.5\ \Delta G_1 - E_p$	$E^{\circ} - 0.5\ \Delta G_1$

For convenience of comparison, the practically used barriers in the unified barrier model are listed in Table 8.9, where the HP catalyst process and the activated low-pressure CVD process were confusedly called the "metal and hydrogen catalyst-assisted catalyst process". In Table 8.9 the original symbols are these used in the original paper of the unified barrier model, such as E_a and G_T^p. That is, $E_a = E^{\circ} - 0.5\ \Delta G_1$, $G_T^p = -\Delta G_1$. According to Table 8.9 only the unreasonable Figs. 8.8 and 8.9 can be drawn. Table 8.9 disagrees also with Fig. 8.4, the original figure in the paper of the unified barrier model. Obviously, in both Figs. 8.8 and 8.9 these unreasonable reaction barriers could not be drawn even by continuous lines. That shows the incorrectness of the unified barrier model.

Table 8.9. Incorrect barriers of the "unified barrier" model

Process	Forward barrier	Reverse barrier
Process	Forward barrier	Reverse barrier
HPHT	$E_a - G_T^p$	$E_a + G_T^p$
Metal and hydrogen catalyst-assisted process (HP catalyst-assisted and activated CVD process)	$E_a + p\Delta V^* - G_T^p$	$E_a + G_T^p$

8.4.3 Analysis of Calculated Results and Conclusions

As early as in 1974 Guan-Yun Zhang et al. proposed an empirical expression for the diamond growth rate in the HP catalyst-assisted process [8.26],

$$r_{\text{dia}} = Ae^{-(E+p\Delta V*)/RT} \tag{8.36}$$

where, r_{dia} represents the diamond growth rate. $E = 1.20\times10^5$ J mol^{-1} and $\Delta V^* = -10.85\times10^{-6}$ m^3 mol^{-1} were taken, A is the pre-exponential coefficient, R is the gas constant, E is the reaction barrier of the forward transformation from graphite to diamond in the high-pressure process without catalyst, ΔV^* was called the activation volume, and has a negative value and $-p\Delta V^*$ represents the decrease of the activation energy due to the existence of the catalyst. For further estimation, the exponential part of the growth rate may be defined as the probability for diamond growth, $f_{\rm dia}$, that is,

$$f_{\rm dia} = e^{-(E+p\Delta V*)/RT}. \tag{8.37}$$

According to Figs. 8.5 and 8.6, and supposing that $E = E^{\rm o} + 0.5\ \Delta G_1$, the possibilities for both diamond growth and graphite growth in different processes are as follows.

1. For the super-high-pressure high-temperature process,

$$f'_{\rm dia} = e^{-(E^{\rm o}+\ 0.5\ \Delta G_1)/RT} \tag{8.38}$$

and

$$f'_{\rm gra} = e^{-(E^{\rm o}-\ 0.5\ \Delta G_1)/RT}. \tag{8.39}$$

2. For the high-pressure catalyst-assisted process,

$$f''_{\rm dia} = e^{-(E^{\rm o}+\ 0.5\ \Delta G_1+\ p\Delta V*)/RT} \tag{8.40}$$

and

$$f''_{\rm gra} = e^{-(E^{\rm o}-\ 0.5\ \Delta G_1+\ p\Delta V*)/RT} \tag{8.41}$$

where, $E^{\rm o}$ is defined as the phase transformation barrier, when the diamond phase and the graphite phase are in equilibrium. ΔG_1 of reaction (1) under high pressure over the Berman–Simon line in the carbon phase diagram has a negative value, as shown in Figs. 8.5 and 8.6. The influence of pressure and temperature on ΔG_1 can be obtained as

$$\begin{aligned}\Delta G_1(\text{J mol}^{-1}) = \{&p(\text{Pa}) - [2.73 \times 10^6 T\ (\text{K}) \\ &+7.23 \times 10^8]\}\Delta V(\text{m}^3\text{mol}^{-1}).\end{aligned} \tag{8.42}$$

Suppose that along the Berman–Simon line $\Delta V \approx -1.77\times10^{-6}$ m^3 mol^{-1} (as an average value) and $E^{\rm o} \approx E = 1.20 \times 10^5$ J mol^{-1} (as a constant), respectively.

In this way, the results of the calculated equal possibility lines for both diamond growth and graphite growth in the unified barrier model should be corrected by the following (8.43)–(8.50).

1. Super-high-pressure high-temperature process

$$f'_{\text{dia}} = 0.01,\ p(\text{Pa}) = 1.363 \times 10^{11} - 4.054 \times 10^{7} T\ (\text{K}) \tag{8.43}$$

$$f'_{\text{dia}} = 0.0001,\ p(\text{Pa}) = 1.363 \times 10^{11} - 8.380 \times 10^{7} T\ (\text{K}) \tag{8.44}$$

and

$$f'_{\text{gra}} = 0.01,\ p(\text{Pa}) = -1.349 \times 10^{11} + 4.600 \times 10^{7} T\ (\text{K}) \tag{8.45}$$

$$f'_{\text{gra}} = 0.0001,\ p(\text{Pa}) = -1.349 \times 10^{11} + 8.926 \times 10^{7} T\ (\text{K}) \tag{8.46}$$

2. High-pressure catalyst-assisted process

$$f''_{\text{dia}} = 0.01,\ p(\text{Pa}) = 1.028 \times 10^{10} - 3.057 \times 10^{6} T\ (\text{K}) \tag{8.47}$$

$$f''_{\text{dia}} = 0.0001,\ p(\text{Pa}) = 1.028 \times 10^{10} - 6.320 \times 10^{6} T\ (\text{K}) \tag{8.48}$$

and

$$f''_{\text{gra}} = 0.01,\ p(\text{Pa}) = 1.198 \times 10^{10} - 4.085 \times 10^{6} T\ (\text{K}) \tag{8.49}$$

$$f''_{\text{gra}} = 0.0001,\ p(\text{Pa}) = 1.198 \times 10^{10} - 7.927 \times 10^{6} T\ (\text{K}). \tag{8.50}$$

These calculated equal possibility lines for both diamond growth and graphite growth can be illustrated for the super-high-pressure high temperature process in Fig. 8.11 and for the high-pressure catalyst-assisted process in Fig. 8.12. From Figs. 8.11 and 8.12 it can be found that both in the super-high-pressure high-temperature process and in the high-pressure catalyst-assisted process all intersection points of diamond and graphite growth equal possibility lines are just on the Berman–Simon line. The intersection points are at a lower temperature in the high-pressure catalyst-assisted process (Fig. 8.12) than that in the super-high-pressure high-temperature process (Fig. 8.11). This means that at lower temperature in the catalyst-assisted process one can get the same diamond growth rate as that at higher temperature in the super-high-pressure process without the catalyst, but equilibrium points (intersection points) are still on the equilibrium (Berman–Simon) line; that is the well-known catalytic effect.

3. Activated low-pressure CVD process
According to the reaction barrier of Fig. 8.7, the calculated equal possibility lines for both diamond growth and graphite growth should be expressed as

$$f'''_{\text{dia}} = e^{-(E^{\text{o}\prime} + 0.5\ \Delta G_1 - E_p)/RT} \tag{8.51}$$

and

$$f'''_{\text{gra}} = e^{-(E^{\text{o}\prime} - 0.5\ \Delta G_1)/RT} \tag{8.52}$$

where, $E^{\text{o}\prime}$ is defined as the phase transformation barrier, when the diamond phase and the activated graphite phase are in a nondissipation transformation state under the activated CVD condition. The concerned values of $E^{\text{o}\prime}$

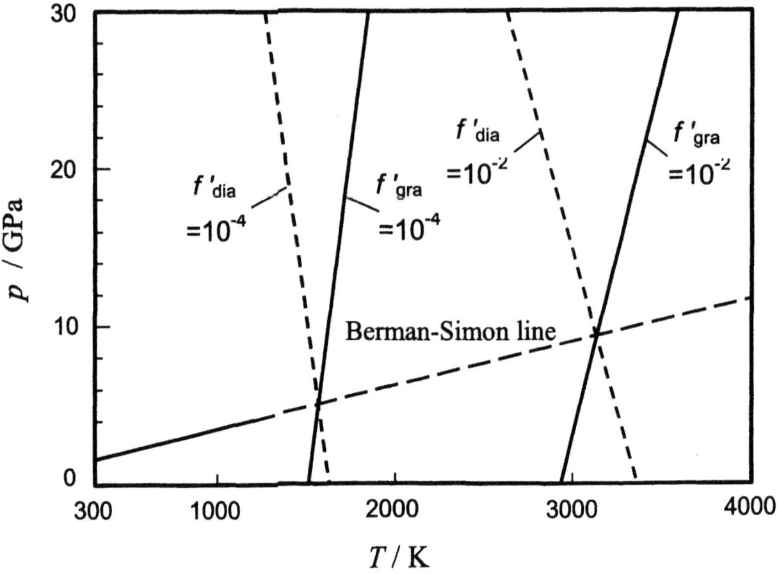

Fig. 8.11. Intersection points of diamond and graphite growth equal possibility lines are just on the Berman–Simon line for the super-high-pressure high-temperature process

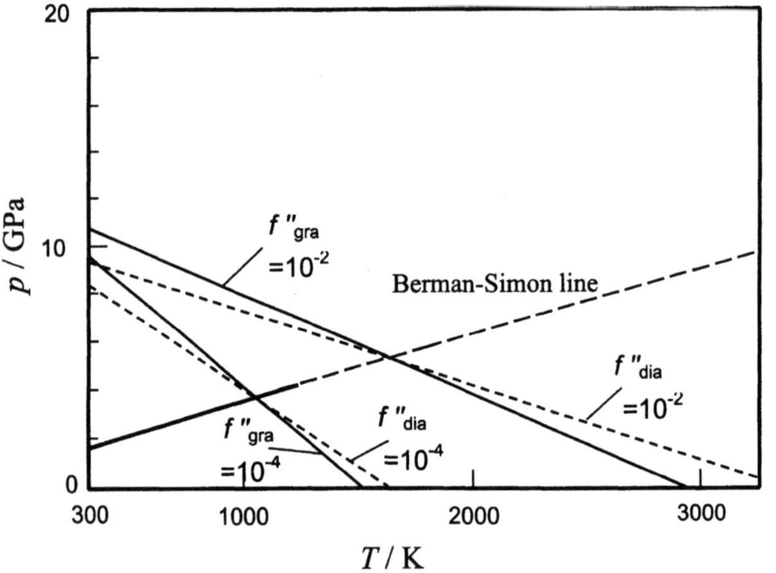

Fig. 8.12. Intersection points of diamond and graphite growth equal possibility lines are just on the Berman–Simon line for the high-pressure catalyst-assisted process

found in the literature are 1.06×10^5 J mol^{-1} (25.2 kcal mol^{-1}) (B. V. Spitsyn, 1995 [8.27]) and 0.84×10^5-1.26×10^5 J mol^{-1} (20-30 kcal mol^{-1}) (W. Zhu et al. 1989 [8.28]). In this paper, $E^{o\prime}$ is still taken to be the same as 1.20×10^5 J mol^{-1}, approximately. E_p is defined as the coupling effect, and equals $-\chi\Delta G_2$. Under the nondissipation transformation condition between activated graphite and diamond, $\chi = \chi^e = -\Delta G_2/\Delta G_2$. The coupling effect should be included only for the forward reaction (from activated graphite to diamond), as shown in reaction (3) and in Fig. 8.7. After the addition of the coupling effect, the graphite (gra) energy level has been changed into a new energy level of activated graphite (gra*). The coupling effect should not be in the reverse reaction. That is the basic principle of the thermodynamic coupling model, and is also a basic and an important difference between the coupling effect and the catalytic effect.

The reaction of the association of superequilibrium atomic hydrogen under low pressure can be treated as an ideal gas system, and then ΔG_2 is nearly independent of the pressure. The relation between ΔG_2 and T can be written as [refer to (1.79)]

$$\begin{aligned}\Delta G_2 = {} & \Delta H_0 - \Delta AT\ln T - \frac{1}{2}\Delta BT^2 \\ & -\frac{1}{2}\Delta CT^{-1} - \frac{1}{6}\Delta DT^3 + \Delta IT\end{aligned} \tag{8.53}$$

where, ΔC_p is the change of heat capacity at constant pressure for reaction (2). ΔH_0 and ΔI are integral constants. Therefore, the detailed expressions are as follows:

(i) For the temperature range 298–1550 K,

$$\begin{aligned}\Delta G_2 = {} & -2.1632\times10^5 + 5.7635\ T\ln T + 6.200\times10^{-4}T^2 \\ & +1.4940\times10^4T^{-1} - 2.2360\times10^{-7}T^3 + 45.0377\ T\end{aligned} \tag{8.54}$$

(ii) For the temperature range of 1550–4000 K,

$$\begin{aligned}\Delta G_2 = {} & -2.1900\times10^5 + 5.9517\ T\ln T - 8.1755\times10^{-4}T^2 \\ & +1.0343\times10^6T^{-1} + 1.7933\times10^{-8}T^3 + 46.6069\ T\end{aligned} \tag{8.55}$$

Under nondissipation transformation condition between activated graphite and diamond, $\chi = \chi^e = -\Delta G_2/\Delta G_2$. Therefore, the relation between χ^e and T can be obtained as listed in Table 8.10.

Table 8.10. Relationship between χ^e and T

T(K)	800	900	1000	1100	1200	1300	1400	1500
χ^e	0.0345	0.0402	0.0466	0.0541	0.0628	0.0731	0.0854	0.0100

Based on (8.51)–(8.55), the calculated equal possibility lines for both diamond growth and graphite growth for the activated low pressure CVD process are listed in the following expressions (8.56)–(8.59) and illustrated in Fig. 8.13.

$$f'''_{\text{dia}} = 0.001,\ \chi^e = 0.0854,$$

$$p(\text{Pa}) = 1.44210 \times 10^9 T^{-1} + 7.527 \times 10^{10} - 5.782 \times 10^7 T\ (\text{K}) \quad (8.56)$$
$$+59.828\ T^2 - 0.022\ T^3 + 5.562 \times 10^5 T \ln T$$

$$f'''_{\text{dia}} = 0.00001,\ \chi^e = 0.0375,$$

$$p(\text{Pa}) = 6.331 \times 10^8 T^{-1} + 8.698 \times 10^{10} - 1.035 \times 10^8 T\ (\text{K}) \quad (8.57)$$
$$+26.271\ T^2 - 0.009\ T^3 + 2.442 \times 10^5 T \ln T$$

and

$$f'''_{\text{gra}} = 0.001,$$

$$p(\text{Pa}) = -9.47 \times 10^{10} + 6.763 \times 10^7 T\ (\text{K}) \quad (8.58)$$

$$f'''_{\text{gra}} = 0.00001,$$

$$p(\text{Pa}) = -9.47 \times 10^{10} + 1.109 \times 10^8 T\ (\text{K}). \quad (8.59)$$

It is obvious that the intersections of each equal possibility line come into the low-pressure region beneath the Berman–Simon line, see Fig. 8.13. That agrees with the well-known facts in the activated CVD processes.

There are different ways of getting artificial diamond, such as the super-high-pressure high-temperature process, the high-pressure catalyst-assisted process, and the activated low-pressure CVD process. Two high-pressure processes run in the diamond stable region of the equilibrium phase diagram of carbon, but the activated low-pressure CVD process runs in the so called "diamond metastable region" of the equilibrium phase diagram of carbon. Correctly speaking, the activated low-pressure process runs under nonequilibrium condition. The activated low-pressure CVD process can be explained by the nonequilibrium thermodynamic coupling theorem, and the concerned nonequilibrium phase diagrams belong to the nonequilibrium nondissipative thermodynamics. The mechanisms and the kinetic barriers are different for these processes. The coupling effect in the activated CVD process is different from the catalytic effect in the high pressure catalyst-assisted process. Confusion between these two effects might lead to the production of the perpetual motion machine.

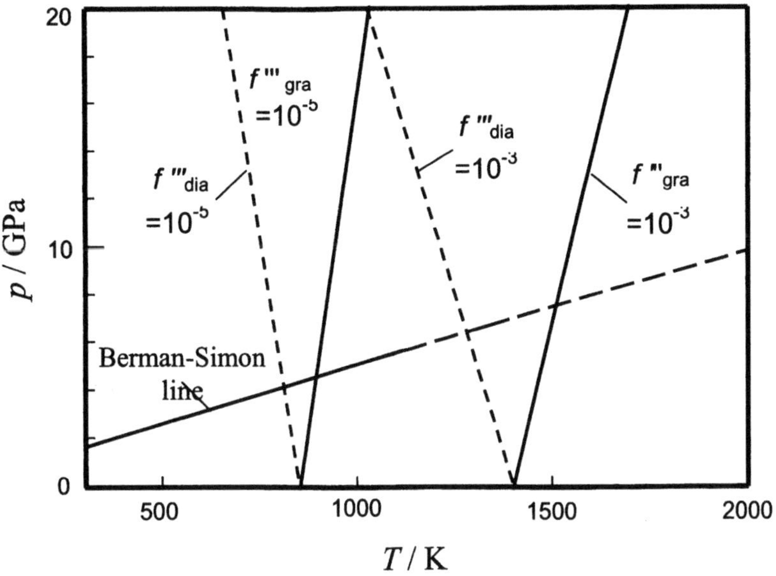

Fig. 8.13. Diamond and graphite growth equal possibility lines for the activated low-pressure CVD process

8.5 Other Thermodynamic Discussions on Activated CVD Diamond

8.5.1 Charged Cluster Model

In 1996, N.M. Hwang et al. proposed a charged cluster model [8.29–8.31]. The basic premise of this model is the great perplexity about activated low-pressure diamond growth from the vapor phase. Hwang thought the well-known experimental observation of "the preferential etching of graphite over diamond" during the existence of superequilibrium atomic hydrogen is a "hypothesis ... , which is contradictory to the well-established stability of graphite with respect to diamond." In particular, he thought another well-confirmed experimental observation that "diamond can be synthesized with graphite as the only source of carbon and diamond is deposited with simultaneous etching of graphite" "has a critical handicap of contradicting the accepted phase diagram of carbon and thereby violating the second law of thermodynamics". Moreover, he wrote: "The experimental observation of diamond deposition with simultaneous graphite etching sounds puzzling in view of the second law of thermodynamics." The activated low-pressure CVD diamond process even seemed to be regarded as a "thermodynamic paradox" perhaps "violating the second law of thermodynamics" in his papers. Correspondingly, Hwang wrote: "Charged carbon nuclei clusters, which are expected to form in the gas phase, are suggested to be responsible for the formation of diamond. When the carbon cluster is sufficiently small, the capillary pressure built up inside the cluster can be high enough to make diamond (should be so called "charged diamond cluster" but not diamond — noted by Wang) more stable than graphite. ... The gas activation process produces charges such as electrons and ions, which are energetically strong heterogeneous nucleation sites for the supersaturated carbon vapor, leading to the formation of charged nuclei clusters ... ." and "Hwang et al. suggested a charged cluster model in which nanometer-sized diamond clusters are formed in the gas phase by charge-induced nucleation." The basic concept of energy levels must be inccorect as shown in Fig. 8.14, and then a confusion between "charged diamond cluster" and bulk diamond was ambiguously introduced for their "theoretical explanation". However, the model proposer might feel there were too many problems in this model, so that Hwang himself also wrote some ambiguous words, such as "Something must be wrong either in interpreting the experimental observation or in applying thermodynamics" and "The thermodynamic treatment here was based on thermal equilibrium, while a real diamond CVD process such as plasma CVD takes place under highly nonequilibrium conditions."

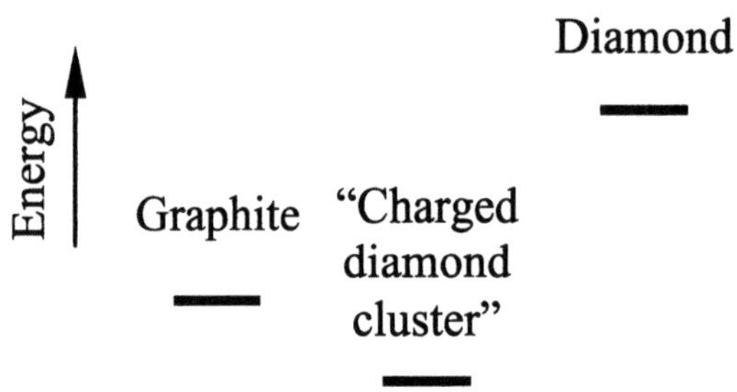

Fig. 8.14. Incorrect concept of energy levels in "charged cluster model" for activated low-pressure CVD process

The present author Wang met Hwang at the *Workshop on Thermodynamics for Gas–Solid Reactions*, Toulouse, France, Sept. 16, 1996. Hwang and Wang exchanged reprints of published papers each other. Hwang presented his model first, and Wang gave an introduction to the nonequilibrium thermodynamic coupling model and presented a series of nonequilibrium stationary phase diagrams, which agree very well with a lot of experimental data reported in the literature. Wang also pointed out the problems in Hwang's model using his own words: "Something must be wrong ... in applying thermodynamics" and "The thermodynamic treatment here was based on thermal equilibrium, while a real diamond CVD process such as plasma CVD takes place under highly nonequilibrium conditions." Wang also said: for example, water flowing toward a lower place is a natural law. However, there is nothing violating the natural law, when we open the tap water in a high building.

There is an invisible pump providing external energy. In the activated diamond CVD process the external energy is also provided, so nonequilibrium thermodynamics must be used for its thermodynamic explanation.

However, recently (during 1998–1999) Hwang insisted on his incorrect logic of the charged cluster model that the prime role of atomic hydrogen and the preferential etching of graphite over diamond during the existence of superequilibrium atomic hydrogen is a "atomic hydrogen hypothesis" but not a experimental observation [8.32–8.34]. Although, he had to say: "This hypothesis is based on the observation that the activated hydrogen etches graphite at rates orders of magnitude faster than diamond" and so on. In this way Hwang arbitrary put the experimental observation and the second law of thermodynamics on both sides of so-called "thermodynamic paradox".

Moreover, Hwang in the "charged cluster model" ambiguously changed the question that diamond (bulk) or graphite (bulk) which is more stable under activated low-pressure diamond growth conditions, into a new question that the charged diamond cluster or graphite (or charged graphite cluster) which is more stable under low pressure diamond growth conditions. His model emphasized that charged diamond clusters were more stable than graphite, as shown in Fig. 8.14. If that were true so as shown in Fig. 8.14, the transformation from charged diamond clusters to diamond would be more difficult than the transformation from graphite to diamond. That would be a more serious "thermodynamics paradox" than the original one. In fact, no scientists are interested in the second question, because both nanometer-sized diamond cluster or nanometer-sized graphite cluster are neither initial reactants nor final products. Thermodynamic study is mainly focused on changes between the initial state (reactant) and the final state (product), but not on intermediate clusters or kinetic mechanisms.

8.5.2 Other Thermodynamic Discussions

Besides the two models mentioned above, there were still other thermodynamic discussions and models in the 1990s. For instance, I. Pinter, A. Tebano, A. Paoletti and P. Paroli [8.35]; N.A. Prijaya, J.C. Angus and P.K. Bachmann [8.36]; N.M.Hwang, J.H. Hahn and G.W. Bahng [8.37]; R. Bachmann, B. Sobisch and W. Kulich [8.38]; I.J. Ford [8.39,8.40], and so on, also discussed or proposed models concerned with thermodynamics. However, these discussions or models were basically qualitative but not quantitative, so they did not really concern the phase diagram or quantitative calculation. Rakov's and Jacob's papers [8.41, 8.42] are concerned with the phase diagram and its calculation. However, Rakov said: "calculations were based on the assumption that diamond cannot be deposited out of the graphite stability domain. Therefore, the lower concentration boundary of the graphite deposition region can be considered at the same time as the lower concentration boundary of the diamond deposition." In fact, Rakov's phase diagram is a typical practical graphite equilibrium phase diagram, calculated by classical equilibrium thermodynamics. Because the Gibbs free energy of diamond is close to the Gibbs free energy of graphite, Rakov explained activated CVD diamond growth by using graphite phase diagrams instead of real diamond phase diagrams. Another problem in Rakov's phase diagram is the loss of the higher concentration boundary between the diamond deposition region and the nondiamond carbon deposition region. Jacob clearly recognized that his calculated diamond growth region is inside the graphite phase region. Jacob had to say: "The deposition region for diamond would be a subset of that for graphite since, at low pressure and moderate temperature, graphite is a stable form of carbon compare to diamond." and "Since graphite is more stable than diamond, it is also expected to codeposit with diamond." No thermodynamic explanation can be found in Rakov's and Jacob's papers for the well-confirmed experimental observation of diamond deposition with the simultaneous graphite etching at activated low pressure.

Recently (1998–1999), W. Piekarczyk was still puzzled by activated CVD diamond processes [8.43, 8.44]. At the beginning of his two papers, *How and why CVD diamond is formed: a solution of the thermodynamics paradox* and *Crystal growth of CVD diamond and some of its peculiarities* he wrote that "Diamond chemical vapor deposition (CVD) thought of as a crystal growth process, is a thermodynamic paradox, because it violates fundamental principles of thermodynamics. The most astonishing violation is the experimental observation that CVD diamond can form in gaseous environments that are carbon under-saturated with respect to diamond." However, the premise does not exist, because no such "experimental observation" could be found in any experiments, except the limitation or prejudice of classical equilibrium thermodynamics. Of course, he hoped to overcome the limitation, so he continued to say in his paper that "it is shown that none of the principles is broken" when the peculiarities of CVD diamond are considered. Maybe because of the influence of Van Rysselberghe's unsuitable speculation of a so called common "coupling substance" in both the coupled reaction and coupling reaction, Piekarczyk did not accept our reaction coupling model at first, but he still listed some similarities between our coupling model and his explanations.

In summary, except for the unified barrier model in which an incorrect derivation made it contradict the basic thermodynamic laws, all other thermodynamic theoretical models or explanations in the 1990s did not break away from the limitations of classical equilibrium thermodynamics.

9. Other Applications and Summary

9.1 Nonequilibrium Phase Diagrams for the Activated CVD cBN Process

Due to the similarities between allotropic pairs of diamond–graphite and cubic boron nitride–hexagonal boron nitride (cBN-hBN), the new kind of phase diagram can also be used for activated CVD cBN processes. Figures 9.1 and 9.2 are projective nonequilibrium phase diagrams calculated on the basis of our coupling model [9.1–9.3] for activated (by hot filament, plasma torch, and/or r.f. plasma, ...) cBN growth from BN-H-Cl and BN-H-F systems, respectively, in the range of total pressure p = 10–10^5 Pa and substrate temperature T = 1200–1500 K. There are three typical phase regions: (i) gas region or no deposition region, (ii) possible cBN growth region or cBN stable region (filled by oblique lines), and (iii) non-cBN deposition region (filled by net-like pattern). One must keep the total composition of the system inside the cBN stable region for growing cBN crystals or films. However, the possible cBN growth region is much narrower in comparison with the diamond growth region in the corresponding phase diagrams for the activated CVD diamond process. The possible cBN growth region seems to be a line in both Figs. 9.1 and 9.2, but it is really a region, as shown amplified in the small circles. That is perhaps the reason why cBN is much more difficult to prepare under low pressure, although activated CVD processes for cBN are very similar to those for diamond. For instance, the hot filament process has been used very easily and stably for the activated CVD diamond process, but a similar hot filament process must be enhanced by a simultaneous r.f. plasma or bias for cBN growth, as reported by F. Zhang, Z. Song, Y. Guo, G. Chen [9.4, 9.5]. Figures 9.1 and 9.2 also show that the cBN growth region from a BN-H system is very close to the apex of H, so cBN can be obtained from B_2H_6-NH_3 gas sources diluted by a large amount of hydrogen (the ratio of $B_2H_6 : H_2$ is about 0.05: 2400, i.e., 2.1×10^{-5} (marked by a open square, and our theoretical value is 1.40×10^{-8}–5.98×10^{-5} in Figs. 9.1 and 9.2) with activation by the hot filament together with the r.f. plasma, as reported by Song et al.. [9.5] Any attempt with a much higher concentration of B_2H_6-NH_3 in H_2 will get hBN or non-cBN instead of cBN, because the gas composition comes into the non-cBN deposition region. That explains Matsumoto's experiment [9.6].

They got non-cBN films from a $B_2H_6 : NH_3 : H_2$ (2.5 : 10 : 1000) mixture (although diluted by Ar), because the concentration of about 0.25% B_2H_6 in H_2 was too high in comparison with our theoretical value of 1.40×10^{-8}–5.98×10^{-5}, and came into the non-cBN deposition region in Figs. 9.1 and 9.2. For the BN-F system, there is no possibility of getting cBN from the BF_3-NF_3 system, because the BN: F ratio is fixed at 1: 6 (marked by $\times$) and comes into the gas region on the BN-F sideline of Fig. 9.2. It had been reported by Matsumoto et al. that "The formation of cubic BN was not confirmed" from the BF_3-NF_3 system even with strong activation of a plasma torch together with another r.f. plasma. If any kind of BN films from a "BF_3-NF_3" gas mixture were obtained, that was due to the reduction of the substrate (Si or Mo), or other reductants inside the reaction chamber. The BN-H-O phase diagram for the activated low-pressure cBN growth wasn't calculated here, although C-H-O phase diagrams are very important for activated low-pressure diamond growth. That is due to the formation of B_2O_3, if any kind of oxygen (including chamber leakage) comes into the reaction chamber. B_2O_3 could form, liquefy and cover the substrate to prevent the growth of cBN.

The existence of the cBN growth region in Figs. 9.1 and 9.2 shows that cBN has become a stable phase under activated low-pressure conditions, while hBN has been activated by superequilibrium atomic hydrogen (denoted by H*) [together with superequilibrium atomic chlorine (Cl*), or superequilibrium atomic fluorine (F*)] into a metastable phase. For instance, on the basis of thermodynamic coupling reactions, as shown in reaction (9.1) (denoted by subscript 1), reaction (9.2) (denoted by subscript 2), and reaction (9.3) (denoted by subscript 3) thermodynamic data for H* activated hBN, denoted by hBN*(H), could be calculated. The calculated data for hBN*(H) with the rate ratio of reaction (9.2) to reaction (9.1) (denoted by χ) are listed in Table 9.1.

$$\text{hBN} = \text{cBN}\ (1400\ \text{K}, 100\ \text{Pa})\ \Delta G_1 > 0 \tag{9.1}$$

$$\text{H}* = 0.5\ \text{H}_2\ (1400\ \text{K}, 100\ \text{Pa}, T_{\text{activation}} = 3000\ \text{K})\ \Delta G_2 < 0 \tag{9.2}$$

(9.3) = (9.1) + ((9.2),

$$\text{hBN} + \chi\ \text{H}* = 0.5\ \chi\ \text{H}_2 + \text{cBN}\ (1400\ \text{K}, 100\ \text{Pa}) \tag{9.3}$$

or

$$\text{hBN} * (\text{H}) = [\text{hBN} + \chi(\text{H}* - 0.5\ \text{H}_2)] = \text{cBN}\ (1400\ \text{K}, 100\ \text{Pa}) \tag{9.4}$$

$$\Delta G_3 = \Delta G_4 < 0,\ \text{ if } \chi \geq 0.28$$

Due to this similarity, thermodynamic data for Cl* and F* activated hBN will not be discussed and listed here. For simplicity, the fixed ratio of B/N = 1 is used during calculation. In practice, a certain excess of N-containing precursors (such as NH_3, NF_3, ...) is usually used.

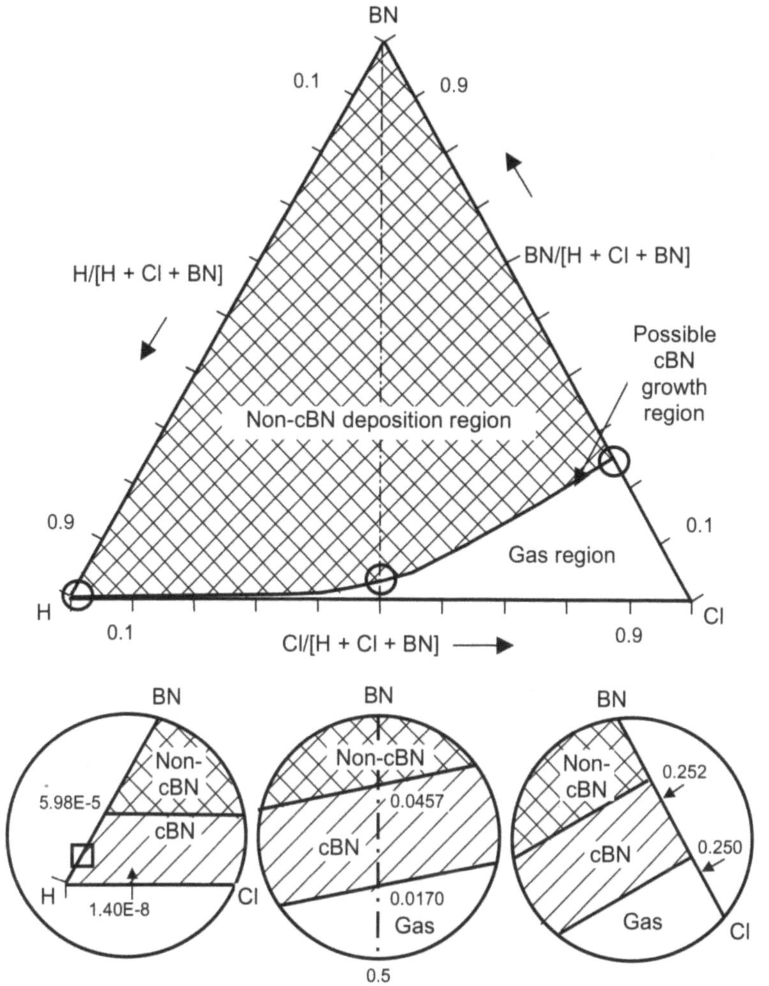

Fig. 9.1. Nonequilibrium phase diagram of BN-H-Cl system for activated CVD of cBN in the range $p = 10$–10^5 Pa and $T = 1200$–1500 K. The highest and lowest BN/(H+Cl+BN) values for cBN growth are shown amplified in the three small circles. The open square mark on the BN-H sideline is Song's experimental point [9.5] inside the theoretical range of BN/(H+BN) = 1.40×10^{-8}–5.98×10^{-5}

It should be noted that all other standard thermodynamic data could be found in ordinary data books, mainly from JANAF [9.7, 9.8], but the entropy (S) and constant pressure heat capacity (C_p) of cBN are taken from another reference (V. L. Solozhenko [9.9, 9.10]). Usually, there is no doubt about the standard Gibbs free energy of hBN (at 298 K and 10^5 Pa, $G^0_{\mathrm{hBN}} = -254.91\pm1.5$ kJ mol^{-1}), but there were different values for the standard Gibbs free energy of cBN (G^0_{cBN}). However, up to now cBN spontaneous crys-

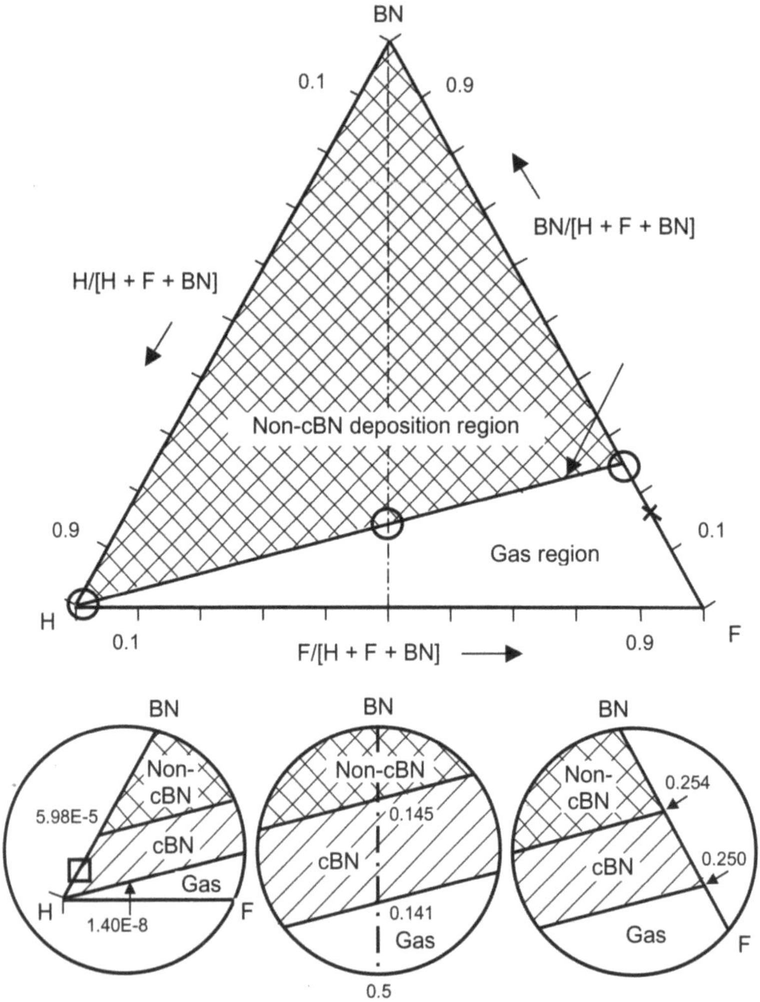

Fig. 9.2. Nonequilibrium phase diagram of BN-H-F system for activated CVD of cBN in the range $p = 10$–10^5 Pa and $T = 1200$–1500 K. The highest and lowest BN/(H+F+BN) values for cBN growth are shown amplified in the three small circles. The open square mark on the BN-H sideline is Song's experimental point [9.5] inside the theoretical range of BN/(H+BN) $= 1.40 \times 10^{-8}$–5.98×10^{-5} for cBN growth; but the mark $\times$ on the BN-F sideline is Matsumoto's unsuccessful experimental point [9.6]

Table 9.1. Thermodynamic data for hBN*(H) with $\chi = 0.28$

$H_{\mathrm{hBN*}}(\mathrm{H}) = -189.464$ kJ mol^{-1}		
$S_{\mathrm{hBN*}}(\mathrm{H}) = 29.143$ J mol^{-1} K^{-1}		
C_p (J mol^{-1} K^{-1})$= A + B10^{-3}T + C10^5T^{-2} + D10^{-6}T^2$		
1550 K	1550–1750 K	1750–3000 K
$A = 19.5121$	$A = 19.5648$	$A = 50.6139$
$B = 40.2709$	$B = 39.4659$	$B = -0.4578$
$C = -8.0777$	$C = -2.3693$	$C = 5.7920$
$D = -13.0917$	$D = -12.6859$	$D = 0.0301$

$T_{\mathrm{activation}}$: 3000 K, Ref. T: 298 K, p range: 10–10^5 Pa
Definitions of B, C and D here are a little different from those in Chap. 1

tallization has not been confirmed below 2.0 GPa. Suppose that the equilibrium pressure between hBN and cBN at 298 K is 1.5 GPa, i.e., the Gibbs free energy change of the phase transformation from hBN to cBN $\Delta G\,(p) = 0$ at 298 K and 1.5 GPa, based on the relation between pressure and the Gibbs free energy, $\left(\frac{\partial \Delta G}{\partial p}\right)_T$, one can get

$$\Delta G(p) - \Delta G(p^0) = \int_{p^0}^{p} \Delta V \mathrm{d}p = \Delta V(p - p^0) \tag{9.5}$$

$$= \left\{\left(\frac{24.83}{3.48} - \frac{24.83}{2.29}\right) \times 10^{-6}\mathrm{m^3mol^{-1}}\right\} \times \left(1.5 \times 10^6 - 101.325\right) \mathrm{kPa}$$
$$= -5.55 \text{ kJ mol}^{-1}.$$

In this way, the standard Gibbs free energy of cBN (G^0_{cBN}) at 298 K equals -249.36 kJ mol^{-1} here. In (9.5), 24.83, 3.48 g cm^{-3} and 2.29 g cm^{-3} are the molecular weight of BN, and the densities of cBN and hBN at 298.15 K, respectively. For χ it is known that in the activated cBN process a much stronger activation condition is needed than that in the activated diamond process, and that χ may be about 0.28 or bigger in the diamond process, so a lower limitation of $\chi = 0.28$ used for calculation here is reasonable.

9.2 Modern Thermodynamics for Belousov–Zhabotinsky Reactions

In the last 40 years the special phenomenon of chemical oscillation had been observed in many chemical reaction systems. The kinetic mechanisms of these reactions are usually very complex and they are difficult to find an explanation in classical thermodynamics. Among these special reactions Belousov–Zhabotinsky reactions (or B–Z reactions, for short) are one of the most interesting reaction systems. Such chemical oscillation reactions was first reported by Belousov in 1958. Later Zhabotinsky reported the ordered picture produced by these reactions. After that much research work was focused on it from the experimental and theoretical aspects. It was difficult to find an explanation from classical thermodynamics, but it is easy to be explained on the basis of modern thermodynamics as self-organization or dissipative structures (see Chap. 2).

For understanding the reaction system we discussed first a simple experiment of B–Z reactions [9.11]. In a beaker with a stirrer, dissolve 4.292 g malonic acid [$CH_2(COOH)_2$] and 0.175 g ammonium cerium (IV) nitrate [$Ce(NH_4)_2(NO_3)_6$] into 150 ml 1 M sulfuric acid. After a short time solution is clear, and 1.415 g sodium bromate ($NaBrO_3$) is added. After a few minutes, the color of the solution will oscillate between yellow and colorless. The period is about 1–2 minute. Such a phenomenon is called chemical oscillation, see Fig. 9.3.

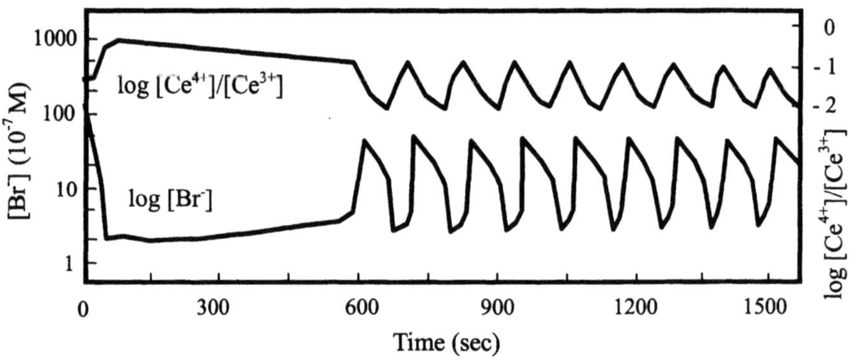

Fig. 9.3. An example of Belousov–Zhabotinsky chemical oscillation reactions. After R.-S. Li [9.11]

The kinetic mechanism of B–Z reactions is usually very complex. The well-accepted mechanism was proposed by Field, Kőrős and Noyes, so the mechanism can be called the Field–Kőrős–Noyes mechanism or simply the FKN mechanism. The FKN mechanism consists of about 10 steps or more. These steps had been summarized into the following three main reaction chains. Those are reaction chain A, reaction chain B and reaction chain C taking place in different intervals of time. The total effects of each reaction chain are as follows.

$$A\text{: } BrO_3^- + 2\ Br^- + 3\ CH_2(COOH)_2 + 3\ H^+ = 3\ BrCH(COOH)_2 + 3\ H_2O \qquad (9.6)$$

$$B\colon \mathrm{BrO_3^-} + 4\ \mathrm{Ce^{3+}} + 5\ \mathrm{H^+} = \mathrm{HOBr} + 4\ \mathrm{Ce^{4+}} + 2\ \mathrm{H_2O} \tag{9.7}$$

$$C\colon \mathrm{HOBr} + 4\ \mathrm{Ce^{4+}} + \mathrm{BrCH(COOH)_2} + \mathrm{H_2O} = 2\ \mathrm{Br^-} + 4\ \mathrm{Ce^{3+}} + 3\ \mathrm{CO_2} + 6\ \mathrm{H^+} \tag{9.8}$$

From viewpoint of thermodynamics [9.12], the whole reaction $W = (A+B+C)$ is an oxidation of an organic compound, which, of course, is an energy-providing or energy-consumption reaction:

$$W\colon 2\mathrm{BrO_3^-} + 3\ \mathrm{CH_2(COOH)_2} + 2\ \mathrm{H^+} = 2\ \mathrm{BrCH(COOH)_2} + 3\ \mathrm{CO_2} + 4\ \mathrm{H_2O} \tag{9.9}$$

The detailed kinetic reaction rate equations including self-catalysis effect must be a set of nonlinear equations for explanation of the chemical oscillation (see references). However, we can also take another simpler way to understanding thermodynamics of the chemical oscillation. All Gibbs free energy changes of these reactions vary with time during the chemical oscillation, because concentrations of all components change with time. For reaction oscillation these must be a set of forward reaction chains A and B together with a set of backward reaction chains $(-A)$ and $(-B)$ taking place in different intervals of time alternately. That is, after A and B $(-A)$ and $(-B)$ must occur and so on. The organic compound (malonic acid) and oxidant (sodium bromate) are in a large excess, so the concentration of these two reagents almost remain constant after each oscillation circle. From the point of view of reaction coupling in modern thermodynamics, reaction C is a coupled reaction consisting of W and $[(-A)+(-B)]$. Here, $(-A)$ and $(-B)$ are backward reactions of A and B, respectively. Roughly speaking, if $\Delta G_A < 0$, $\Delta G_B < 0$, $\Delta G_C < 0$ and $\Delta G_W < 0$ are taken, then $\Delta G_{[(-A)+(-B)]} > 0$, $\Delta G_W < 0$ and the coupled reaction $\Delta G_C < 0$. This strictly agrees with the criterion of reaction coupling in modern thermodynamics, i.e., $\Delta G_1 > 0$, $\Delta G_2 < 0$ and the overall coupled reaction $\Delta G_C = [\Delta G_1 + \Delta G_2] \leq 0$, if reaction $[(-A)+(-B)]$ and reaction W are taken as reaction 1 and reaction 2, respectively. There is nothing violating the second law of thermodynamics in chemical oscillations, in self-organization or in dissipative structures.

9.3 A Characteristic of Life: Drawing Negative Entropy from its Environment

Erwin Schrödinger in his book *What is life?* [9.13] wrote: "... a living organism continually ... produces positive entropy ... and thus tends to approach the dangerous state of maximum entropy, which is death. It can only keep ... alive by continually drawing from its environment negative entropy ... [The] essential thing in metabolism is that the organism succeeds in freeing itself from all the entropy it cannot help producing while alive." However, in these famous words there was no clear definition of "negative entropy". There was no discussion of how the living organism can continually draw negative entropy from its environment for keeping alive either. Therefore, I. N. Levine cited the words in his book *Physical Chemistry* [9.14] and said: "As the preceding analysis shows, there is no reason to believe that living organisms violate the second law. Of course, a quantitative demonstration of this (by measurement of $\Delta S_{syst} + \Delta S_{surr}$ for an organism and its surroundings) would be a difficult task. In the absence of such a quantitative demonstration, there have been occasional speculations that organisms violate the second law of thermodynamics." From these two citations it can also be seen that negative entropy is very important for the description of biological phenomena. Since then, sometimes roughly speaking: What is life? Life is negative entropy. In this way, the concept of negative entropy has also been clouded by a deep "mystique".

After the quantitative verification of reaction coupling in the activated low-pressure diamond process, the negative entropy production reaction was also demonstrated [9.15–9.18], because in the criterion of reaction coupling [i.e., $d_iS_1 < 0$, $d_iS_2 > 0$ and $d_iS \geq 0$] it was already confirmed that a negative entropy production of reaction 1 (also called a coupled reaction) can be compensated by another simultaneous positive entropy production of reaction 2 (also called a coupling reaction) in the same system. That is, the coupled reaction is a nonspontaneous reaction and cannot take place alone. The coupled reaction can go forward to the right side only due to another coupling reaction with positive entropy production coupled with it. For further understanding of both negative entropy production reactions in this book and Schrődinger's "negative entropy", let's go back to Figs. 8.1 and 8.2. If the hot filament method is only taken as an example and the part around the gas solid interface is only taken as the system under consideration, then an open system could be drawn as shown in Fig. 9.4a, or the mechanism of the overall reaction coupling can be drawn as shown in Fig. 9.4b in brief.

The overall reaction is the reaction between graphite and superequilibrium atomic hydrogen to produce diamond and hydrogen molecules. Figures 9.4, 8.1a and 8.2a are the system for the hot filament method of the activated diamond process with three different system boundaries selected. The common point is the reaction coupling taking place inside the same system. Correspondingly, no difficulty has been met during the application of the modern formulation of the second law. These systems quantitatively agree with the second law of thermodynamics.

It is known that the boundary of a system may be chosen somewhat arbitrarily. If the system in Fig. 9.4 is split into two smaller systems and the gas–solid interface is the boundary between two smaller systems, then the reaction coupling cannot be seen as an overall reaction taking place inside the same system, but at the gas–solid interface between the two systems.

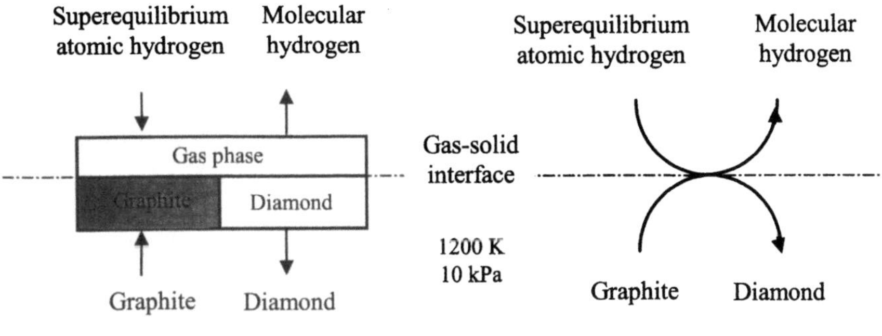

Fig. 9.4. (**a**) An open system around the interface and (**b**) mechanism of reaction coupling for activated low-pressure diamond growth from the vapor phase

Discussions on multiphase reaction coupling are rarely found in the literature. Quantitative calculations on multiphase reaction coupling have not been found either. In the following discussion the solid phase part in Fig. 9.5 is taken as system I, while the gas phase part is taken as system II, or all the others except system I are taken as the surroundings of system I.

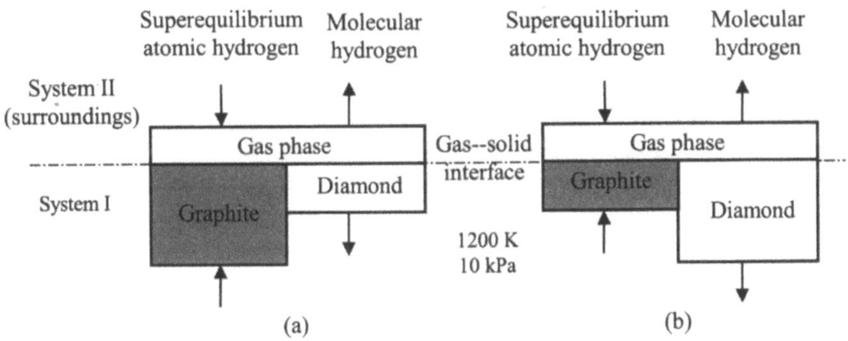

Fig. 9.5. The change of the solid system in the activated low-pressure diamond process from (**a**) the initial state to (**b**) the final state

The choice of the solid phase system in Fig. 9.5 is quite different from the choices of systems in Figs. 8.1a, 8.2a and 9.4. The system chosen in Fig. 9.5 is more naturally based on visual representation. In this way, the region in which the reaction coupling takes place is not inside system I, but at the boundary between system I and system II (or between system I and its surroundings), i.e., at the gas–solid interface in Fig. 9.5. None of the calculation results of the entropy flow d_eS and the entropy production d_iS for Figs. 8.1, 8.2 and 9.4 will be suitable here. According to the definition of entropy flow d_eS, i.e., that part of the entropy change due to the interaction between system I and its surroundings, the reaction coupling at the boundary (gas–solid interface) is only an entropy flow with respect to system I, while there is no irreversible process inside system I, i.e., $\Delta_i S^{\mathrm{I}} = 0$. Suppose that the total change from the initial state to the final state is 1 mole transformation from graphite to diamond. Now it is very easy to calculate the total entropy change of system I, that is, the total entropy change of system I, ΔS^{I} at 1200 K and 10 kPa, is the entropy difference between diamond and graphite, i.e., $\Delta S^{\mathrm{I}} = 23.734 - 28.506 = -4.772$ J mol^{-1} K^{-1} (at 10 kPa, 1200 K). Data for graphite and diamond and at 1200 K are taken from Tables 8.2 and 8.3 independent of pressure, respectively. According to $\Delta S = \Delta_e S + \Delta_i S$, it can be seen that $\Delta_e S^{\mathrm{I}} = -4.772$ J mol^{-1} K^{-1}, that is to say, in the activated low-pressure CVD diamond process, the solid phase (system I) successfully draws from its environment (gas phase) negative entropy to realize the negative entropy change from graphite to diamond. That may seem at first sight to quantitatively demonstrate that a negative entropy change of the system can be realized by a reaction coupling (not only by heat, work or exchange of matter) at the boundary of the system.

It is well known that a living organism is usually separated from its exterior by several kinds of membranes, and that a lot of biological coupling reactions are taking place on biological membranes (such as cell membranes, inner mitochondrial membranes, ...). Therefore, these biological coupled reactions on membranes can be treated as multiphase coupled reactions. Correspondingly, continuously drawing negative entropy from its environment is still possible and reasonable.

Life is really very complex. Incubation of eggs can also be taken as an example. Suppose there are two eggs in an incubator. One of the eggs is covered by an air-proof film before incubation. After incubation the covered egg is dead and decayed, while the other egg becomes a chicken, as shown in Fig. 9.6. Usually, incubation from an egg to a chicken is seen as a negative entropy process of ordering, while the decay of an egg is seen as a positive entropy process of disordering. In fact, the chicken has evolved from only the largest part of the egg, while the substance of the smaller part of the egg has been dissipated in the metabolism of the living egg. Meanwhile, the absorption of oxygen and the exhaustion of carbon dioxide take place continuously during incubation. Because the system discussed is not an isolated system, the classical entropy increase principle could not directly be used here. The concepts of positive entropy, "increase in entropy (surroundings)" and negative entropy or "decrease in entropy (system)" are still cloudy. Figure 9.6b is redrawn after a guide book of biochemistry written by P. van Eikeren [9.19].

From this it is known that there are a lot of thermodynamic coupling processes, especially reaction coupling (including multiphase reaction coupling in membranes) processes in biological systems. Negative entropy production processes can take place in the interior of a living body. Negative entropy

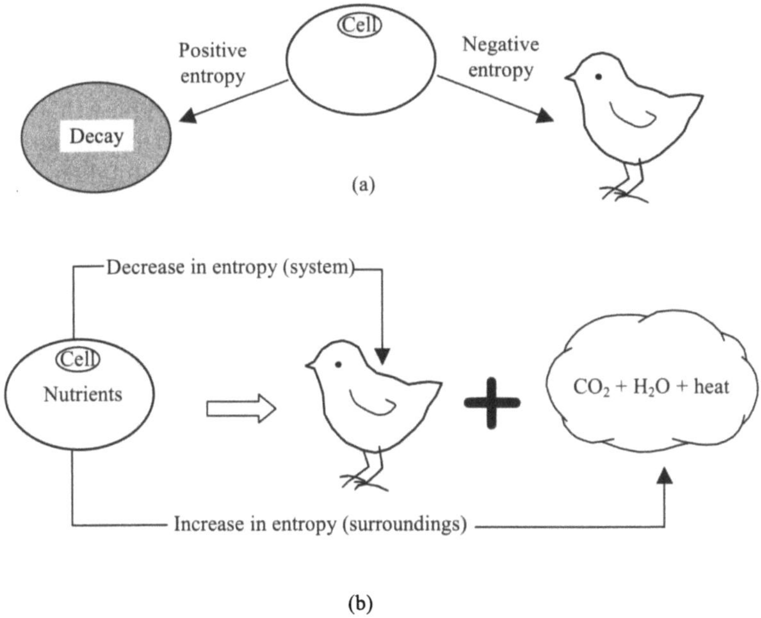

Fig. 9.6. (**a**) and (**b**) two popular viewpoints of entropy change during incubation of eggs. (b) after P. van Eikeren [9.19]

change processes of biological systems can also take place due to drawing from its environment negative entropy flow. The most important thing is that the different concepts of negative entropy production reaction, negative entropy flow and negative entropy change of system come from different sides of the reaction coupling (especial multiphase reaction coupling) process. The different terms are dependent on the choice of the boundary of the system. Therefore, on the basis of a deep understanding the general term of negative entropy can still be used. To sum up, when a living body draws from its environment negative entropy more than the positive entropy produced for keeping alive, then the living body increases its ordering during its growth and development stage. When a living body draws from its environment negative entropy equal to the positive entropy produced for keeping alive, then the living body is in its steady and mature stage. When it draws from its environment negative entropy less than the positive entropy produced for keeping alive, then the living body is in the decline of life. Of course, when it has never drawn from its environment negative entropy, that is death. That is, one of the characteristics of life is continually drawing from its environment negative entropy, but that should not be the peculiar character of life. Some inanimate systems, which can also continually draw from their environment negative entropy, may not be alive. The solid-phase system of activated low-pressure diamond growth is an example of that. Sometimes briefly speaking, life means negative entropy (negative entropy change of system, or negative entropy production of reaction or process), but by contrast negative entropy may not be life.

9.4 Some Similarities of Reaction Coupling in Biological and Inanimate Systems

In bioenergetics ATP (adenosine triphosphate) plays a very important role in living bodies. The function of ATP is to store the available energy when food is metabolized for supplying energy on demand to a wide variety of processes, including muscular contraction, reproduction, and vision. A popular view is that ATP is a special energy source that can be directly used, so that ATP is seen as a "gasoline fuel". Without gasoline a car cannot be driven. ATP in the human body is synthesized from ADP (adenosine diphosphate) and inorganic phosphoric acid (Pi). The synthesis of ATP from ADP is not a spontaneous process. The synthesis reaction can progress only under conditions accompanied by the input of other energy. Human cells contain hundreds of subcellular organelles called mitochondria. These organelles are approximately the size of small bacteria. Their double membrane encloses a matrix and a convoluted structure of inner membranes that contains a complex array of enzymes and proteins. The major function of mitochondria is to convert food energy to the chemical energy of the cell (ATP). Most ATP is synthesized in mitochondria by a series of reactions that are driven by the energy released during metabolic oxidation. That is called oxidative phosphorylation. The mechanisms of oxidative phosphorylation are already rather clearly understood. The key effect of oxidative phosphorylation is the proton (H^+) gradient across the inner membrane and the special structure of the inner membranes. A chemiosmotic hypothesis can explain how the free energy generated by the electron transport chain is used to produce ATP from ADP + Pi. The members of the electron transport chain are arrayed asymmetrically in the inner mitochondrial membrane, forming loops, as illustrated in Fig. 9.7. The chemiosmotic hypothesis states that these loops consist of alternating hydrogen atoms and electron carriers that transfer protons H^+ from the mitochondrial matrix (inside) to the cytoplasmic side (outside) of the inner mitochondrial membrane. In other words, the electron transport chain is an active proton transport system. During the oxidation protons are pumped from the matrix side to the cytoplasmic side of the inner mitochondrial membrane. This process creates across the inner mitochondrial membrane a H^+ concentration gradient (the H^+ concentration is higher on the cytoplasmic side, i.e., outside), and an electric potential with that side positive is generated, see Fig. 9.8.

The energy inherent in this gradient is sufficient to drive ATP synthesis. The free energy inherent in the concentration gradient of the proton is called

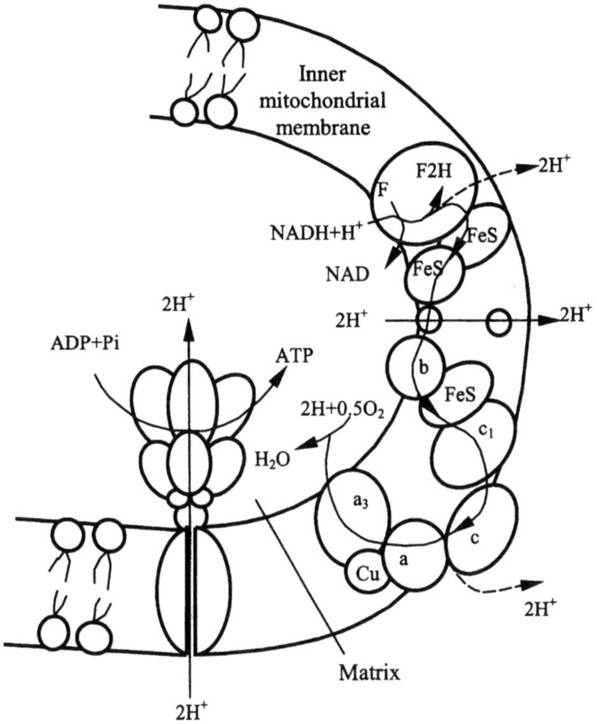

Fig. 9.7. The electron transport chain in the inner mitochondrial membrane and mechanisms of oxidative phosphorylation

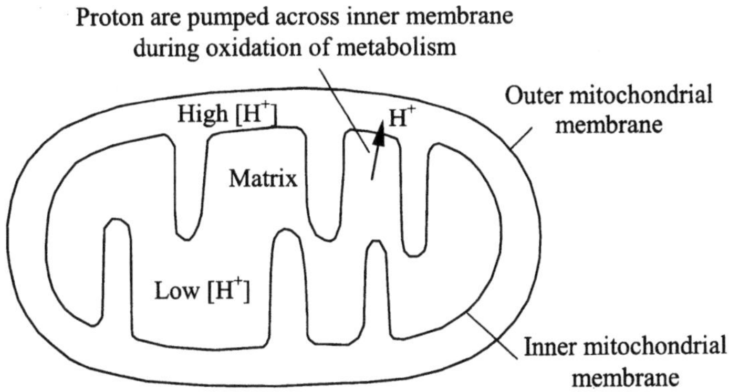

Fig. 9.8. Structure of mitochondrion and the proton (H^+) gradient across the inner membrane

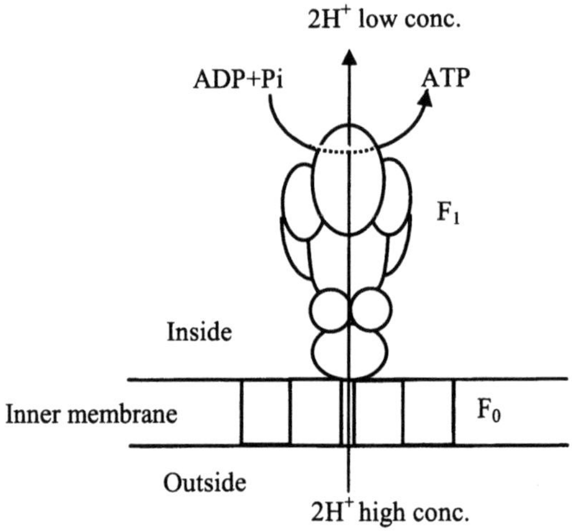

Fig. 9.9. H^+ concentration gradient across the inner mitochondrial membrane drives ATP synthesis. F_1/F_2-ATPase is the enzyme complex that synthesizes ATP

the driving force of the protons. When protons reenter the mitochondrial matrix (inside of the inner mitochondrial membrane) then pass through a channel in the F_1/F_2-ATPase molecule, resulting at the same time in the synthesis of ATP from ADP + Pi, see Fig. 9.9. Many experimental results confirm these mechanisms. For instance, once the completeness of the inner mitochondrial membrane is destroyed, ATP will not be able to synthesize. As another example, the existence of 2,4-dinitro-phenol (a reagent) allows free passage of protons through the inner mitochondrial membrane, and the concentration gradient of the protons disappears. That results in the inhibition of ATP synthesis. The most powerful evidence is the ATP synthesis realized by a man-made concentration gradient, and therefore this mechanism became generally acknowledged. The chemiosmotic hypothesis was proposed by the English biochemist Peter Mitchell in 1961 [9.20]. Other scientists hardly believed it at that time. After many years scientists searched diligently, but unsuccessfully, to find other explanations, and at last Mitchell got the Nobel Prize in 1978. Further detailed description of oxidative phosphorylation and other bioprocesses can be found in many books on biochemistry.

However, it is interesting that there are some similarities of multiphase thermodynamic coupling between biological systems and inanimate systems [9.18]. Figure 9.10a illustrates Mitchell's chemiosmotic hypothesis for the biosynthesis of ATP. Figure 9.10b illustrates the author's coupling explanation for Spitsyn's high (temperature) gradient chemical transport reaction (HGCTR) synthetic diamond process (see also Fig. 4.11). In Fig. 9.10a ATP synthesis is coupled by proton flow across the inner mitochondrial membrane, while in Fig. 9.10b diamond synthesis is coupled by heat flow. The mechanisms or principles of reaction coupling are thermodynamically similar. Both of them are a negative entropy production reaction coupled with a positive entropy production process together with positive total entropy production. None of them violates the second law of thermodynamics. In Fig. 9.10a the proton flow from high concentration to low concentration across the membrane is a positive entropy production process, while in Fig. 9.10b the heat flow from high temperature to low temperature across the gas phase layer is a positive entropy production process. Both of them can be seen as multiphase coupled reactions. Therefore, drawing from the environment negative entropy to achieve a negative entropy change of the system by a multiphase reaction coupling completely agrees with the second law of thermodynamics.

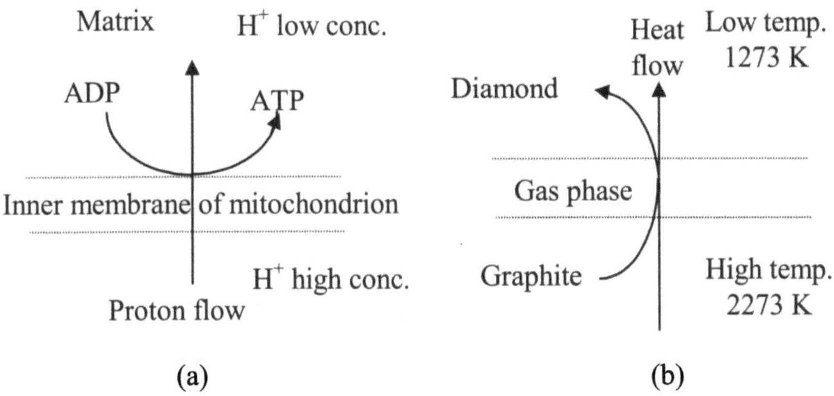

Fig. 9.10. Illustration of the similarity between reaction coupling mechanisms in biological and inanimate systems. (**a**) Mitchell's chemiosmotic hypothesis for ATP biosysthesis. (**b**) Coupling model for Spitsyn's HGCTR diamond process

For another example, human brain cells have a high concentration of K^+ and a low concentration of Na^+ relative to the external medium. These ionic gradients are generated by a special transport system that is called the Na^+, K^+-ATP pump because the movement of these ions is linked. The pump is a special protein present in the cell membrane. The active transport of Na^+ and K^+ ions is of great physiological significance. There is also a kind of valve-like channel. If a signal stimulates the brain cell, the valve-like channel suddenly opens. The inside K^+ ions flow toward the outside of the cell membrane, while the outside Na^+ ions flow toward the inside. Thus the brain cell is excited, because the electrolyte is out of balance. The Na^+, K^+-ATP pump starts working automatically. The inside Na^+ ions are pumped opposite to the concentration gradient to the outside of the cell membrane, while the outside K^+ ions are pumped to the inside. The active transport of Na^+ and K^+ ions is driven by the energy released during hydrolysis from ATP to ADP. Once the normal state with a higher concentration of K^+ inside the cell and a higher concentration of Na^+ outside the cell is recovered, the Na^+, K^+-

ATP pump stops working. Correspondingly, the brain cell comes back to its normal quiet state. All these pump effects are very similar to the chemical pump effect (coupling effect) in activated low-pressure diamond synthesis discussed in the previous chapters.

9.5 Some Conclusions, Further Development and Prospects

Now some conclusions and comments about further development and prospects can be clearly obtained as follows [9.3, 9.21–9.32].

1. Importance of reaction coupling in modern thermodynamics
Reaction coupling plays a very important role in modern thermodynamics. As mentioned in Chap.2, "the term of reaction may sometime be of extended meanings". That is, the term of reaction coupling may include process coupling, such as thermodiffusion. Reaction coupling is concerned with not only the author's nonequilibrium phase diagrams but also the Onsager's reciprocity relation and the Prigogine's dissipative structures. Thermodiffusion is a typical and well-known example of linear dissipative thermodynamics, in which the Onsager's reciprocity relation and the Prigogine's entropy production minimization principle are valid. Thermodiffusion is also a typical example of reaction or process coupling between spontaneous heat flow (from high temperature to low temperature) and nonspontaneous diffusion (from homogeneous to inhomogeneous), as discussed in Chap.2. Chemical oscillation is an example of dissipative structures, and is also a typical reaction coupling between spontaneous and nonspontaneous reactions, as discussed in this chapter. Confusions or mistakes on reaction coupling had seriously hindered the completion of the systematization of modern thermodynamics and the general acceptance of modern thermodynamics.

2. Coupling systems and noncoupling systems
The definitions of "coupling" and "noncoupling" systems here are systems including and not including reaction coupling, respectively. Research objects of classical thermodynamics are noncoupling systems, while those of modern thermodynamics can be extended to any system including coupling and noncoupling systems. If classical thermodynamics is unsuitable to be used for coupling systems, then the results and conclusions should be checked and corrected by modern thermodynamics, such as examples for activated low-pressure diamond synthesis, and many thermodynamic problems in the life sciences. All results obtained from modern thermodynamics must be the same as those obtained from classical thermodynamics for noncoupling systems. The contents of classical thermodynamics in textbooks of thermodynamics, physical chemistry for undergraduate and graduate students are still correct,

if it is pointed out that noncoupling systems only are concerned, i.e., "within the framework of the application of its basic concepts" as said by Einstain. However, some basic concepts of modern thermodynamics or part of them, such as written in Chap. 2 of this book, should be introduced for new generations of scientists as soon as possible.

3. Different stages of thermodynamics

Both classical thermodynamics and modern thermodynamics are thermodynamics in its different stages of its development. In other words, the framework of the application of classical thermodynamics and modern thermodynamics are different. The framework of the application of classical thermodynamics is much narrower than that of modern thermodynamics. In general, modern thermodynamics does not contradict classical thermodynamics, but extends the framework of the application of the basic concepts of thermodynamics. One can say that classical thermodynamics is only a special case of modern thermodynamics. If in concerned systems driving forces $[(-\Delta G)_{T,p}]$ of all reaction (or processes) are positive or zero, i.e., there is no reaction coupling, modern thermodynamics is degraded into classical thermodynamics.

4. Systematization of modern thermodynamics

In the Preface and in Fig. 9.11 here, the systematization of modern thermodynamics based on Gibbs free energy changes is listed [9.3,9.31,9.32]. Maybe it will be easier for most scientists to accept it at the present time, because many of them are not familiar with the term "entropy production". However, in classical thermodynamics the Gibbs free energy change can be used as a criterion only for isobaric and isothermal closed systems. For multicomponent systems the Gibbs free energy change can be expressed as (1.60), i.e.,

$$\mathrm{d}G = -S\mathrm{d}T + V\mathrm{d}p + \sum_i \mu_i \mathrm{d}n_i$$

The term of $\sum_i \mu_i dn_i$ can be further divided into two parts, i.e.,

$$\mathrm{d}G = -S\mathrm{d}T + V\mathrm{d}p + \sum_i \mu_i \mathrm{d}n_i + \sum_i \nu_i \mu_i \mathrm{d}\xi \tag{9.10}$$

where, μ_i, n_i, ν_i and ξ are chemical potential, mole number, chemical stoichiometric coefficient of component i, and the extent of reaction, respectively. The third term of (9.10) is the Gibbs free energy change due to matter flow between system and its surroundings, while the fourth term of (9.10) is due to chemical reaction inside the system. Therefore, if the third term is deleted, then Gibbs free energy change, $(\mathrm{d}G)_{T,p}$ can still be used as a criterion of irreversibility for open systems. However, such a systematization of modern thermodynamics, as shown in Preface and in Fig. 9.11 here can only be used for isothermal and isobaric systems. Therefore, in the main part of the book the general systematization of modern thermodynamics, which can be used for any system (i.e., any isolated, closed, or open system) as shown in Fig. 2.2, is illustrated on the basis of the entropy production expression [9.23, 9.24, 9.27, 9.28]. The importance of entropy production in modern thermodynamics is something similar to entropy function in classical thermodynamics.

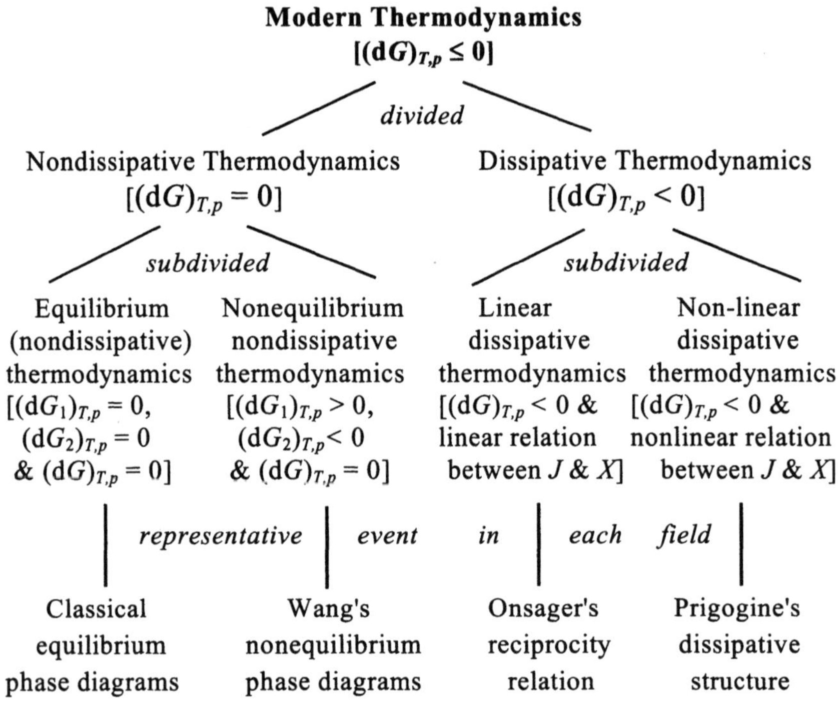

Fig. 9.11. Systematization of modern thermodynamics for isothermal and isobaric systems

Note that a more complete systematization of modern thermodynamics may be obtained by the following three steps.

1. The first order division is based on the second law of thermodynamics. This is the same as mentioned above and in Chap. 2.
2. The second order division is based on reaction coupling. After the second order division nondissipative thermodynamics is divided into equilibrium nondissipative thermodynamics and nonequilibrium nondissipative thermodynamics. This is also the same as above and in Chap. 2. However, after the second order division dissipative thermodynamics is divided into classsical nonequilibrium thermodynamics and (modern) dissipative thermodynamics.
3. The third order division is based on linearity, and the third order division is only used for modern dissipative thermodynamics. After the third order division modern

dissipative thermodynamics is divide into linear dissipative thermodynamics and nonlinear dissipative thermodynamics.

However, classical nonequilibrium thermodynamics is less important than other subfields. Classical nonequilibrium thermodynamics may only be regarded as a part of dissipative thermodynamics, and is not listed separatively as a subfield, as done in Preface, Fig. 2.2 and Fig. 9.11.

5. Limitation of classical thermodynamics

Now it is obvious that the framework of the application of basic concepts of classical thermodynamics is limited for noncoupling systems. That is the limitation of classical thermodynamics. The sencond order subdivision in systematization of modern thermodynamics are mainly based on rection coupling. Both Osanger's reciprocity relation and Prigogine's dissipative structure are concerned with reaction coupling. If noncoupling systems are only under consideration, the second order subdivision is not necessary at all. The systematization of modern thermodynamics, as shown in Fig. 9.11, will automatically be degraded into the systematization of classical thermodynamics. The nomenclature of equilibrium thermodynamics and nonequilibrium thermodynamics can still be used, as shown in Fig. 9.12. Now it is also very clear that the systematization of traditional thermodynamics (equilibrium thermodynamics, linear nonequilibrium thermodynamics and nonlinear nonequilibrium thermodynamics), as shown in Fig. 2.1 is an incomplete systematization of modern thermodynamics. In comparison between Fig. 9.12 and Fig. 2.2, one can find that research in modern thermodynamics is deep into the individual reaction level of reaction coupling, and research in classical thermodynamics is mainly at the global system level. If there is no reaction coupling, multireaction systems can be treated by classical thermodynamics, as done in equilibrium phase diagram calculation. However, all these multireactions are treated as a whole toward the spontaneous direction. That is the reason why chemical oscillations, activated low-pressure CVD diamond processes and many processes (such as ATP biosynthesis) in the life science had met skepticism on the basis of classical thermodynamics.

6. A sufficient and necessary condition for an equilibrium system

After the second-order subdivision in Fig. 9.11 equilibrium nondissipative thermodynamics can still be called equilibrium thermodynamics, because even in the new systematization for any system it is still the unique field really concerned with equilibrium states. However, the criterion of equilibrium thermodynamics has already change from only $[(\mathrm{d}G)_{T,p} = 0]$ for the system into $[(\mathrm{d}G_1)_{T,p} = 0$, $(\mathrm{d}G_2)_{T,p} = 0$ and $(\mathrm{d}G)_{T,p} = 0]$ for system and all reactions. The sufficient and necessary condition for a system being in equilibrium is not only the equality of the second law of thermodynamics but also all driving forces of reactions inside the system equal to zero. Therefore, in the first-order division the equality part of the second law should no longer

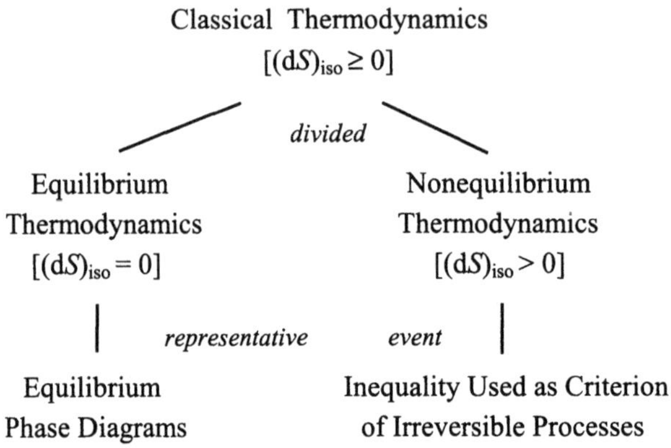

Fig. 9.12. Systematization of classical thermodynamics for isolated systems

be called equilibrium thermodynamics for any system including coupling and noncoupling systems, and the inequality part of the second law should not be called nonequilibrium thermodynamics, either. Maybe it is difficult to think that such a basic concept in classical thermodynamics should be changed, but it is necessary to do so for the new developments out of the framework of applications of classical thermodynamics. More basically speaking, the equality of the second law is not a sufficient and necessary condition for a system being in equilibrium, however an incorrect basic concept in classical and traditional thermodynamics had lasted for about 150 years.

7. Equilibrium thermodynamics and nonequilibrium thermodynamics

The systematization of modern thermodynamics is mainly on the basis of the second law of thermodynamics. If one prefers to make a classification on the basis of the assumption of local equilibrium, then one could say only equilibrium nondissipative thermodynamics (or classical equilibrium thermodynamics) belongs to equilibrium thermodynamics without the assumption; all other three fields (i.e., nonequilibrium nondissipative thermodynamics, linear dissipative thermodynamics and nonlinear dissipative thermodynamics) belong to nonequilibrium thermodynamics with the necessity of the assumption of local equilibrium. However, such a classification in modern thermodynamics is not based on the second law of thermodynamics — a basic law of thermodynamics, and thus it is less meaningful.

8. Nonequilibrium phase diagrams

An important characteristic of nonequilibrium nondissipative thermodynamics is its relation to the equality of the second law of thermodynamics. That makes possible the calculation of phase diagrams (CALPHAD) for both equilibrium nondissipative thermodynamics and nonequilibrium nondissipative thermodynamics, as shown in Fig. 9.13 [9.23, 9.25, 9.29, 9.30]. equilibrium nondissipative thermodynamics results in equilibrium phase diagrams, while nonequilibrium nondissipative thermodynamics results in nonequilibrium phase diagrams. As mentioned in Chap. 5, it should be keep in mind that the difference between equilibrium phase diagrams and nonequilibrium phase diagrams is different from that between stable phase diagrams and metastable phase diagrams. No confusion should arise between these differences. In particular, it should be careful to distinguish nonequilibrium phase diagrams from metastable equilibrium phase diagrams. In any corresponding system of nonequilibrium phase diagrams there is at least one macroscopic process, while in the corresponding system of metastable equilibrium phase diagrams there is no macroscopic process and there is a lack of stable phase or there is a hindrance for a chemical reaction.

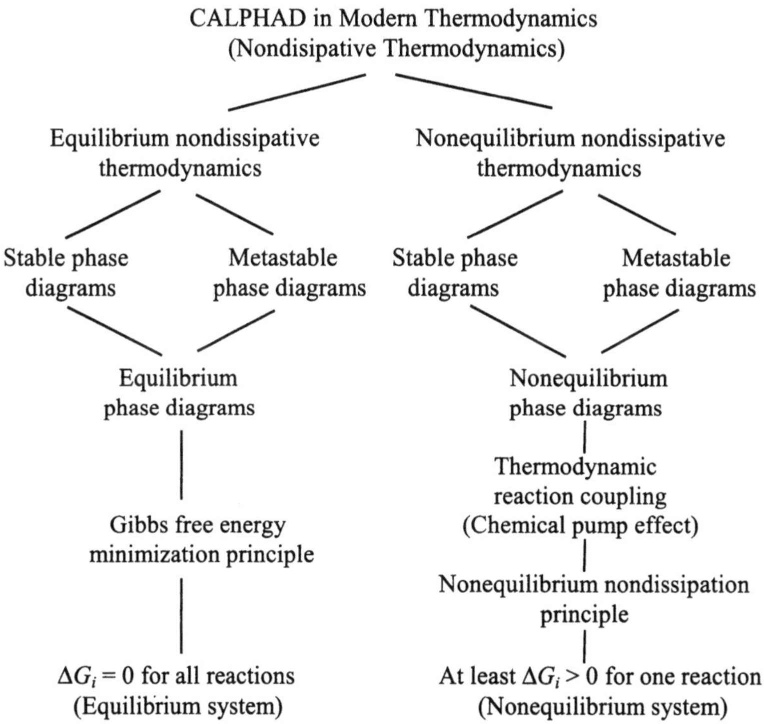

Fig. 9.13. An overview of the calculation of phase diagrams (CALPHAD) in modern thermodynamics

9. Nonequilibrium nondissipation principle

After quantitative verification of reaction coupling in activated low-pressure diamond synthesis, nonequilibrium phase diagrams can be calculated on the basis of nonequilibrium nondissipation principle, i.e., the criterion of nonequilibrium nondissipative thermodynamics [such as $(\mathrm{d}G_1)_{T,p}$ ¿ 0, $(\mathrm{d}G_2)_{T,p}$ ¡ 0 and $(\mathrm{d}G)_{T,p} = 0$]. The equality part of reaction coupling is also an extremum of reaction coupling, but not a practical situation for running reaction coupling. Any practical reaction coupling must have at least a small negative value of $(\mathrm{d}G)_{T,p}$. However, extremum in thermodynamics is very important for calculations of both equilibrium phase diagrams and nonequilibrium phase diagram. The nonequilibrium nondissipation principle may also be regarded as a special case [such as $(\mathrm{d}G)_{T,p} = 0$] of the Prigogine's entropy production minimization principle.

10. Application of nonequilibrium phase diagrams
The calculation of nonequilibrium phase diagrams, which this book has discussed in detail, is one of the most important applications in the new field of nonequilibrium nondissipative thermodynamics. It is well known that classical equilibrium phase diagram is a very active field in equilibrium thermodynamics. Equilibrium phase diagrams play an important role in fields of physics, chemistry, materials science, metallurgy, and so on. In particular, in metallurgy the discipline of metallography has formed. In research and development of salt lake products, ceramics, superconductors and semiconductors, phase diagrams are always needed. All of these systems are treated as equilibrium systems, even though the change rate in an equilibrium system approaches zero. However, activated low-pressure diamond growth together with simultaneous graphite etching can happen only under typical nonequilibrium conditions, i.e. the existence of superequilibrium atomic hydrogen. The equilibrium phase diagram of carbon or a carbon–hydrogen system should no longer be used. That has resulted in the appearance of nonequilibrium nondissipative thermodynamics and a series of nonequilibrium phase diagrams for activated low-pressure diamond (or c-BN) growth from the vapor phase. Nonequilibrium phase diagrams not only excellently agree with experimental data but also provide a quantitative theoretical guide for designing, optimizing and improving the practical processes for diamond (or c-BN) synthesis. Astronomers have observed diamond dust in space for a long time [9.33], but they found it difficult to give a theoretical explanation. In fact, all the conditions, such as hydrocarbons, hydrogen, plasma, high-energy particles, or cosmic rays generally exist in space. In particular, under weightless conditions in space it may be very favorable to grow big diamond crystals. It should be possible in someday to grow or catch diamond crystals in space. It is also predicted that nonequilibrium phase diagrams may be used for the synthesis of carbon nitride, fullerenes, carbon nanotube and other new materials.

11. Prospects in life science
As discussed in Sects. 9.3 and 9.4 the life sciences (including biology, biochem-

istry, bioenergetics, ...) should be one of the most potential and important applications of nonequilibrium nondissipative thermodynamics and other dissipative thermodynamics, because systems concerned in the life sciences are mainly coupling systems, in which many nonspontaneous reactions or processes are included. However, there is a severe handicap due to the confusion between the concepts of the standard Gibbs free energy change ΔG^0 and the practical Gibbs free energy change ΔG in most present textbooks in biology, biochemistry and bioenergetics. In these textbooks the incorrect criterion of reaction coupling, i.e., $[(\mathrm{d}G_1^0)_{T,p}$ ¿ 0, $(\mathrm{d}G_2^0)_{T,p}$ ¡ 0 and $(\mathrm{d}G^0)_{T,p} \leq 0]$, was used instead of the correct criterion $[(\mathrm{d}G_1)_{T,p}$ ¿ 0, $(\mathrm{d}G_2)_{T,p}$ ¡ 0 and $(\mathrm{d}G)_{T,p} \leq 0]$. For instance, in some textbooks this was explained on the "assumption" that "Usually, ΔG for biologic reactions has the same sign as ΔG^0, and the two often do not differ greatly in value." [9.34] or that "the actual free-energy change under cellular conditions is about the same as this (the standard free-energy change)" [9.35]. However, such an "assumption" is still incorrect for most biological reactions, in particular, for shuttle reactions in living cells. Figure 9.14 shows the mechanism of the malic acid – oxaloacetic acid shuttle reaction system taking place via α-ketoglutaric acid and malic acid through the inner mitochondrial membrane [9.36].

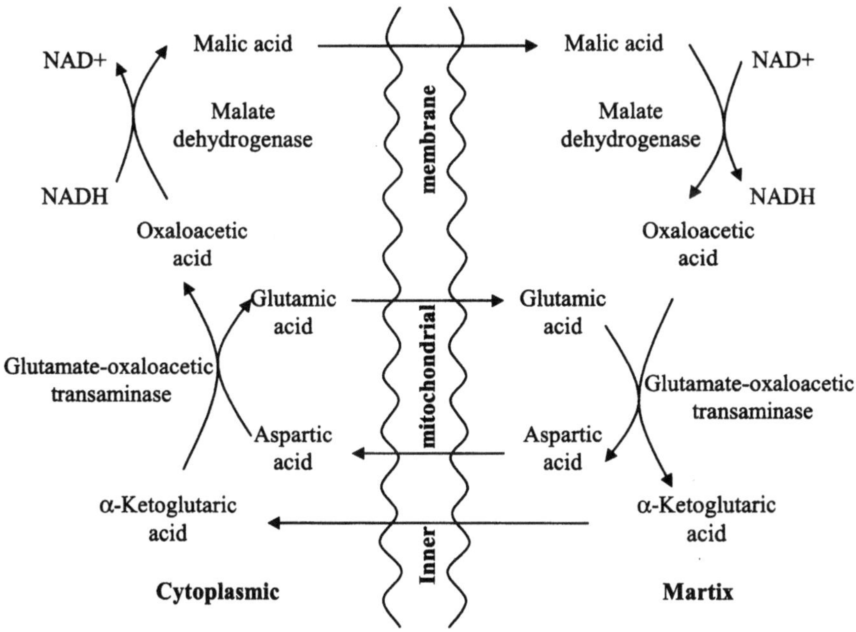

Fig. 9.14. An example of the mechanism for a shuttle reaction system in a mitochondrion. After R.-Q. Shen [9.36]

The directions of the reactions taking place on both sides of the inner mitochondrial membrane are just opposite to each other. That must be due to the different practical concentrations of these components on the different sides, and therefore the different signs and different values of ΔG must be connected with these opposite reactions. If the "criterion" in standard Gibbs free energy change form (ΔG^0) were used, all values and signs for these opposite reactions would be the same and independent of any practical concentrations, and how could even a qualitative explanation for shuttle reactions in opposite directions be found?

12. Modern thermodynamics in social science and philosophy

Modern thermodynamics in macroscopic region (but not for the Universe) may be "the only physical theory of universal content", as Einstein said, so it should be used not only for the natural sciences but also for the social sciences and even philosophy. Many economic and social activities are subordinate to statistical rules and therefore also to thermodynamic rules. Any substantial progress in thermodynamics might affect social science. For instance, an equilibrium state has been regarded as an ideal state for getting the highest efficiency for energy transformation, but in any equilibrium state there is no macroscopic process run. Based on modern thermodynamics mentioned in this book, the nondissipative state is of more extended meaning than the equilibrium state. In an ideal nonequilibrium nondissipative state, macroscopic process(es) might still be going on without any free energy dissipation, i.e., with the highest efficiency. If such a basic new idea were widely accepted that will promotes human society developing with higher efficiency.

Finally, as we know the entropy increase principle is a basic principle of classical thermodynamics, so classical thermodynamics may be called thermodynamics of disordering. In a narrow sense of modern thermodynamics it mainly deals with negative entropy production process(es), so modern thermodynamics may be called thermodynamics of ordering. This is also the reason why modern thermodynamics is more important for biological evolution, social development and emerging science and technology. To summarize this book in one sentence, the book is devoted to the development of the second law of thermodynamics and promote understanding of the second law of thermodynamics to the deepest level, which had never been reached before.

References

Chapter 1

1.1 J.-T. Wang, D. W. Zhang, Z.-J. Liu: *Thermodynamic Coupling Model for Low Pressure Diamond Growth from the Vapor Phase, 175 pages* (Science Press, Beijing 1998 and reprinted in 2000) p. 43
1.2 J.-T. Wang: *Phase Diagrams of Stationary Nonequilibrium States Thermodynamics for Low Pressure Diamond Growth from the Vapor Phase, 212 pages* (Science Press, Beijing, 2000) p. 8
1.3 D. Kondepudi, I. Prigogine: *Modern Thermodynamics — From Heat Engines to Dissipative Structures*(John Wiley & Sons, New York 1998) p.3, also for Preface

Further Reading

A. Einstein: *The World As I See It* translated by A. Harris (The Wisdom Library, A Division of Philosophical Library, New York 1949) reference for preface
X.-C. Fu, W.-X. Shen, T.-Y. Yao: *Physical Chemistry*, 4th edn. (High Education Press, Beijing 1990)
D. Kondepudi, I. Prigogine: *Modern Thermodynamics — From Heat Engines to Dissipative Structures*(John Wiley & Sons, New York 1998)
K. Wark: *Thermodynamics*, 4th edn.(McGraw-Hill Book Co., New York 1983)

Chapter 2

2.1 J. H. Keenan: *Notes for Special Summer Program in Modern Thermodynamics, MIT June 29–July 17* (Technology Press of MIT, Cambridge 1959)
2.2 D. Kondepudi, I. Prigogine: *Modern Thermodynamics — From Heat Engines to Dissipative Structures* (John Wiley & Sons, New York 1998) http://www.wiley.com/
2.3 I. Prigogine: *Introduction to Thermodynamics of Irreversible Processes* 3rd edn.(Interscience Publishers, John Wiley & Sons, New York 1967)
2.4 R.-S. Li: *Nonequilibrium Thermodynamics and Dissipative Structure* (Tsinghua University Press, Beijing 1986)
2.5 X.-C. Fu, W.-X. Shen, T.-Y. Yao: *Physical Chemistry*, 4th edn. (High Education Press, Beijing 1990) p. 172
2.6 J.-T. Wang: Physics **29**(9), 524–530 (2000)

2.7 J.-T. Wang: 'Modern thermodynamics in CVD of hard materials. *The Science of Hard Materials, 7th International Conference, at Ixtapa, Mexico, March 5-9, 2001*, (PDVSA-Intevep, Venezuela, 2001) pp. 105–106; J.-T. Wang, J.-Y. Shen, D.W. Zhang: J. Refractory Metals and Hard Materials, 19(4-6), 461–466 (2001)
2.8 Th. De Donder, P. Van Rysselberghe: *Thermodynamic Theory of Affinity — A book of principles*, 2nd edn. (Stanford University Press, Stanford 1936)
2.9 D. Burk: J. Phys. Chem. **35**, 432 (1931)
2.10 J.-T. Wang: 'Modern thermodynamics for phase diagrams of stationary states. In: *Phase Diagram Calculation (CALPHAD XXIX) (Abstracts), 29th International Conference at MIT, Cambridge, Massachusetts, USA, June 18–23, 2000* p. 8
2.11 J.-T. Wang: 'Systematization of modern thermodynamics'. In: *Chemical Thermodynamics (ICCT-2000), abstracts of 16th IUPAC Internat Conf at Halifax, Canada, Aug 6–11, 2000* p. 150
2.12 J.-T. Wang: 'Modern thermodynamics for advanced materials research. *Advanced Materials, vol. 1, International Conference on Engineering & Technical Science 2000, Session 3, at Beijing, China, Oct. 2000*, ed. by J. Song, R.-Y. Yin (New World Press, Beijing, 2000) pp. 864–870
2.13 J.-T. Wang: 'Modern thermodynamics for activated chemical vapor deposition (CVD) processes. *Chemical Vapor Deposition (extended abstract), 2nd Asian Conference, at Gyeongju, Korea, May 28-30, 2001*, (Ponhang University of Science and Technology, Korea, 2001) pp. 63–66
2.14 J.-T. Wang: 'Nonequilibrium nondissipative thermodynamics – and its application to diamond film deposition. *Solid-State and Integrated-Circuit Technology, 6th International Conference, at Shanghai, China, October 22–25, 2001*, (IEEE Press, People's Posts and Telecommunication Publishing House, Beijing, 2001) vol.2, pp. 1421–1426
2.15 J.-F. Deng, K.-N. Fan: *Physical Chemistry* (High Education Press, Beijing 1993)
2.16 I. Prigogine: Acad. Roy. Belg., Bull. Classe Sci. **31**, 600 (1945)
2.17 S. Chandrasekhar: *Hydrodynamic and Hydromagnetic Stability* 3rd press (Dover Publication, Inc., New York 1981; Clarendon Press, Oxford 1961) Chap. II, pp.10–12
2.18 R. Mayer-Spasche: *Pattern Formation in Viscous Flows* (Birkhauser Verlag, Switzerland 1999) Fig.4.1 on page 149.
2.19 M.D. van Dyke: *An Album of Fluid Motion* (The Parabolic Press, Stanford 1982) Fig.142.
2.20 E.L. Koschmieder, Adv. Chem. Phys., **26**, 177 (1974)
2.21 X.-F. Shen, G. Hu, L. Jiang: *Theory of Dissipative Structures* (Shanghai people's Publishing House, Shanghai 1987) pp. 36–45

Chapter 3

3.1 Shi-Zhen Li: *Encyclopedia of Herb & Medicines* (published in about 1594, reprinted by People's Health Press, Beijing, 1957) p. 633
3.2 X.-S. Lu: 'Preface'. In: *Crystal Growth.* ed. by K.-C. Zhang, L.-H. Zhang (Science Press, Beijing 1981)
3.3 J.-T. Wang: 'Chemical vapor deposition in China'. In: *Chemical Vapor Deposition, 13th International Conf. at Los Angeles, May 5–10, 1996*, ed. by T.M. Besmann, M.D. Allendorf, McD. Robinson et al. (The Electrochem. Soc., Inc., 1994) **PV96-5**, 651–655

3.4 C.P. Powell, J.H. Oxley, J.M. Blocher, Jr.: *Vapor Deposition*, (John Wiley & Sons Inc., New York 1966)
3.5 Y.-G. Cui: Scientific Chinese No.5, 48–50 (1995)
3.6 J.B. Hanney: Nature **22**, 255, (1880);
3.7 F.A. Bannister, K. Lonsdale: Nature **151**, 334 (1943)
3.8 Rayleigh: Nature **151**, 394 (1943)
3.9 C.H. Desch: Nature **152**, 148 (1943)
3.10 C. Persons, Philos. Trans. Roy. Soc. **A220**, 67(1920)
3.11 F.D. Rossini, R.S. Jessup: J. Res. Nat. Bur. Stand. **21**, 491(1938)
3.12 R. Berman, F. Simon: Z. Electrochem. **59**, 333 (1955)
3.13 J.P. Bromberg: *Physical Chemistry*, 2nd edn., (Allyn and Bacon, Inc., Boston 1984) pp.178–179
3.14 X.-C. Fu, W.-X. Shen, T.-Y. Yao: *Physical Chemistry*, 4th edn. (High Education Press, Beijing 1990) vol. 1, 421
3.15 F.P. Bundy, H.T. Hall, H.M. Strong, R.H. Wentorf: Nature **176**, 51–54 (1955)
3.16 W.J. Moore: *Physical Chemistry*, 5th edn. (Longman, London, 1976) Chap 6, p. 223–226
3.17 X.-H. Fang: *Basis and Standard of Superhard Materials* (China Building Materials Industry Press, Beijing 1998) p. 7
3.18 Physical Chemistry Teaching Group of Nanjing Univ.: *Physical Chemistry*, 1st edn. (Prople's Education Press, Beijing 1961) vol.1, 68
3.19 B. Lewic: in *Crystal Growth*, 2nd ed., ed. by B. R. Pamplin, 1980, 23
3.20 J.-T. Wang, M.-D. Liu: *Semiconductor Materials* (High Education Press, Beijing 1990), 638 pages
3.21 W.G. Eversole: US Patent Nos. 3 030 187 and 3 030 188. Filed July 23, 1958
3.22 W.G. Eversole: Can. Patent No. 628567. Filed 1959
3.23 B.V. Spitsyn, B.V. Deryagin: USSR Author(s Certificate No. 399 134, Appls. 716358/23-26 and 964957/23-26. Filed July 10, 1956. Patent issued in 1980
3.24 R.C. DeVries: Ann. Rev. Mater. Sci. **17**, 161–187 (1987)
3.25 J.C. Angus, H.A. Will, S. Stanko: J. Appl. Phys. **39**, 2915–2922 (1968)

Chapter 4

4.1 B.V. Deryagin, D.V. Fedoseev, V.M. Lukyanovich, B.V. Spitsyn, V.A. Ryabov, A.V. Lavrentev: J. Cryst. Growth **2**, 380–384 (1968)
4.2 B.V. Spitsyn, et al.: (USSR Author's Certificate No 987912 (Filed April 21,1971)
4.3 Inst. of Physical Chemistry et al.: *Physicochemical Synthesis of Diamond from the Gas Phase*, ("Technik" Press, Kiev 1971) p.18
4.4 B.E. Dzevitskii, B.V. Spitsyn et al: US Patent No. 3 705 937 (1972)
4.5 B.V. Deryagin, D.V. Fedoseev, K.S. Uspenskaya: Zurnal Fisicheskoy Khimii **47**, 24 (1973)
4.6 B.V. Deryagin, D.V. Fedoseev: Carbon **11**, 299–308 (1973)
4.7 B. V. Deryagin and B. V. Spitsyn et al.: Dokl. Akad. Nauk., **231**, 333 (1976)
4.8 B.V. Spitsyn, L.L. Bouilov, B.V. Deryagin: J. Cryst. Growth **52**, 219–226 (1981)
4.9 B.V. Spitsyn, 'The Roads to Metastable Diamond Growth'. In: *Diamond Materials, 3rd International Symposium at Honolulu, Hawaii, May 16–21, 1993*, ed. by J.P. Dismukas, K.V. Ravi, K.E. Spear, B. Lux, N. Setaka (The Electrochemical Society, Inc., Pennington, New Jersey 1993) **PV93-17**, 345–350

4.10 B.V. Spitsyn: 'Chap. 10 Growth of Diamond Films from the Vapor Phase'. In: *Handbook of Crystal Growth, vol.1*, ed. by D.T.J. Hurle (Elsevier Science, Amsterdam, NL 1994) pp. 401–456
4.11 B.V. Spitsyn, 'Crystallization of Diamond by the Chemical Transport Reaction: Thermodynamics and Kinetics'. In: *Diamond Materials, 4th International Symposium at Reno, Nevada, May 21–26, 1995*, ed. by K.V. Ravi & J.P. Dismukas (The Electrochemical Society, Inc., Pennington, New Jersey 1995) **PV95-4**, pp.61–72
4.12 S. Matsumoto, Y. Sato, M. Kamo, N. Setaka: J. Appl. Phys. **21**(4), L183–L185 (1982)
4.13 S. Matsumoto, Y. Sato, M. Tsutsumi, N. Setaka: J. Mater. Sci. **17**, 3106–3112 (1982)
4.14 M. Kamo, Y. Sato, S. Matsumoto, N. Setaka: J. Cryst. Growth **62**, 642–644 (1983)
4.15 S. Matsumoto, M. Hino, Kobayashi: Appl. Phys. Lett. **51**(10), 737–739 (1987)
4.16 N. Setaka: J. Materials Research. **4**(3), 664–670 (1989)
4.17 N. Setaka: 'One Proposal on Diamond Growth Mechanisum from Vapor Phase'. In: *Diamond Materials, 3rd International Symposium at Honolulu, Hawaii, May 16–21, 1993*, ed. by J.P. Dismukas, K.V. Ravi, K.E. Spear, B. Lux, N. Setaka (The Electrochemical Society, Inc., Pennington, New Jersey 1993) **PV93-17**, pp. 1–8
4.18 Z.-S. Jin et at.: Acta Scientiarum Naturatium Universitatis Jilinensis **2**, 127 (1987)
4.19 M. Murakawa, S. Takeuchi, Hirose: Surface and Coatings Tech. **39/40**, 235–240 (1989)
4.20 K.E. Spear: Earth and Mineral Science **56**(4), 53–59 (1987)
4.21 S. Jou, H.J. Doerr, R.F. Bunshah: Thin Solid Films **253**, 95–102 (1994)
4.22 J.C. Angus: Science **241**, 913–921 (1988)
4.23 J.C. Angus: 'History and current status of diamond growth at metastable conditions'. In: *Diamond and Diamond-like Films, 1st International Symp., 1989* ed. by J.P. Dismukes, A.J. Purdes, K.E. Spear et al. (The Electrochem. Soc., Inc., Pennington, New Jersey, 1989) **PV89-12**, 1–13
4.24 J.C. Angus: MRS Bulletin 38–47 (October, 1989)
4.25 J.-T. Wang, C.-B. Cao, P.-J. Zheng, J. Electrochem. Soc. **141**(1), 278–281 (1994)
4.26 M. Sommer, K. Mui, F.W. Smith: Solid State Communication **69**(7), 775–778 (1989)
4.27 J.C. Batty and R.R. Stickney: J. Chemical Physics **51** (10), 4475–4484 (1969)
4.28 W. A. Yarbrough: MRS Fall Meeting, Boston Nov 28, 1989, Paper F 13
4.29 W.A. Yarbrough: J. Electronic Materials **20**(2), 133–139 (1991)
4.30 Y. Bar-Yam, T.D. Moustakas: Nature **342**, 786–787 (1989)
4.31 W.A. Yarbrough, R. Missier: Science **247**, 688–696 (1990)
4.32 J.-T. Wang, J.-O. Carlsson: Surface and Coatings Technology **43/44**, 1–9 (1990); presented at 8th Internat Conf on Thin Films, April 2–6, San Diego

Chapter 5

5.1 J.-T. Wang, J.-O. Carlsson: Surface and Coatings Technology, **43/44**, 1–9 (1990)
5.2 J.-T. Wang: Thin Film Science and Technology, **4**(3), 54–67 (1991)

5.3 J.-T. Wang: 'Three Theoretical Models for Chemical Vapor Depositon'. In: *Thin Solid Films, 3rd National Conference at Guiling, China, Nov. 9–14, 1992*, ed. by Chinese Electronics Society (Guangxi Normal University, Guiling 1992) pp. 10–13

5.4 J.-T. Wang, P.-J. Zheng, C.-B. Cao, G.-Y. Li: 'A chemical pump model for diamond growth from the vapor phase'. In: *Diamond Materials, 3rd International Symposium at Honolulu, Hawaii, May 16–21, 1993*, ed. by J.P. Dismukas, K.V. Ravi, K.E. Spear, B. Lux, N. Setaka (The Electrochemical Society, Inc., Pennington, New Jersey 1993) **PV93-17**, pp. 962–967

5.5 J.-T. Wang, P.-J. Zheng: 'Thermodynamic coupling model for low pressure diamond syntheses'. In: *Proc. of International Union of Materials Research Society, International Conference in Asia, Supplementary Issue of Chinese J. of Mater. Res. (I), at Wuhan–Chongqing, Sept. 6–10, 1993*, ed. by Y.-F. Han, Y. Huang, J.-M. Tian, P.-X. Wang, (Supp. Issue of Chinese J. of Mater. Res., 1994, vol.1) pp. 363–368

5.6 J.-T. Wang, C.-B. Cao, P.-J. Zheng: J. Electrochem. Soc. **141**(1), 278–281 (1994)

5.7 B. Nolang: Thesis, University of Uppsala, Faculty of Science, Uppsala (1983)

5.8 D. R. Stull et al., *JANAF Thermochemical Table*, 2nd ed. (US Government Printing Office, Washington 1971)

5.9 M.W. Chase Jr., C.A. Davies, J.R. Downey Jr, D.J. Frurip, R.A. McDonald, A.N. Syverud: *JANAF Thermochemical Tables, American Chemical Society and American Institute of Physics*, 3rd edn. (1986)

5.10 S. Matsumoto, Y. Sato, M. Tsutsumi, N. Setaka: J Materials Sci. **17**, 3106–3112(1982)

5.11 J.-T. Wang, H. Wang, Q.-H. Yang, P.-J. Zheng: 'Phase diagrams of stationary nonequilibrium states for low pressure diamond syntheses'. In: *Materials Research Soc., Advanced Materials and Surface Techniques vol. III-3, '94 Materials Conference of Chinese Materials Research Society at Beijing, Nov. 11–15, 1994* ed. by G.-T. Zou, J.-J. Liu, S.-Z. Li (Chemical Industry Press, Beijing 1995) pp. 190–195

5.12 J.-T. Wang: 'Nonequilibrium thermodynamics for materials research'. In: *Materials Research Soc., New Techniques for Materials Processing and Research vol. IV-2, '94 Materials Conference of Chinese Materials Research Society at Beijing, Nov. 11–15, 1994* ed. by X.-Y. Yuan, B.-L. Zhou, C.-X. Shi, H.-D. Li (Chemical Industry Press, Beijing 1995) pp. 496–499

5.13 J.-T. Wang, P.-J. Zheng: Chinese Science Bulletin, English edn. **40**(13), 1141–1143 (1995); Chinese edn. **40**(11), 1056 (1995)

5.14 J.-T. Wang, P.-J. Zheng, Q.-H. Yang, H. Wang: 'Phase diagram calculations of non-equilibrium stationary states for CVD diamond growth'. In: *Diamond Materials, 4th International Symposium at Reno, Nevada, May 21–26, 1995*, ed. by K.V. Ravi & J.P. Dismukas (The Electrochemical Society, Inc., Pennington, New Jersey 1995) **PV95-4**, pp.13–19

5.15 J.-T. Wang: Thin Film Science and Technology **8**(3), 275 (1995)

5.16 J.-T. Wang, Z.-Q. Huang, Q.-H. Yang, D.W. Zhang, Y.-Z. Wan: 'Non-equilibrium thermodynamic coupling model for activated CVD diamond growth'. In: *Chemical Vapor Deposition, 13th International Conference at Los Angeles, May 5–10, 1996*, ed. by T.M. Besmann, M.D. Allendorf, McD. Robinson, R.K. Ulrich (The Electrochemical Society, Inc., Pennington, New Jersey 1996) **PV96-5**, pp.727–732

5.17 J.-T. Wang, D.W. Zhang, Y.-Z. Wan, Z.-Q. Huang: 'Calculated Phase Diagrams for Activated Low Pressure Diamond Growth from C-H Projective and

cross sectional phase diagrams for activated CVD diamond growth'. In: *Materials Research, Functional Materials vol. I-3, '96 Fall Meeting of Chinese Materials Research Society, Section C, at Beijing, 1996* (Chemical Industry Press, Beijing 1997) pp. 552–555
5.18 J.-T. Wang, Z.-Q. Huang, Y.-Z. Wan, D.W. Zhang, Q.-H. Yang: Chinese Science Bulletin, English edn. **42**(11): 967–968 (1997); Chinese edn. **42**(7): 783–784 (1997)
5.19 J.-T. Wang, Y.-Z. Wan, D.W. Zhang, Z.-Q. Huang, P.-J. Zheng: Progress in Nature Science, English edn. **7**(3), 265–271 (1997); Chinese edn. **7**(4), 392–396 (1997)
5.20 J.-T. Wang, D. W. Zhang, Z.-J. Liu: *Thermodynamic Coupling Model for Low Pressure Diamond Growth from the Vapor Phase, 175 pages* (Science Press, Beijing 1998 and reprinted in 2000)
5.21 S.-X. Zou, J.-T. Wang: 'Chap. 42 Plasma Enhanced Chemical Vapor Deposition (PCVD)'. In: *Handbook of Materials Surface Technology and Applications.* ed. By M.-G. Qian (Mechanical Industry Press, Beijing 1998) pp. 757–775
5.22 J.-T. Wang: 'Chap. 2 Nonequilibrium Thermodynamics of Diamond Film Growth'. In: *Some New Film Materials.* ed. by J.-L. Wu, Q.-D. Wu (Beijing University Press, Beijing 1999) p. 24–38
5.23 J.-T. Wang, D.W. Zhang, Y.-Z. Wan: 'Chapter 5 Thermodynamic Calculations for the Low Pressure Diamond Growth from the Vapor Phase'. In: *Computerized Chemistry of Metallurgy and Materials.* ed. by Z.-Y. Qiao, Z.-H. Xu, H.-L. Liu (Metallurgical Industry Press, Beijing 1999) pp. 129–146
5.24 J.-T. Wang: 'Chap. 9 Chemical Vapor Deposition and Related Theories in Inorganic Synthesis and Materials Preparation'. In: *Inorganic Synthesis and Preparation Chemistry.* ed. by R.-R. Xu, W.-Q. Peng (Higher Education Press, Beijing, 2001) pp. 257–285
5.25 D. Burk: J. Phys. Chem. **35**, 432 (1931)
5.26 Th. De Donder, P. Van Rysselberghe: *Thermodynamic Theory of Affinity — A book of principles*, 2nd edn. (Stanford University Press, Stanford 1936)
5.27 S. Glasstone: *Textbook of Physical Chemistry*, 2nd edn. (1946) p.466
5.28 D. Feng: *Dictionary of Solid Physics* (High Education Press, Beijing 1995) p.893
5.29 D. Jou, J. Casas-Vázquez, M. Criado-Sancho: *Thermodynamics of fluids under flow* (Springer, Berlin 2000)
5.30 Y. Sato Y, et al.: Surface and Coatings Tech. **39/40**, 183–198 (1989)
5.31 P. Southworth et al.: 'In: *Diamond Materials, 3rd Intern'l Symp. at Honolulu, Hawaii 1993*, pp. 87–94
5.32 J.-Y. Zhang, P.-F. Wang, S.-J. Ding, D.W. Zhang, J.-T. Wang, Z.-J. Liu: Thin Solid Films **368**(2), 266–268 (2000)
5.33 K. Kobashi, K. Nishimura, Y. Kawate et al.: Phys. Rev. B. **38**, 4067 (1988)
5.34 B.V. Spitsyn: 'Chap. 10 Growth of Diamond Films from the Vapor Phase'. In: *Handbook of Crystal Growth, vol.1*, ed. by D.T.J. Hurle (Elsevier Science, Amsterdam 1994) pp. 403–456

Chapter 6

6.1 J.-T. Wang, H. Wang, Q.-H. Yang, P.-J. Zheng: 'Phase diagrams of stationary nonequilibrium states for low pressure diamond syntheses'. In: *Materials Research Soc., Advanced Materials and Surface Techniques vol. III-3, '94*

Materials Conference of Chinese Materials Research Society at Beijing, Nov. 11–15, 1994 ed. by G.-T. Zou, J.-J. Liu, S.-Z. Li (Chemical Industry Press, Beijing 1995) pp. 190–195

6.2 J.-T. Wang: 'Nonequilibrium thermodynamics for materials research'. In: *Materials Research Soc., New Techniques for Materials Processing and Research vol. IV-2, '94 Materials Conference of Chinese Materials Research Society at Beijing, Nov. 11–15, 1994* ed. by X.-Y. Yuan, B.-L. Zhou, C.-X. Shi, H.-D. Li (Chemical Industry Press, Beijing 1995) pp. 496–499

6.3 J.-T. Wang, P.-J. Zheng: Chinese Science Bulletin, English edn. **40**(13), 1141–1143 (1995); Chinese edn. **40**(11), 1056 (1995)

6.4 J.-T. Wang, P.-J. Zheng, Q.-H. Yang, H. Wang: 'Phase diagram calculations of non-equilibrium stationary states for CVD diamond growth'. In: *Diamond Materials, 4th International Symposium at Reno, Nevada, May 21–26, 1995*, ed. by K.V. Ravi & J.P. Dismukas (The Electrochemical Society, Inc., Pennington, New Jersey 1995) **PV95-4**, pp.13–19

6.5 C.-X. Gao, G.-T. Zou, T.-C. Zhang, Q.-C. Cui, X.-Y. Lu, Z.-S. Jin.: 'Epitaxial Diamond Growth on Cubic Boron Nitride Surface'. In: *Abstracts, vol.1, '94 Meeting of CMRS at Beijing, Nov. 11–15, 1994* (Chinese Materials Research Soc., Beijing 1994) p. 382

6.6 M.-Y. Mao, T.-Y. Wang, J.-F. Xie, S.-S. Tan, W.-Y. Wang: 'Nucleation and Growth of Highly Orientational Diamond Films'. In: *Abstracts, vol.1, '94 Meeting of CMRS at Beijing, Nov. 11–15, 1994* (Chinese Materials Research Soc., Beijing 1994) p. 383

6.7 Q.-Q. Yang, S.-L. Li, Y.-C. Su, B.-J. Chen, X.-Z. Wang, J.-C. Jin: 'Diamond Film Growth with High Rate by Hot-filament CVD'. In: *Abstracts, vol.1, '94 Meeting of CMRS at Beijing, Nov. 11–15, 1994* (Chinese Materials Research Soc., Beijing 1994), p. 387

6.8 Y. Wang, Y.-B. Xie, H. Wang: 'Diamond Films Deposited on Aluminum Oxide Ceramic Substrate by MW Plasma CVD'. In: *Abstracts, vol.1, '94 Meeting of CMRS at Beijing, Nov. 11–15, 1994* (Chinese Materials Research Soc., Beijing 1994) p. 401

6.9 Y.-Z. Bai, Z.-S. Jin, X.-Y. Lu, C.-L. Wang, G.-T. Zou.: 'Diamond Films Deposited by MW PCVD'. In: *Abstracts, vol.1, '94 Meeting of CMRS at Beijing, Nov. 11–15, 1994* (Chinese Materials Research Soc., Beijing 1994) p. 415

6.10 Z.-G. Zhang et al.: In: *Abstracts, vol.1, '94 Meeting of CMRS at Beijing, Nov. 11–15, 1994* (Chinese Materials Research Soc., Beijing 1994) p. 385

6.11 Z.-D. Lin, K.-A. Feng, B.-W. Sun, J. Yang, Q.-J. Chen, Y. Chen: 'Atomic Scale Mechanism on Diamond Films Growth'. In: *Abstracts, '96 Meeting of CMRS at Beijing, Nov. 1996* (Chinese Materials Research Soc., Beijing 1996) Session I, p. 9

6.12 D.W. Zhang, Y.-Z. Wan, J.-T. Wang: Progress in Nature Science, English edn. **7**(4): 478–482 (1997); Chinese edn. **7**(2): 170–174 (1997)

6.13 D.W. Zhang, Y.-Z. Wan, J.-T. Wang: Progress in Nature Science, English edn. **7**(4), 506–508 (1997); Chinese edn. **7**(5), 634–637 (1997)

6.14 D.W. Zhang, Y.-Z. Wan, J.-T. Wang: J. Crystal Growth **177**(1/2), 171–173 (1997)

6.15 D.W. Zhang, Y.-Z. Wan, J.-T. Wang: Chinese J. of Inorganic Materials **12**(3): 331–335 (1997)

6.16 D.W. Zhang, Y.-Z. Wan, J.-T. Wang: Chinese J. of Inorganic Materials **12**(4), 617–619 (1997)

6.17 D.W. Zhang, Y.-Z. Wan, Z.-J. Liu, J.-T. Wang: Acta Metallurgica Sinica **33**(11), 1189–1193 (1997)

6.18 D.W. Zhang, J.-T. Wang, Y.-Z. Wan: Chinese J. of Physics **46**(6): 1237–1242 (1997)
6.19 D.W. Zhang, Z.-J. Liu, Y.-Z. Wan, J.-T. Wang: Applied Physics A **66**(1), 49–51 (1998)
6.20 Z.-J. Liu, D.W. Zhang, Y.-Z. Wan, J.-T. Wang: Chinese J. of Inorganic Materials **13**(3), 432–434 (1998)
6.21 Z.-J. Liu, D.W. Zhang, Y.-Z. Wan, J.-T. Wang: Chinese J. of Inorganic Chemistry **14**(4): 412–417 (1998)
6.22 Z.-J. Liu, D.W. Zhang, Y.-Z. Wan, J.-T. Wang: Microfabrication Technology **5**, 69–73 (1998)
6.23 Z.-J. Liu, Y.-Z. Wan, D.W. Zhang, J.-T. Wang: Chemical J. of Chinese Universities **19**(7), 1140–1143 (1998)
6.24 Z.-J. Liu, D.W. Zhang, Y.-Z. Wan, J.-T. Wang: Functional Materials **29** (Suppl), 1025–1027 (1998)
6.25 D.W. Zhang, Y.-Z. Wan, Z.-J. Liu, J.-T. Wang: J. Materials Science Lett. **16**, 1349–1351 (1997)

Chapter 7

7.1 P.K. Bachmenn, D. Leers, H. Lydtin: Diamond and Related Materials **1**, 1–12(1991)
7.2 P.K. Bachmenn, W. van Enckevort: Diamond and Related Materials **1**, 1021–1034 (1992)
7.3 M. Marinelli, E. Milani, M. Montuori et al.: J. Appl. Phys. **76**(10), 5702–5705 (1994)
7.4 P.K. Bachmenn, H.J. Hagemann, H. Lade, D. Leers, F. Picht, D.U. Wiechert, H. Wilson: 'Diamond Chemical Vapor Deposition: Gas Compositions and Film Properties'. In: *Diamond, SiC and Nitride Wide Bandgap Semiconductor, Spring Meeting of Materials Research Society at San Francisco, California 1994*, ed. by C.H. Carter, Jr. G. Gildenblat, et al. (1994) pp. 267–277
7.5 P.K. Bachmenn, H.J. Hagemann, H. Lade et al.: Diamond and Related Materials **4**, 820–826(1995)
7.6 J.-T. Wang, Z.-Q. Huang, Q.-H. Yang, D.W. Zhang, Y.-Z. Wan: 'Non-equilibrium thermodynamic coupling model for activated CVD diamond growth'. In: *Chemical Vapor Deposition, 13th International Conference at Los Angeles, May 5–10, 1996*, ed. by T.M. Besmann, M.D. Allendorf, McD. Robinson, R.K. Ulrich (The Electrochemical Society, Inc., Pennington, New Jersey 1996) **PV96-5**, pp.727–732
7.7 J.-T. Wang, Y.-Z. Wan, D.W. Zhang, Z.-J. Liu, Z.-Q. Huang: J. of Materials Research **12**(12), 3250–3253 (1997)
7.8 J.-T. Wang, Y.-Z. Wan, D.W. Zhang, Z.-Q. Huang, P.-J. Zheng: Progress in Nature Science, English edn. **7**(3), 265–271 (1997); Chinese edn, **7**(4), 392–396 (1997)
7.9 J.-T. Wang, Z.-Q. Huang, Y.-Z. Wan, D.W. Zhang, Q.-H. Yang: Chinese Science Bulletin, English edn. **42**(11): 967–968 (1997); Chinese edn. **42**(7): 783–784 (1997)
7.10 M.-Y. Mao, X.-F. Jin, T.-P. Wang et al.: (1995) Appl. Phys. Lett. **66**, 16
7.11 Y.-Z. Wan, D.W. Zhang, J.-T. Wang, Z.-Q. Huang: ' Projective and cross sectional phase diagrams for activated CVD diamond growth'. In: *Materials Research, Functional Materials vol. I-3, '96 Meeting of Chinese Materials Research Society, International Section C, at Beijing, 1996* (Chemical Industry Press, Beijing 1997) pp. 552–555

7.12 Y.-Z. Wan, D.W. Zhang, Z.-J. Liu, J.-T. Wang: High Technol. Lett. **3**(2), 80–83 (1997)
7.13 Y.-Z. Wan, Z.-J. Liu, D.W. Zhang, J.-T. Wang: Microfabrication Technology **4**, 43–48 (1997)
7.14 Y.-Z. Wan, D.W. Zhang, Z.-J. Liu, J.-T. Wang: Applied Physics A **67**(2), 225–231 (1998)
7.15 J.-T. Wang, Y.-Z. Wan, Z.-J. Liu, H. Wang, D.W. Zhang, Z.-Q. Huang: Mater. Lett. **33**(4): 311–314 (1998)
7.16 J.-T. Wang: Physics (Beijing) **27**(2), 77–83 (1998)
7.17 Z.-J. Liu, D.W. Zhang, Y.-Z. Wan, J.-T. Wang: Functional Materials **29**(5), 506–508 (1998)
7.18 Z.-J. Liu, D.W. Zhang, Y.-Z. Wan, J.-Y. Zhang, J.-T. Wang: Chinese Chemistry Bulletin **1999**(2), 59–62 (1999)
7.19 Z.-J. Liu, D.W. Zhang, P.-F. Wang, S.-J. Ding, J.-Y. Zhang, J.-T. Wang: Thin Solid Films 368(2), 253–256 (2000)
7.20 R. Ramesham, M.F. Rose, R.F. Askew: In: *Diamond Materials, 4th International Symposium at Reno, Nevada, May 21–26, 1995*, ed. by K.V. Ravi & J.P. Dismukas (The Electrochemical Society, Inc., Pennington, New Jersey 1995) **PV95-4**, pp.182–187
7.21 W.D. Cassidy, P.W. Morrison Jr., J.C. Angus: In: *Diamond Materials, 4th International Symposium at Reno, Nevada, May 21–26, 1995*, ed. by K.V. Ravi & J.P. Dismukas (The Electrochemical Society, Inc., Pennington, New Jersey 1995) **PV95-4**, pp.85–94
7.22 Y.-Z. Wan, D.W. Zhang, J.-T. Wang, Z.-Q. Huang: Chinese J. of Applied Science **16**(2), 185–190 (1998)
7.23 Y.-Z. Wan, D.W. Zhang, Z.-J. Liu, J.-T. Wang: Chinese J. of Synthetic Crystals **27**(3), 109–113 (1998)
7.24 J.-T. Wang, Y.-Z. Wan, Z.-J. Liu, D.W. Zhang, J.-Y. Zhang: 'Low pressure growth of diamond films under fluorine addition'. In: *Solid-state Integrated Circuit Technol., 5th International Conference at Beijing, Oct. 1998* (Publishing House of Electronics Industry, Beijing 1998) pp. 800–803
7.25 J.-T. Wang, Z.-J. Liu, D.W. Zhang, J.-Y. Zhang, Y.-Z. Wan: 'Thin film growth conditions for CVD diamond under low pressure'. In: *Solid-state Integrated Circuit Technol., 5th International Conference at Beijing, 21–23 Oct. 1998* Ed. by M. Zhang, K.-N. Tu (IEEE Press, Publishing House of Electronic Industry, Beijing 1998) pp. 819–822
7.26 Y.-Z. Wan, D.W. Zhang, Z.-J. Liu, J.-Y. Zhang, J.-T. Wang: Materials Chemistry and Physics **56**(3), 275–279 (1998)
7.27 Y.-Z. Wan, J.-Y. Zhang, Z.-J. Liu, D.W. Zhang, J.-T. Wang: Chinese J. of Materials Research **12**(6), 604–609 (1998)
7.28 Z.-J. Liu, D.W. Zhang, Y.-Z. Wan, J.-T. Wang: Chinese J. of Inorganic Materials **13**(4), 634–636 (1998)
7.29 Z.-J. Liu, D.W. Zhang, Y.-Z. Wan, J.-T. Wang: High Technol. Lett. **8**(6), 43–46 (1998)
7.30 Z.-J. Liu, D.W. Zhang, Y.-Z. Wan, J.-T. Wang: J. of Synthetic Crystals **27**(3), 201–205 (1998)
7.31 Z.-J. Liu, D.W. Zhang, Y.-Z. Wan, J.-Y. Zhang, J.-T. Wang, C.-B. Cao, H.-S. Zhu: High Technol. Lett. **8**(8), 36–39 (1998)
7.32 Z.-J. Liu, D.W. Zhang, J.-Y. Zhang, Y.-Z. Wan, J.-T. Wang, C.-B. Cao, H.-S. Zhu: Functional Materials **29**(Suppl), 1002–1003 (1998)
7.33 Z.-J. Liu, D.W. Zhang, J.-Y. Zhang, Y.-Z. Wan, J.-T. Wang, C.-B. Cao, H.-S. Zhu: Chemical J. of Chinese Universities **20**(1), 111–114 (1999)

7.34 Z.-J. Liu, D.W. Zhang, J.-Y. Zhang, S.-J. Ding, P.-F. Wang, J.-T. Wang: Chinese J of Inorganic Chemistry **15**(1), 132–134 (1999)
7.35 Z.-J. Liu, D.W. Zhang, J.-Y. Zhang, Y.-Z. Wan, J.-T. Wang: Progress in Nature Science, Chinese edn. **9**(4), 392–396 (1999)
7.36 Z.-J. Liu, D.W. Zhang, Y-Z Wan, J.-T. Wang: Appl. Phys. A **68**(3), 359–362 (1999)
7.37 Z.-J. Liu, D.W. Zhang, J.-Y. Zhang, Y.-Z. Wan, J.-T. Wang: Thin Solid Films **342**, 42–46 (1999)
7.38 Z.-J. Liu, S.-J. Ding, P.-F. Wang, D.W. Zhang, J.-Y. Zhang, J.-T. Wang: Thin Solid Films **368**(2), 208–210 (2000)

Chapter 8

8.1 D. Burk: J. Phys. Chem. **35**, 432–455 (1931)
8.2 Th. De Donder, P. Van Rysselberghe: *Thermodynamic Theory of Affinity — A book of principles*, 2nd edn. (Stanford University Press, Stanford 1936)
8.3 Ruhland, Johrbuch. Wiss. Bot., **63**, 321 (1924)
8.4 P. Van Rysselberghe: Bull. Ac. Roy. Belg. (Cl. Sc.) **22**, 1330 (1936)
8.5 P. Van Rysselberghe: Bull. Ac. Roy. Belg. (Cl. Sc.) **23**, 416 (1937)
8.6 P. Van Rysselberghe: J. Phys. Chem. **41**, 787–796 (1937)
8.7 I. Prigogine, R. Defay: *Chemical Thermodynamics*, translated by D.H. Everett (Longmans Green and Co, Inc, London 1954) p. 42
8.8 W.M. Southerland: *Biochemistry* (Churchill Livingstone, New York 1990) 53–55
8.9 G.W. Castellan: *Physical Chemistry*, 3rd edn. (Addison–Wesley Pub. Co., London 1983)
8.10 X.-C. Fu, W.-X. Shen, T.-Y. Yao: *Physical Chemistry*, 4th edn. (High Education Press, Beijing 1990) vol 1, pp. 427–431, pp. 437–444
8.11 K.J. Leidler, J.H. Meiser: *Physical Chemistry*, (Benjamin/ Cummings Publishing Inc, 1982) p.154
8.12 Z.-Z. Zheng: *Introduction to Thermodynamics for Irreversible Processes and Modern Reaction Kinetics* (High Education Press, Beijing 1983) p. 50–51
8.13 M. Boudart: J. Phys. Chem. **87**, 2786–2789 (1983)
8.14 I. Prigogine: *Introduction to Thermodynamics of Irreversible Processes*, 3rd edn. (Interscience Publishers, John Wiley & Sons, New York 1967)
8.15 N.N. Semenov, "Some Problems in Chemical Kinetics and Reactivity", vol.II, translated by M. Boudart, Princeton Univ. Press, Princeton, N.J. 1959, p.151
8.16 D. Kondepudi, I. Prigogine: *Modern Thermodynamics — From Heat Engines to Dissipative Structures*(John Wiley & Sons, New York 1998) pp.398–399
8.17 R.-S. Li: *Nonequilibrium Thermodynamics and Dissipative Structure* (Tsinghua University Press, Beijing 1986) pp. 257–259
8.18 I.M. Koltz, R.M. Rosenberg: *Chemical Thermodynamics*, (Benjamin,New York 1972) p.150
8.19 J.-T. Wang, D.W. Zhang: Chinese J. of Fudan University **38**(1), 75–79 (1999)
8.20 J.-T. Wang: *Phase Diagrams of Stationary Nonequilibrium States Thermodynamics for Low Pressure Diamond Growth from the Vapor Phase, 212 pages* (Science Press, Beijing, 2000)
8.21 Y.-F. Zhang, F.-Q. Zhang, G.-H. Chen: J. Materials Research **9**(11), 2845–2849 (1994)
8.22 W.J. Moore: *Physical Chemistry*, 4th edn. (Longman, New Jersey 1976) Chap, 9, 408

8.23 J.-T. Wang, Z.-Q. Huang, D.W. Zhang, Y.-Z. Wan, Z.-J. Liu: 'Reaction barrier during the activated CVD diamond growth'. In: *Functional Materials vol. I-2, '96 Materials Conference of Chinese Materials Research Society, Section I at Beijing, Nov.17–21, 1996* (Chemical Industry Press, Beijing 1997) pp. 365–369
8.24 J.-T. Wang, Z.-Q. Huang, Y.-Z. Wan, D.W. Zhang, H.-Y. Jia: J. of Materials Research 12(6), 1530–1535 (1997)
8.25 I.N. Levine: *Physical Chemistry*, 2nd edn. (McGraw-Hill, Inc, 1983) translated in Chinese by D.-Y. Zhu et al: (Beijing University Press, Beijing) vol. 2, p.179
8.26 G.-Y. Zhang, Z.-H. Li, J.-Z. Wu, Physics **3**(4), 241 (1974)
8.27 B.V. Spitsyn: 'Crystallization of Diamond by the Chemical Transport Reaction: Thermodynamics and Kinetics'. In: *Diamond Materials, 4th International Symposium at Reno, Nevada, May 21–26, 1995*, ed. by K.V. Ravi & J.P. Dismukas (The Electrochemical Society, Inc., Pennington, New Jersey 1995) **PV95-4**, pp. 61–72
8.28 W. Zhu, R. Mewwier, A.R. Badian: In: *Diamond and Diamond-Like Films, 1st Internat. Symp., 1989*, ed. by J.P. Dismukas) p. 61
8.29 N.M. Hwang, J.H. Hahn, D.Y. Yoon: J. Cryst. Growth **160**, 87–97 (1996)
8.30 N.M. Hwang, D.Y. Yoon: J. Cryst. Growth **160**, 98–103 (1996)
8.31 N.M. Hwang, J.H. Hahn, D.Y. Yoon, J. Cryst. Growth **162**, 55–68 (1996)
8.32 N.M. Hwang, J. Crystal Growth **204**, 85–90 (1999)
8.33 H.M. Jiang, N.M. Hwang, J. Materials Research **13**(12), 3527–3535 (1998)
8.34 H.M. Jiang, N.M. Hwang, J. Materials Research **13**(12), 3536–3549 (1998)
8.35 I. Pinter, A. Tebano, A. Paoletti, P. Paroli: Diamond and Relat. Mater. **3**, 126–128 (1993)
8.36 N.A. Prijaya, J.C. Angus, P.K. Bachmann: Diamond and Relat. Mater. **3** 129–136 (1993)
8.37 N.M. Hwang, J.H. Hahn, G.W. Bahng: Diamond and Relat. Mater. **3**, 163–167 (1993)
8.38 R. Bachmann, B. Sobisch, W. Kulich: Diamond and Relat. Mater. **4**, 256–260 (1995)
8.39 I.J. Ford: J. Appl. Phys. **78**(1), 510–513 (1995)
8.40 I.J. Ford: J. Phys D: Appl. Phys. **29**, 2229–2234 (1996)
8.41 E.G. Rakov: Appl. Phys. Lett. **69**(16), 2370–2372 (1996)
8.42 K.T. Jacob, N. Gundiah, A.G. Menon: High Temperature Materials and Processes **15**(4), 223–235 (1996)
8.43 W. Piekarczyk: J. Mater. Sci. **33**, 3443–3453 (1998)
8.44 W. Piekarczyk: Cryst. Res. Technol. **34**, 553–563 (1999)

Chapter 9

9.1 J.-T. Wang, J.-Y. Zhang, S.-J. Ding, D.W. Zhang, P.-F. Wang: 'Phase diagrams for activated CVD diamond and cubic BN'. In: *Chemical Thermodynamics, Abstracts of the 16th IUPAC International Conference (ICCT-2000) at Dalhousie University, Halifax, Canada, Aug 6–11, 2000*, p. 43
9.2 J.-T. Wang, S.-J. Ding, D.W. Zhang, J.-Y. Zhang, P.-F. Wang: 'Modern Thermodynamics for Activated Cubic Boron Nitride Growth from the Vapor Phase'. In: *Session 3 Advanced Materials, vol. 1, International Conference on Engineering & Technical Science at Beijing, China, Oct. 2000*, ed. by J. Song, R.-Y. Yin (New World Press, Beijing, 2000) pp. 857–863

9.3 J.-T. Wang, D. W. Zhang, J.-Y. Shen: J. Refractory Metals & Hand Materials, **19**(4-6), 461–466 (2001); J.-T. Wang: 'Modern thermodynamics in CVD of hard materials'. In: *Science of Hard Materials (Extended abstracts), 7th International Conference at Ixtapa, Mexico, March 5–9, 2001*, (PDVSA-Intevep, Venezuela, 2001) pp. 105–106

9.4 F. Zhang, Y. Guo, Z. Song, G. Chen: Appl. Phys. Lett. **65**(8), 971 (1994)

9.5 Z. Song, F. Zhang, Y. Guo, G. Chen: Appl. Phys. Lett. **65**(21), 2669 (1994)

9.6 S. Matsumoto, N. Nishida, K. Akashi, K. Sugai: J. of Mater. Sci. **31**, 713 (1996)

9.7 D. R. Stull et al.: *JANAF Thermochemical Tables*, 2nd ed., US Government Printing Office (1971)

9.8 M.W. Chase, Jr., C.A. Davies, J.R. Downey, Jr. et al.: *JANAF Thermochemical Tables* 3rd ed. J. Phys. Chem. Ref. Data **14** Suppl. (1985)

9.9 L. Solozhenko: Diamond and Related Materials **4**, 1–4 (1994)

9.10 L. Solozhenko: J. Hard Materials **6**, 51–65 (1995)

9.11 R.-S. Li: *Nonequilibrium Thermodynamics and Dissipative Structure* (Tsinghua University Press, Beijing 1986) p.259

9.12 S.-J. Ding , J.-T. Wang, D.W. Zhang: 'Reaction coupling in modern thermodynamics'. In: *Chemical Thermodynamics, Abstracts of the 16th IUPAC International Conference (ICCT-2000) at Dalhousie University, Halifax, Canada, Aug 6–11, 2000*, p. 151

9.13 E. Schrödinger: *What is life*, 2nd edn. (Cambridge University Press, Doubleday 1956) p.71

9.14 I.N. Levine: *Physical Chemistry*, 2nd edn. (McGraw-Hill, Inc., 1983) translated into Chinese by D.-Y. Zhu, Z.-F. Li, Y.-F. Zhang, (Beijing University Press, Beijing 1987) vol. 2, p. 203

9.15 J.-T. Wang, P.-J. Zheng, D.W. Zhang, Z.-J. Liu, Y.-Z. Wan, J.-Y. Zhang, C.-B. Cao: 'Negative entropy reactions for preparation of materials'. In: *Calculation of Phase Diagrams and its Application(CALPHAD), 27th International Conference at Beijing, May, 1998*

9.16 J.-T. Wang: Physics (Beijing) **27**(9), 571–572 (1998)

9.17 J.-T. Wang, D.W. Zhang: Chinese J. of Fudan University **38**(1), 75–79 (1999)

9.18 J.-T. Wang, D.W. Zhang: 'From activated CVD diamond to the life science'. In: *Chemical Vapor Deposition, Abstracts of 1999 Asian Conference at Shanghai, May 10–13, 1999*, paper no 1-6 (1999)

9.19 P. van Eikeren: *Guide to Lehninger's Principles of Biochemistry* (Worth Publisher, Inc., New York 1985) p. 194

9.20 P. Mitchell, Science **206**,1148–1159 (1979)

9.21 J.-T. Wang, D.W. Zhang, Z.-J. Liu, J.-Y. Zhang: 'Structure stability in plasma chemistry'. In: *Plasma Deposition and Treatment of Polymer, 1998 Full Meeting of Materials Research Society, at Boston, Nov. 30–Dec. 2, 1998*, ed. by W.W. Lee, R. d'Agostino, M.R. Wertheimer (Materials Research Society, Warrendale, Pennsylvania) Symp. Proc. Vol. **544**, 101–108

9.22 J.-T. Wang: 'Low Pressure Growth of Diamond Films and Modern Thermodynamics. In: *Thin Films (TFC99'), Chinese National Conference in Shanghai, Oct. 1999* pp. 75–76

9.23 J.-T. Wang: 'Modern thermodynamics for phase diagrams of stationary states. In: *Phase Diagram Calculation (CALPHAD XXIX) (Abstracts), 29th International Conference at MIT, Cambridge, Massachusetts, USA, June 18–23, 2000* p. 8

9.24 J.-T. Wang: 'Systematization of modern thermodynamics'. In: *Chemical Thermodynamics, Abstracts of the 16th IUPAC International Conference (ICCT-2000) at Dalhousie University, Halifax, Canada, Aug 6–11, 2000*, p. 150

9.25 J.-T. Wang, S.-J. Ding, D.W. Zhang, J.-Y. Zhang, P.-F. Wang: 'Calculation of phase diagrams of stationary states'. In: *Chemical Thermodynamics, Abstracts of the 16th IUPAC International Conference (ICCT-2000) at Dalhousie University, Halifax, Canada, Aug 6–11, 2000*, p49
9.26 J.-T. Wang, S.-J. Ding, D.W. Zhang, J.-Y. Zhang, P.-F. Wang: 'Modern Thermodynamics for shuttle reactions in biological systems'. In: *Chemical Thermodynamics, Abstracts of the 16th IUPAC International Conference (ICCT-2000) at Dalhousie University, Halifax, Canada, Aug 6–11, 2000*, p. 84
9.27 J.-T. Wang: Physics (Beijing) **29**(9), 524–530 (2000)
9.28 J.-T. Wang: 'Modern Thermodynamics for Advanced Materials Research. In: *Advanced Materials, vol. 1, International Conference on Engineering & Technical Science 2000, Session 3, at Beijing, China, Oct. 2000*, ed. by J. Song, R.-Y. Yin (New World Press, Beijing, 2000) pp 864–870
9.29 J.-T. Wang, D.W. Zhang, S.-J. Ding, P.-F. Wang: CALPHAD (J. of CALculated PHAse Diagrams) **24**(4), 427–434 (2000); J.-T. Wang:'A new field of phase diagrams of stationary nonequilibrium states'. In: *CALPHAD, Abstracts of 27th International Conference at Beijing, May 17–22, 1998*, p 66
9.30 J.-T. Wang: 'Modern Thermodynamics for Phase Diagrams of Stationary States. In: *Phase Diagram, 10th National Symp. on, Chinese Physical Society, Wuhu, Anhui, Sept 25–28, 2000*, pp 135–138
9.31 J.-T. Wang, D.W. Zhang: 'Modern thermodynamics for activated CVD processes'. In: *Chemical Vapor Deposition (CVD) (extended abstracts), 2nd Asian Conference at Gyeongju, Korea, May 28–30, 2001* pp. 63–66
9.32 J.-T. Wang: 'Nonequilibrium nondissipative thermodynamics – and its application to diamond film deposition. *Solid-State and Integrated-Circuit Technology, 6th International Conference, at Shanghai, China, October 22–25, 2001*, (IEEE Press, People's Posts and Telecommunication Publishing House, Beijing, 2001) vol.2, pp. 1421–1426
9.33 L. P. Wright, Nature **360**, 20 (1992)
9.34 W.M. Southerland: *Biochemistry* (Churchill Livingstone, New York 1990) p. 55.
9.35 G. Zubay: *Biochemistry* (Addison–Wesley Publishing Company, Massacusetts 1983) p.395
9.36 R.-Q. Shen, Q.-M. Gu: *Textbook of Biochemistry* (High Education Press, Beijing 1993) p. 328

Index

Springer Series in Chemical Physics

Editors: Vitalii I. Goldanskii Fritz P. Schäfer J. Peter Toennies

1 **Atomic Spectra and Radiative Transitions** By I. I. Sobelman
2 **Surface Crystallography by LEED** Theory, Computation and Structural Results. By M. A. Van Hove, S. Y. Tong
3 **Advances in Laser Chemistry** Editor: A. H. Zewail
4 **Picosecond Phenomena** Editors: C. V. Shank, E. P. Ippen, S. L. Shapiro
5 **Laser Spectroscopy** Basic Concepts and Instrumentation By W. Demtröder 3rd Printing
6 **Laser-Induced Processes in Molecules** Physics and Chemistry Editors: K. L. Kompa, S. D. Smith
7 **Excitation of Atoms and Broadening of Spectral Lines** By I. I. Sobelman, L. A. Vainshtein, E. A. Yukov
8 **Spin Exchange** Principles and Applications in Chemistry and Biology By Yu. N. Molin, K. M. Salikhov, K. I. Zamaraev
9 **Secondary Ion Mass Spectrometry SIMS II** Editors: A. Benninghoven, C. A. Evans, Jr., R. A. Powell, R. Shimizu, H. A. Storms
10 **Lasers and Chemical Change** By A. Ben-Shaul, Y. Haas, K. L. Kompa, R. D. Levine
11 **Liquid Crystals of One- and Two-Dimensional Order** Editors: W. Helfrich, G. Heppke
12 **Gasdynamic Laser** By S. A. Losev
13 **Atomic Many-Body Theory** By I. Lindgren, J. Morrison
14 **Picosecond Phenomena II** Editors: R. M. Hochstrasser, W. Kaiser, C. V. Shank
15 **Vibrational Spectroscopy of Adsorbates** Editor: R. F. Willis
16 **Spectroscopy of Molecular Excitions** By V. L. Broude, E. I. Rashba, E. F. Sheka
17 **Inelastic Particle-Surface Collisions** Editors: E. Taglauer, W. Heiland
18 **Modelling of Chemical Reaction Systems** Editors: K. H. Ebert, P. Deuflhard, W. Jäger
19 **Secondary Ion Mass Spectrometry SIMS III** Editors: A. Benninghoven, J. Giber, J. László, M. Riedel, H. W. Werner
20 **Chemistry and Physics of Solid Surfaces IV** Editors: R. Vanselow, R. Howe
21 **Dynamics of Gas-Surface Interaction** Editors: G. Benedek, U. Valbusa
22 **Nonlinear Laser Chemistry** Multiple-Photon Excitation By V. S. Letokhov
23 **Picosecond Phenomena III** Editors: K. B. Eisenthal, R. M. Hochstrasser, W. Kaiser, A. Laubereau
24 **Desorption Induced by Electronic Transitions DIET I** Editors: N. H. Tolk, M. M. Traum, J. C. Tully, T. E. Madey
25 **Ion Formation from Organic Solids** Editor: A. Benninghoven
26 **Semiclassical Theories of Molecular Scattering** By B. C. Eu
27 **EXAFS and Near Edge Structures** Editors: A. Bianconi, L. Incoccia, S. Stipcich
28 **Atoms in Strong Light Fields** By N. B. Delone, V. P. Krainov
29 **Gas Flow in Nozzles** By U. G. Pirumov, G. S. Roslyakov
30 **Theory of Slow Atomic Collisions** By E. E. Nikitin, S. Ya. Umanskii
31 **Reference Data on Atoms, Molecules, and Ions** By A. A. Radzig, B. M. Smirnov
32 **Adsorption Processes on Semiconductor and Dielectric Surfaces I** By V. F. Kiselev, O. V. Krylov
33 **Surface Studies with Lasers** Editors: F. R. Aussenegg, A. Leitner, M. E. Lippitsch
34 **Inert Gases** Potentials, Dynamics, and Energy Transfer in Doped Crystals. Editor: M. L. Klein
35 **Chemistry and Physics of Solid Surfaces V** Editors: R. Vanselow, R. Howe
36 **Secondary Ion Mass Spectrometry, SIMS IV** Editors: A. Benninghoven, J. Okano, R. Shimizu, H. W. Werner
37 **X-Ray Spectra and Chemical Binding** By A. Meisel, G. Leonhardt, R. Szargan
38 **Ultrafast Phenomena IV** By D. H. Auston, K. B. Eisenthal
39 **Laser Processing and Diagnostics** Editor: D. Bäuerle

Springer Series in Chemical Physics

Editors: Vitalii I. Goldanskii Fritz P. Schäfer J. Peter Toennies

Managing Editor: H. K. V. Lotsch

40 **High-Resolution Spectroscopy of Transient Molecules**
By E. Hirota

41 **High Resolution Spectral Atlas of Nitrogen Dioxide 559–597 nm**
By K. Uehara and H. Sasada

42 **Antennas and Reaction Centers of Photosynthetic Bacteria**
Structure, Interactions, and Dynamics
Editor: M. E. Michel-Beyerle

43 **The Atom-Atom Potential Method**
Applications to Organic Molecular Solids
By A. J. Pertsin and A. I. Kitaigorodsky

44 **Secondary Ion Mass Spectrometry SIMS V**
Editors: A. Benninghoven, R. J. Colton, D. S. Simons, and H. W. Werner

45 **Thermotropic Liquid Crystals, Fundamentals**
By G. Vertogen and W. H. de Jeu

46 **Ultrafast Phenomena V**
Editors: G. R. Fleming and A. E. Siegman

47 **Complex Chemical Reaction Systems**
Mathematical Modelling and Simulation
Editors: J. Warnatz and W. Jäger

48 **Ultrafast Phenomena VI**
Editors: T. Yajima, K. Yoshihara, C. B. Harris, and S. Shionoya

49 **Vibronic Interactions in Molecules and Crystals**
By I. B. Bersuker and V. Z. Polinger

50 **Molecular and Laser Spectroscopy**
By Zu-Geng Wang and Hui-Rong Xia

51 **Space-Time Organization in Macro molecular Fluids**
Editors: F. Tanaka, M. Doi, and T. Ohta

52 **Multiple-Photon Laser Chemistry**
By. R. V. Ambartzumian, C. D. Cantrell, and A. Puretzky

53 **Ultrafast Phenomena VII**
Editors: C. B. Harris, E. P. Ippen, G. A. Mourou, and A. H. Zewail

54 **Physics of Ion Impact Phenomena**
Editor: D. Mathur

55 **Ultrafast Phenomena VIII**
Editors: J.-L. Martin, A. Migus, G. A. Mourou, and A. H. Zewail

56 **Clusters of Atoms and Molecules**
Solvation and Chemistry of Free Clusters, and Embedded, Supported and Compressed Clusters
Editor: H. Haberland

57 **Radiationless Transitions in Polyatomic Molecules**
By E. S. Medvedev and V. I. Osherov

58 **Positron Annihilation in Chemistry**
By O. E. Mogensen

59 **Soot Formation in Combustion**
Mechanisms and Models
Editor: H. Bockhorn

60 **Ultrafast Phenomena IX**
Editors: P. Barbara, W. H. Knox, G. A. Mourou, and A. H. Zewail

61 **Gas Phase Chemical Reaction Systems**
Experiments and Models 100 Years After Max Bodenstein
Editors: J. Wolfrum, H.-R. Volpp, R. Rannacher, and J. Warnatz

62 **Ultrafast Phenomena X**
Editors: P. F. Barbara, J. G. Fujimoto, W. H. Knox, and W. Zinth

MIX
Papier aus verantwortungsvollen Quellen
Paper from responsible sources
FSC® C105338

If you have any concerns about our products,
you can contact us on
ProductSafety@springernature.com

In case Publisher is established outside the EU,
the EU authorized representative is:
Springer Nature Customer Service Center GmbH
Europaplatz 3, 69115 Heidelberg, Germany

Printed by Libri Plureos GmbH
in Hamburg, Germany